W0230145

PROCESS CONTROL
Designing Processes and Control Systems

PROCESS CONTROL
Designing Processes and Control Systems

Contributors :
Ravendra Singh,
Fernando J. Muzzio, *et al.*

AURIS REFERENCE LTD.
London, UK

Process Control: Designing Processes and Control Systems
Contributors : Ravendra Singh *and* Fernando J. Muzzio, *et al.*

Auris Reference Ltd., UK

www.aurisreference.com

United Kingdom

Copyright 2016

Printed in 2017 for Sale in the Indian Subcontinent

Notice

Process Control: Designing Processes and Control Systems

ISBN: 978-1-78154-504-1

British Library Cataloguing in Publication Data
A CIP record for this book is available from the British Library

Exclusively distributed by CBS Publishers & Distributors Pvt. Ltd.

Sales & Distribution Rights only for India, Pakistan, Bangladesh, Sri Lanka, Nepal and Bhutan.This book is not to be sold outside these territories.

PREFACE

Process Control: Designing Processes and Control Systems deals with the process of designing a new control system, or improving an existing one. That is, how can an individual engineer or a team of engineers and project managers, tackle the design in a systematic way. With respect to the traditional views on the design process, this book adds an extra point of view — the design context of each particular system — to increase insight in the various interactions between the traditional design steps, and their relative importance. In many places, the book provides a motivated list of things to think of, do's and don'ts and best practices in the design of a system from individual components.

Although this book focuses on Designing Processes and Control Systems with an important amount of mechanical subsystems, the design steps can probably also be applied to design problems in most other areas of engineering.

This page left intentionally blank.

CONTENTS

This page left intentionally blank.

List of Contributors

Ravendra Singh

Engineering Research Center for Structured Organic Particulate Systems (C-SOPS), Department of Chemical and Biochemical Engineering, Rutgers, The State University of New Jersey, Piscataway, NJ 08854, USA; E-Mails: fjmuzzio@yahoo.com (F.J.M.); marianth@soemail.rutgers.edu (M.I.)

Fernando J. Muzzio

Engineering Research Center for Structured Organic Particulate Systems (C-SOPS), Department of Chemical and Biochemical Engineering, Rutgers, The State University of New Jersey, Piscataway, NJ 08854, USA; E-Mails: fjmuzzio@yahoo.com (F.J.M.); marianth@soemail.rutgers.edu (M.I.)

Marianthi Ierapetritou

Engineering Research Center for Structured Organic Particulate Systems (C-SOPS), Department of Chemical and Biochemical Engineering, Rutgers, The State University of New Jersey, Piscataway, NJ 08854, USA; E-Mails: fjmuzzio@yahoo.com (F.J.M.); marianth@soemail.rutgers.edu (M.I.)

Rohit Ramachandran

Engineering Research Center for Structured Organic Particulate Systems (C-SOPS), Department of Chemical and Biochemical Engineering, Rutgers, The State University of New Jersey, Piscataway, NJ 08854, USA; E-Mails: fjmuzzio@yahoo.com (F.J.M.); marianth@soemail.rutgers.edu (M.I.)

This page left intentionally blank.

Chapter 1

INTRODUCTION TO PROCESS CONTROL

Control is a science that is used in many engineering disciplines such as chemical, electrical and mechanical engineering and it is applied to a wide range of physical systems from chemical processes to electrical circuits to guided missiles to robots.

Control: to maintain desired condition in a physical system by adjusting a selected variable in the system.

Controlled variable: Is the system variable that represents the desired condition.

Manipulated variable: The selected adjustable variable.

Control System Examples

- Toilet Float
- Driving a car
- Room air conditioner
- Heated Tank

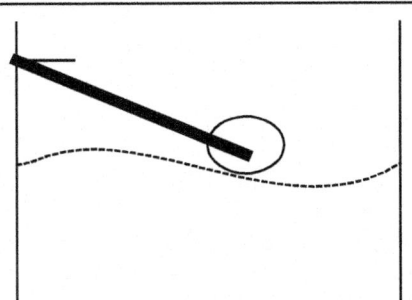 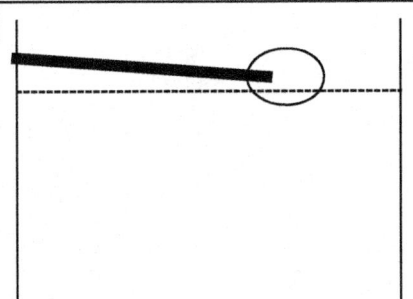

Fig. : Toilet float.

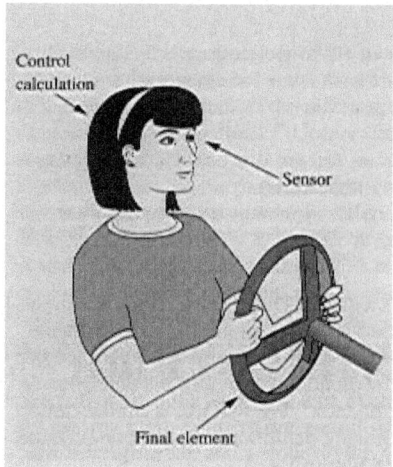

Fig. : Driving a car example.

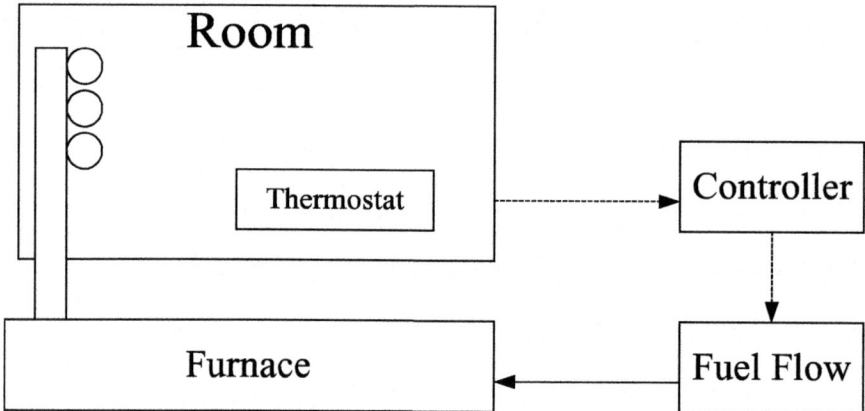

Fig. : Feedback control for room temperature.

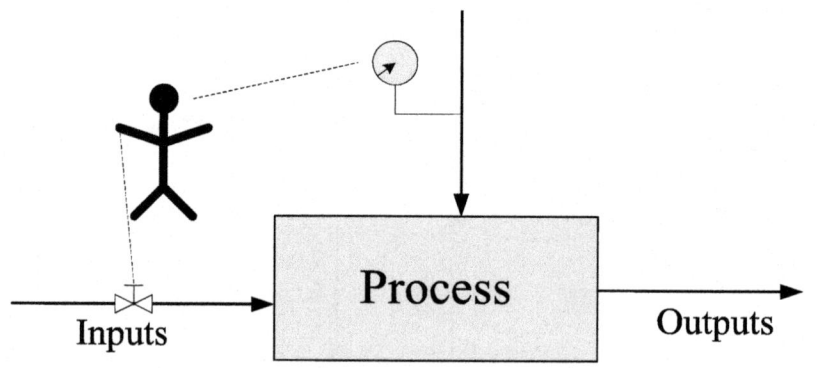

Fig. :Manual Control Concept.

Control System Elements

The major components of a control system:

- Process (system)
- Instrumentation (sensors, final control element, transmission lines)
- Control algorithm

These elements can be used to represent many control systems.

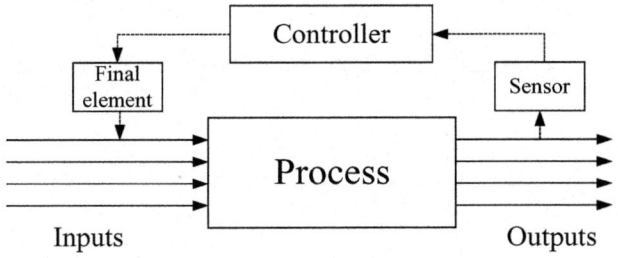

Fig. : Schematic diagram of a general feedback control system.

Sensors ➜ transducers

Final control element ➜ actuator: valves, pump motor

Transmission line: pneumatic, electrical, digital

Controller: manual → on/off → mechanical elements → electronic devices → computer programs

Control Implementation

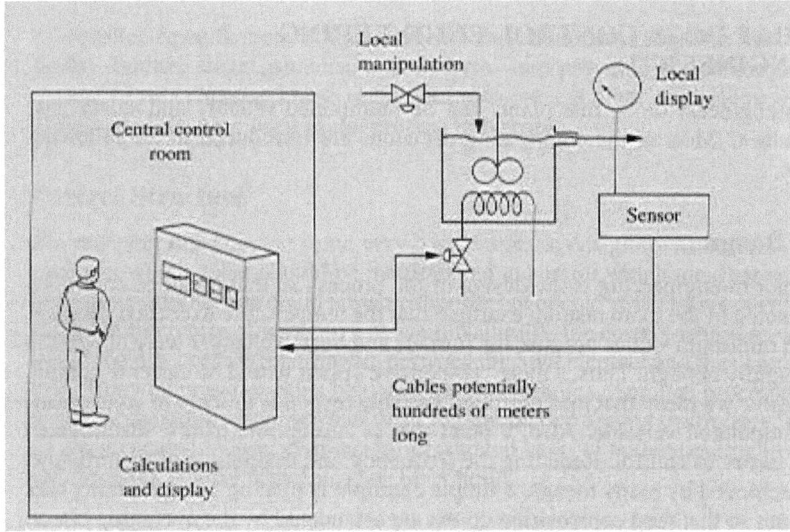

Fig. : Control room.

Commercial Distributed Control Systems

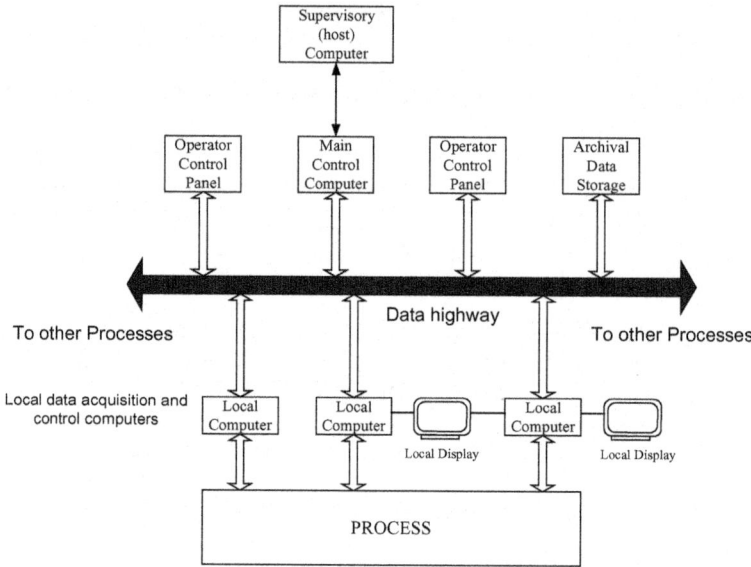

Fig. : The elements of a commercial distributed control system network.

CONTROL SYMBOLS AND DRAWINGS

Display Symbol

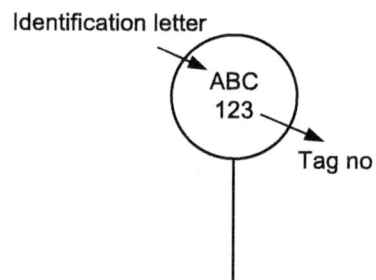

Identification letter

	First letter	*Successive letter*
A	Analyzer	Alarm
C		Control
F	Flow	Ratio
I		Indicator
L	Level	
P	Pressure	
T	Temperature	Transmitter

Transmission Line

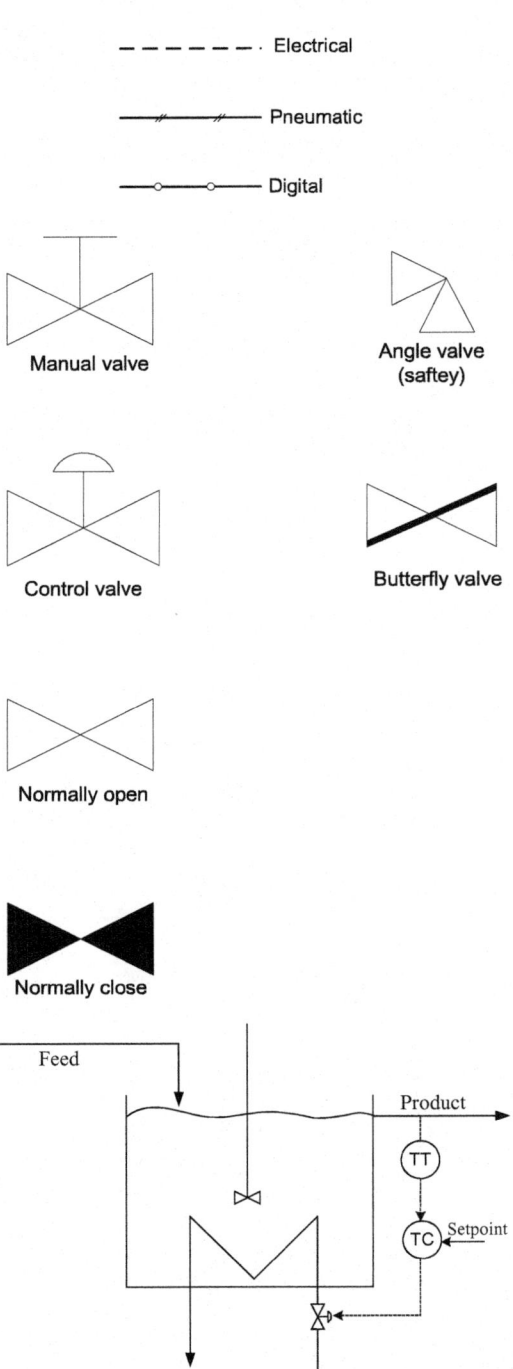

------- Electrical

———//———//——— Pneumatic

———o———o——— Digital

Manual valve

Angle valve
(saftey)

Control valve

Butterfly valve

Normally open

Normally close

Feed

Product

TT

TC ← Setpoint

Heating oil

Fig. : A stirred tank heater system.

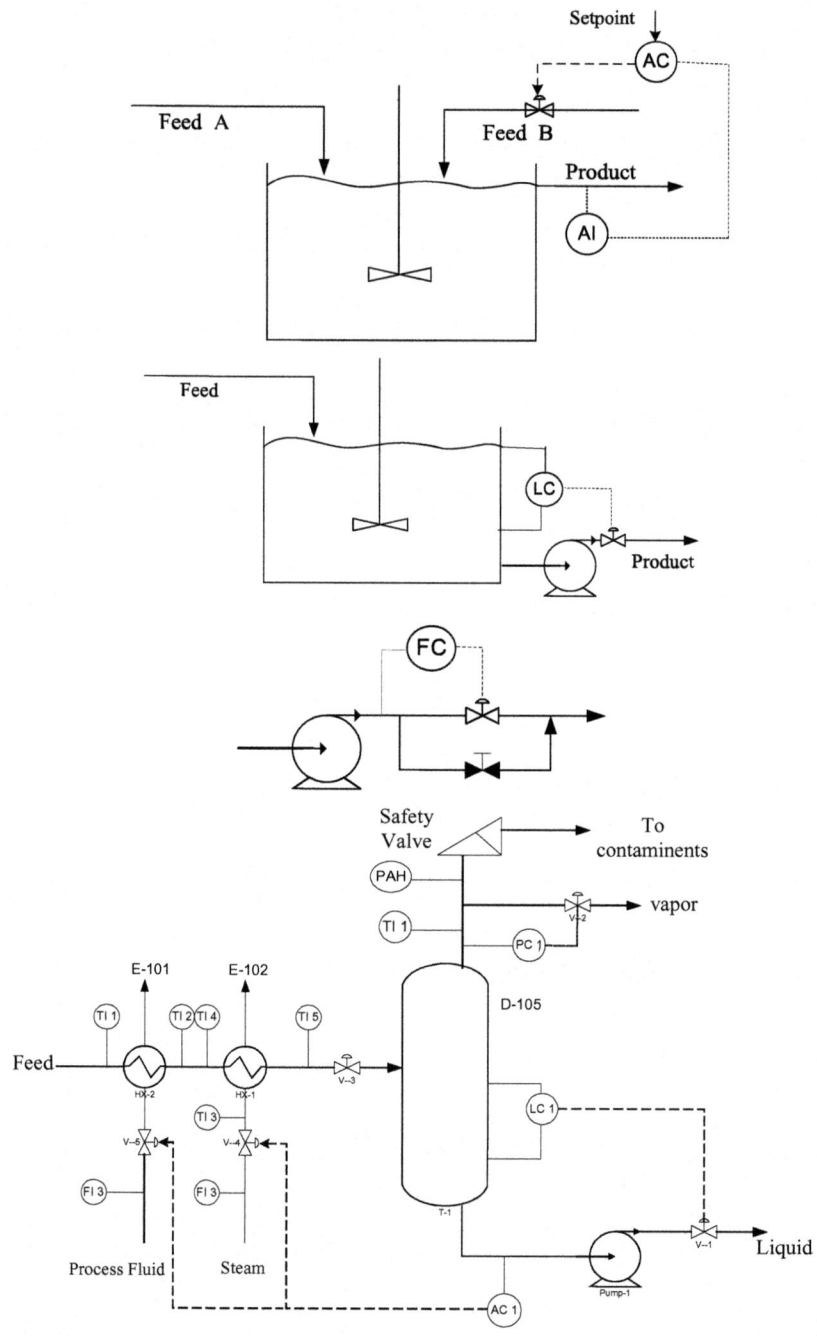

PAH : high pressure alarm.
PC : Pressue controller
AC: Analyzer controller
LC: Level controller
TI : temperature indicator (sensor)
FI: Flow indicator (sensor)

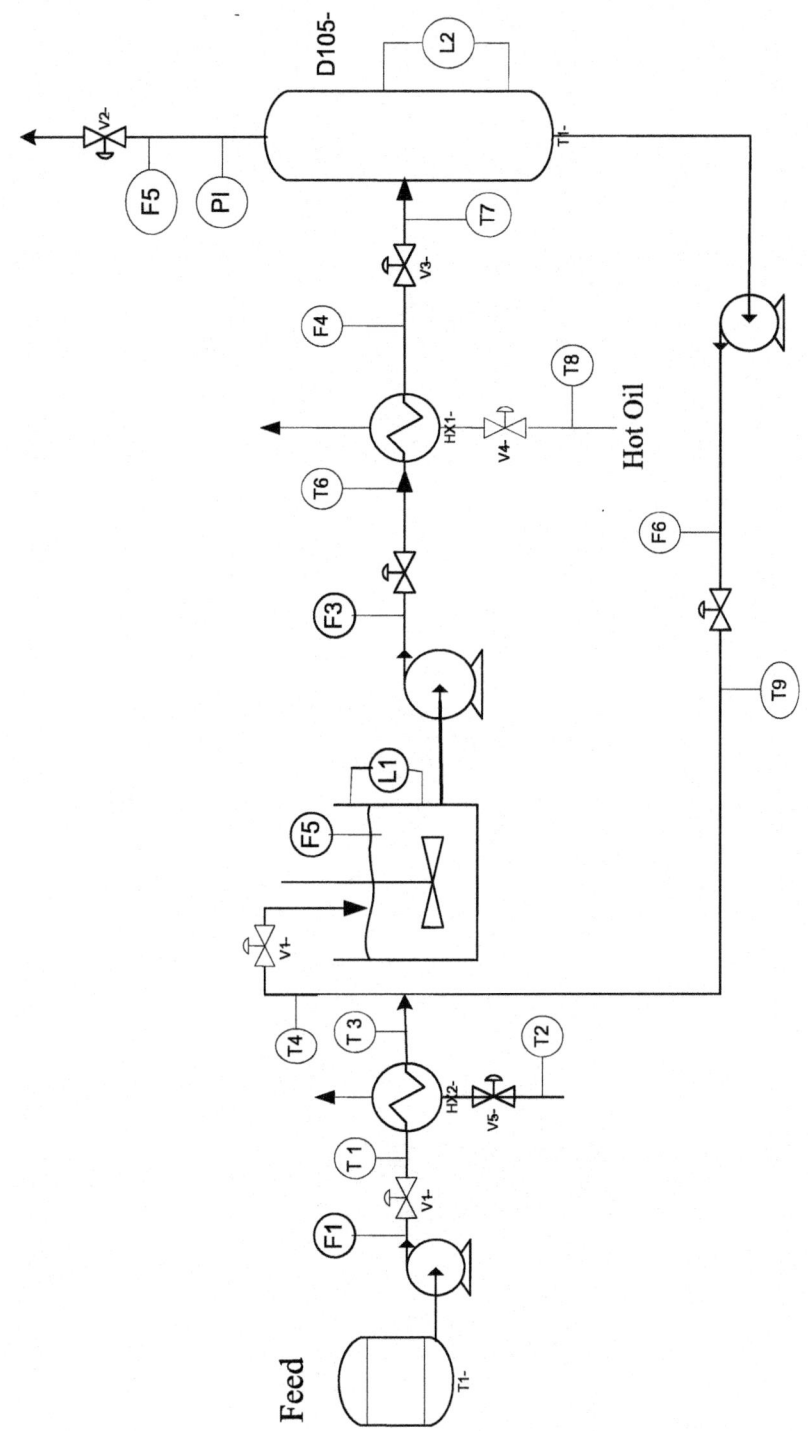

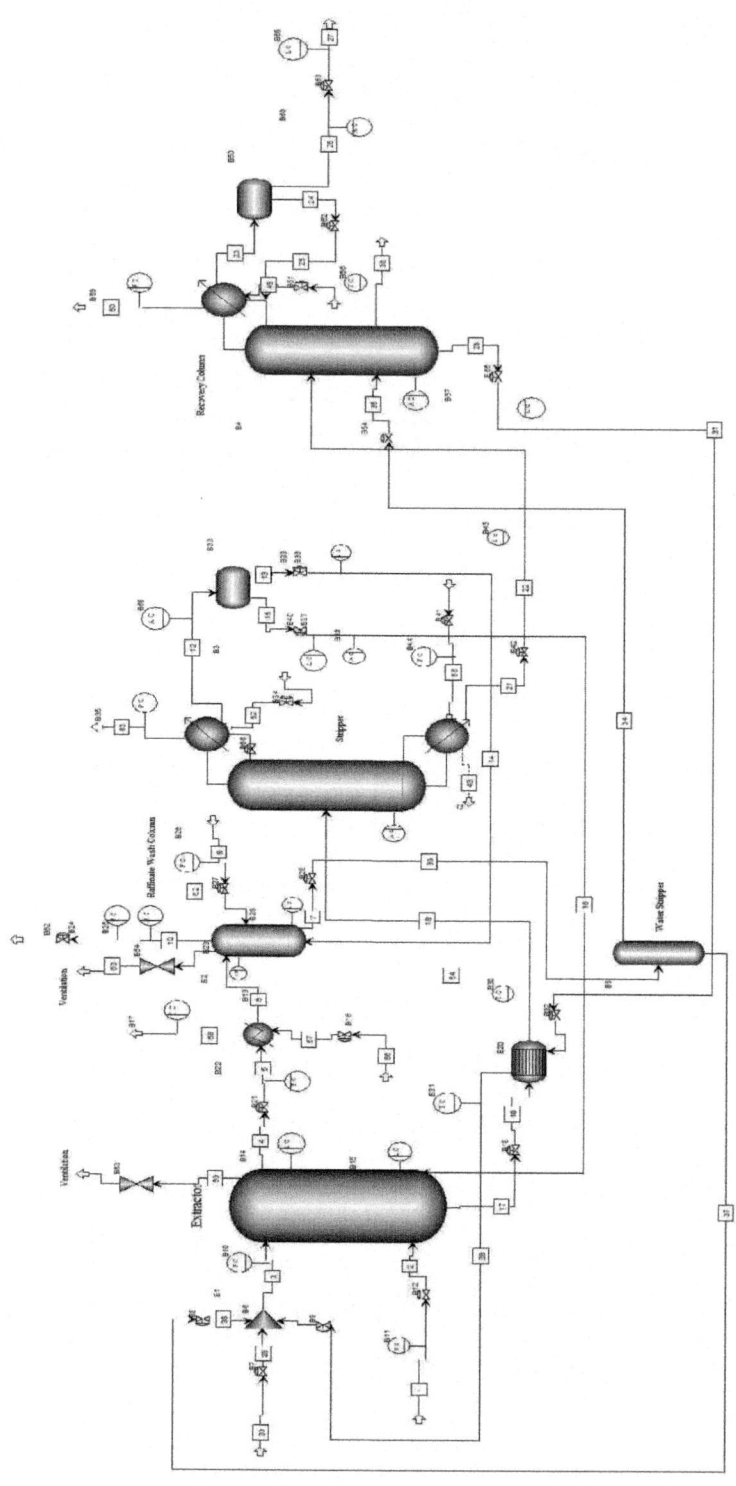

Equipment	Manipulated Variables	Controlled Variables	Action
Extractor	St.1 Flow rate	Flow rate	Control Valve (B12)
	St.1 Flow rate	Flow rate	Control Valve (B7) Control Valve (B7) Control Valve (B9)
	St.4 Flow rate	High level control	Control Valve (B21)
	St.17 Flow rate	Low level control	Control Valve (B18)
Stripper	Reflux flow rate	Overhead compositions	Control Valve (B66)
	St.55 Flow rate	Bottom compositions	Control Valve (B41)
	St.13 Flow rate	High level control	Control Valve (B36)
	St.15 Flow rate	High level control	Control Valve (B37)
	St.15 Flow rate	Taking light Aliphatics	Control Valve (B37)
	St.21 Flow rate	Low level control	Control Valve (B42)
	St.52 Flow rate	Condenser pressure	Control Valve (B34)
	St.55 Flow rate	Reboiler temperature	Control Valve (B41)
Recovery Column	St.24 Flow rate	Overhead compositions	Control Valve (B52)
	St.35 Flow rate	Bottom compositions	Control Valve (B54)
	St.35 Flow rate	Column temperature	Control Valve (B54)
	St.26 Flow rate	High level control	Control Valve (B53)
	St.28 Flow rate	Low level control	Control Valve (B55)
	St.49 Flow rate	Condenser pressure	Control Valve (B51)
Raffinate Wash Column	St.5 Flow rate	Flow rate	Control Valve (B21)
	St.5 Flow rate	High level control	Control Valve (B21)
	St.7 Flow rate	Low level control	Control Valve (B28)

Variable	Control priority	Cases
Level	Mandatory	All vessels containing liquids
Pressure	High	High pressure vessels, processes involving gases. Flash drum, distillation columns
Temperature	medium	May become important for reactors with runaway reactions, operations sensitive to temperature.
Composition	Low/ as needed	May be required for the final product to maintain high quality products.
Flow	Common	All flow are regulated

STATISTICAL PROCESS CONTROL

Statistical process control (SPC) is a method of quality control which uses statistical methods. SPC is applied in order to monitor and control a process. Monitoring and controlling the process ensures that it operates at its full potential. At its full potential, the process can make as much conforming product as possible with a minimum (if not an elimination) of waste (rework or scrap). SPC can be applied to any process where the "conforming product" (product meeting specifications) output can be measured. Key tools used in SPC includecontrol charts; a focus on continuous improvement; and the design of experiments. An example of a process where SPC is applied is manufacturing lines.

Overview

Objective Analysis of Variation

SPC must be practiced in 2 phases: The first phase is the initial establishment of the process, and the second phase is the regular production use of the process. In the second phase, a decision of the period to be examined must be made, depending upon the change in 4 - M conditions (Man, Machine, Material, Method) and wear rate of parts used in the manufacturing process (machine parts, jigs, and fixture).

Emphasis on Early Detection

An advantage of SPC over other methods of quality control, such as "inspection", is that it emphasizes early detection and prevention of problems, rather than the correction of problems after they have occurred.

Increasing Rate of Production

In addition to reducing waste, SPC can lead to a reduction in the time required to produce the product. SPC makes it less likely the finished product will need to be reworked.

Limitations

SPC is applied to reduce or eliminate process waste. This, in turn, eliminates the need for the process step of post-manufacture inspection. The success of SPC relies not only on the skill with which it is applied, but also on how suitable or amenable the process is to SPC. In some cases, it may be difficult to judge when the application of SPC is appropriate.

History

SPC was pioneered by Walter A. Shewhart at Bell Laboratories in the early 1920s. Shewhart developed the control chart in 1924 and the concept of a state of statistical control. Statistical control is equivalent to the concept of exchangeability developed by logician William Ernest Johnson also in 1924 in his book *Logic, Part III: The Logical Foundations of Science*. Along with a gifted team at AT&T that included Harold Dodge and Harry Romig he worked to put sampling inspection on a rational statistical basis as well. Shewhart consulted with Colonel Leslie E. Simon in the application of control charts to munitions manufacture at the Army's Picatinny Arsenal in 1934. That successful application helped convince Army Ordnance to engage AT&T's George Edwards to consult on the use of statistical quality control among its divisions and contractors at the outbreak of World War II.

W. Edwards Deming invited Shewhart to speak at the Graduate School of the U.S. Department of Agriculture, and served as the editor of Shewhart's book *Statistical Method from the Viewpoint of Quality Control* (1939) which was the result of that lecture. Deming was an important architect of the quality control short courses that trained American industry in the new techniques during WWII. The graduates of these wartime courses formed a new professional society in 1945, the American Society for Quality Control, which elected Edwards as its first president. Deming traveled to Japan during the Allied Occupation and met with the Union of Japanese Scientists and Engineers (JUSE) in an effort to introduce SPC methods to Japanese industry.

"Common" and "Special" Sources of Variation

Shewhart read the new statistical theories coming out of Britain, especially the work of William Sealy Gosset, Karl Pearson, and Ronald Fisher. However, he understood that data from physical processes seldom produced a "normal distribution curve"; that is, a Gaussian distribution or "bell curve". He discovered that data from measurements of variation in manufacturing did not always behave the way as data from measurements of natural phenomena (for example, Brownian motion of particles). Shewhart concluded that while every process displays variation, some processes display variation that is natural to the process ("common" sources of variation)- these processes were described as in (statistical) control. Other processes additionally display variation that is not present in the causal system of the process at all times ("special" sources of variation), and these were described as 'not in control'.

Application to Non-manufacturing Processes

In 1988, the Software Engineering Institute suggested that SPC could be applied to non-manufacturing processes, such as software engineering processes, in the Capability Maturity Model (CMM). The Level 4 and Level 5 practices of the Capability Maturity Model Integration (CMMI) use this concept.

The notion that SPC is a useful tool when applied to non-repetitive, knowledge-intensive processes such as research and development or systems engineering has encountered skepticism and remains controversial.

Variation in Manufacturing

In manufacturing, quality is defined as conformance to specification. However, no two products or characteristics are ever exactly the same, because any process contains many sources of variability. In mass-manufacturing, traditionally, the quality of a finished chapter is ensured by post-manufacturing inspection of the product. Each chapter (or a sample of articles from a production lot) may be accepted or rejected according to how well it meets its design specifications. In contrast, SPC uses statistical tools to observe the performance of the production process in order to detect significant variations before they result in the production of a sub-standard chapter. Any source of variation at any point of time in a process will fall into one of two classes.

1. "Common Causes" - sometimes referred to as nonassignable, normal sources of variation. It refers to many sources of variation that consistently acts on process. These types of causes produce a stable and repeatable distribution over time.'

2. "Special Causes" - sometimes referred to as assignable sources of variation. It refers to any factor causing variation that affects only some of the process output. They are often intermittent and unpredictable.

Most processes have many sources of variation; most of them are minor and may be ignored. If the dominant sources of variation are identified, however, resources for change can be focused on them. If the dominant assignable sources of variation are detected, potentially they can be identified and removed. Once removed, the process is said to be "stable". When a process is stable, its variation should remain within a known set of limits.

That is, at least, until another assignable source of variation occurs. For example, a breakfast cereal packaging line may be designed to fill each cereal box with 500 grams of cereal. Some boxes will have slightly more than 500 grams, and some will have slightly less. When the package weights are measured, the data will demonstrate a distribution of net weights. If the production process, its inputs, or its environment (for example, the machines on the line) change, the distribution of the data will change.

For example, as the cams and pulleys of the machinery wear, the cereal filling machine may put more than the specified amount of cereal into each box. Although this might benefit the customer, from the manufacturer's point of view, this is wasteful and increases the cost of production. If the manufacturer finds the change and its source in a timely manner, the change can be corrected (for example, the cams and pulleys replaced).

Application of SPC

The application of SPC involves three main phases of activity:

- Understanding the process and the specification limits.
- Eliminating assignable (special) sources of variation, so that the process is stable.
- Monitoring the ongoing production process, assisted by the use of control charts, to detect significant changes of mean or variation.

Control Charts

The data from measurements of variations at points on the process map is monitored using control charts. Control charts attempt to differentiate "assignable" ("special") sources of variation from "common" sources. "Common" sources, because they are an expected part of the process, are of much less concern to the manufacturer than "assignable" sources. Using control charts is a continuous activity, ongoing over time.

Stable Process

When the process does not trigger any of the control chart "detection rules" for the control chart, it is said to be "stable". A process capability analysis may be performed on a stable process to predict the ability of the process to produce "conforming product" in the future.

Excessive Variation

When the process triggers any of the control chart "detection rules", (or alternatively, the process capability is low), other activities may be performed to identify the source of the excessive variation. The tools used in these extra activities include: Ishikawa diagrams, designed experiments, and Pareto charts. Designed experiments are a means of objectively quantifying the relative importance (strength) of sources of variation. Once the sources of variation have been quantified, actions may be taken to reduce or eliminate them. Methods of eliminating a source of variation might include: development of standards; staff training; error-proofing and changes to the process itself or its inputs.

Mathematics of Control Charts

Digital control charts use logic based rules that determine "derived values" which signal the need for correction. For example,

$$\text{derived value} = \text{last value} + \text{average absolute difference between the last N numbers.}$$

USING GLOBAL PROCESS CONTROL IN SEMICONDUCTOR FABS TO ACHIEVE APC

As the semiconductor industry experiences rapid technological changes, new products and processes are continuously developed while technology becomes ever more complicated and precise. These changes give rise to the need for increasing numbers of manufacturing steps. In addition, with the introduction of 300-mm production, human intervention alone can no longer control semiconductor processes. Consequently, automation has become widespread in IC fabs, expanding new technology horizons.

Nevertheless, it is still necessary to reduce manufacturing costs and increase return on investment. Among other things, the semiconductor industry must decrease the use of nonproduct wafers, increase overall equipment effectiveness, reduce defects, improve product quality and yields, meet specifications for reduced feature sizes, improve production efficiency, and achieve faster process development and yield learning.

Achieving these and other objectives requires the implementation of process and equipment modeling, monitoring, and control, which are used widely in many industries and are beginning to take hold in the semiconductor industry, where such methods are commonly called advanced process control (APC) and advanced equipment control.

As device features shrink to 0.15 μm and below and 300-mm fabs proliferate, APC systems are becoming increasingly important. From the outset of 300-mm manufacturing, productivity improvements and fab effectiveness have been key considerations. To ensure the profitability of these new fabs, productivity gains must reach 30–35%.

Diagnosis, monitoring, and APC are needed to predict when tools require maintenance and to avoid unscheduled downtime. These control strategies also should reduce the mean time to repair. APC enables tools to perform with greater consistency, repeatability, and precision, leading fabs to increase production efficiency and reduce their total cost of tool ownership.

Even 200-mm fabs are discovering that the implementation of such control strategies is necessary to extend the lifetime of their process tools. In 200-mm fabs, implementing APC allows manufacturers to integrate new components such as sensors so that they can ensure that their tools keep pace with new technologies.

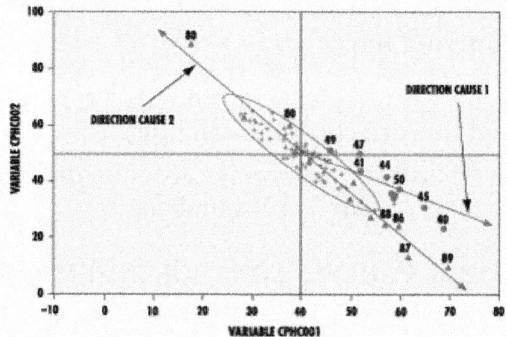

Fig. : A two-dimensional T2 control chart showing error signatures for two detected errors with two variables, one of which is a function of the other.

Given the complexity of semiconductor processes and the need to adapt rapidly to technological changes, it is critical to choose optimal process and equipment modeling, monitoring, and control methodologies.

Comparing Process Control Methods

There are two major types of APC systems: statistical process control (SPC) and model-based process control (MBPC). SPC can be broken down into univariate and multivariate fault detection methods.

Fault Detection through Univariate SPC.

Univariate SPC systems are based on the idea that variations of a controlled variable (variations that affect a process all the time and are essentially unavoidable within that process) have a common cause. SPC allows the normal (common-cause) variation of a controlled variable within acceptable limits and detects any assignable cause of abnormal variation of the controlled variable as soon as possible. The technique is characterized by the creation of control charts with a target value and upper and lower control limits that plot individual monitored variables.

Because univariate SPC leads to the creation of as many control charts as monitored variables, it is of no value at the fab level, where a huge number of variables must be monitored, and of little value at the tool level because an engineer can monitor a maximum of only four charts at the same time. Moreover, of all control methodologies, univariate SPC results in the most undetected errors and false alarms, leading to wasted time. Also, the monitored variables must theoretically be independent, which is usually not the case in the semiconductor industry.

Fault Detection through Multivariate SPC

Multivariate SPC is superior to univariate SPC. Rather than create separate control charts for each variable, multivariate SPC uses global control charts that represent as many variables as needed. Moreover, multivariate SPC is much more precise than univariate SPC in detecting errors and avoiding false alarms. Finally, multivariate SPC takes into account the dependencies among monitored variables.

Multivariate methodologies have their limitations, however. While they generate almost no false alarms or undetected errors, they cannot indicate the origin of errors and are thus unable to automatically manage error occurrences. Detecting error origins with these methodologies can be so complex that engineers either require a very high level of expertise and intuition to use them or cannot use them at all.

Model-Based Process Control

MBPC directly links product variables and process variables to models. This control method tries to determine the variables of a process from the characteristics of a desired output. Inversely, it also tries to predict product output from the process variables. These two applications are the basis for feed-forward, feed-backward, and real-time control. The power of the models is their ability to compare desired and real output variables.

Building models for MBPC requires a good knowledge of equipment and processes. While performing MBPC on one type of equipment may provide good results, doing so on another may provide unsatisfactory results. In all cases, performing MBPC requires that customers learn the method or that suppliers learn the processes, resulting in slow, complicated, and relatively inflexible implementation. But the main limitation of MBPC is theoretical: it acts on measured variables to obtain a desired output but gives no indication about the underlying causes of observed deviations. Finally, MBPC corrects the symptoms of errors but not the errors themselves, resulting in fewer out-of-control outputs but more-significant problems.

Global Process Control

An optimal control methodology must be able to perform automatic fault identification and single out anomalies when detected errors do not fit existing classifications. Few available tools indicate error types, let alone errors themselves, and those that do require that engineers analyze monitored profiles when faults are detected. That procedure can lead to misjudgments because profiles are not always obvious. Moreover, engineer intervention is time-consuming and ineffective when new faults appear.

To identify faults, some methods associate a particular signature with an out-of-control event. Each out-of-control event must be compared with a set of signatures to identify its origin. Considering the great number of candidate signatures, it is difficult to apply this method in real time. Engineer subjectivity can also be a source of errors, because the relationship between the signature and a multidimensional observation often is not clear.

Because no single control solution provides adequate results, a combination of different solutions is needed. Real-time control (starting with run-to-run and wafer-to-wafer controls) requires automatic forward or backward error correction. But an automatic correction system presupposes the existence of automatic fault detection and classification. Fault detection must be able to function as a plug-in module that fits the central APC system and is easy to reuse. A modeling application necessarily has limited relevance unless an efficient fault identification system is established. On the other hand, only MBPC can efficiently automate the correction of anomalies.

Global process control was designed to carry out these multiple tasks. A subtle combination of multivariate SPC, MBPC, and mathematical procedures, GPC combines the advantages of all these methodologies while avoiding their inherent drawbacks.

The use of multivariate control charts for fault detection is essential for extracting relevant information from the large amounts of data generated by equipment and processes. This method reduces the frequency of irrelevant alerts and detects faults that are invisible on common SPC charts. Moreover, multivariate methods solve the problem of fault identification. Because the origin of a fault is identified in real time, fab personnel can take corrective action quickly and limit rejected product.

GPC is performed by following a sequence of steps. First, a GPC error detection chart is created on the basis of Hotelling's T2 multivariate SPC methodology. This chart enables engineers to monitor all error occurrences in a single chart and takes new measurements into account.

Next, the GPC methodology adds signatures to already known errors. These signatures are linear tracks on the T2 chart. The establishment of signatures makes it possible to determine whether an error fits an already known error type or whether it is a new type. The error signatures in a two-dimensional T2 control chart for two detected errors. The x- and y-axes depict two different variables, both of which have absolute values. The red circles represent one type of error, the red triangles the other; the blue crosses correspond to controlled measurements. The ellipse is the graphical representation of the T2 control limit. The numbered plots represent products.

To show the intensity and the variability of an error, the next GPC step is to create a specific error control chart for each error signature, which enables engineers to monitor error types that have already been encountered.

GPC then uses a factorial methodology to determine the correlation of each variable to these signatures (or the impact of these variables on the signatures). This step has two major advantages: it automatically determines the origin or cause of errors and also provides a model to determine the impact of each variable on the errors, enabling engineers to determine which parameters have to be modified to maintain a controlled process.

In the event of a new error type, the linear track or signature of the newly encountered error is detected automatically. As an associated specific error control chart is then created for this new error signature. The demonstrates that new measurements will either be in control (inside the ellipse) or out of control (outside the ellipse) and that the out-of-control measurements will appear either on the existing linear tracks or on other linear tracks. Each linear track corresponds to an error type.

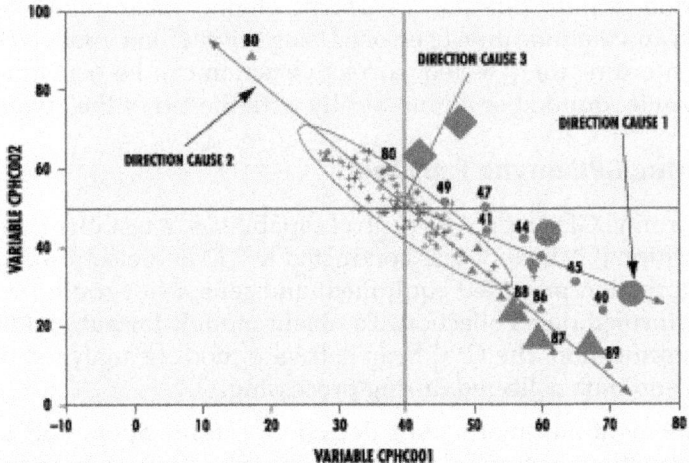

Fig. : Two-dimensional chart extending the data. The new (larger) measurements can either be in control (inside the ellipse) or out of control (outside the ellipse). Out-of-control measurements appear on the existing linear tracks and on other linear tracks; each linear track corresponds to an error type.

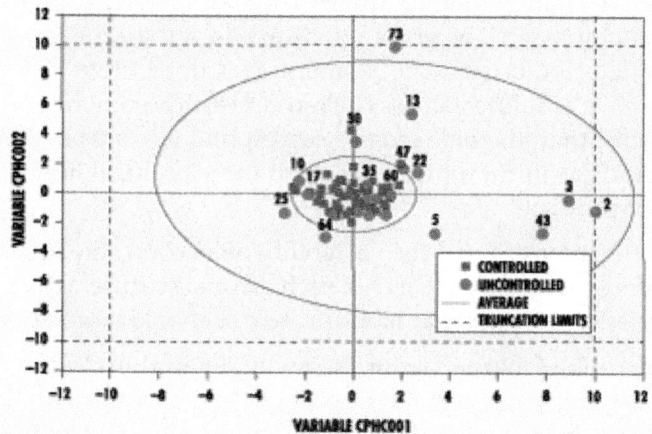

Fig.: Control chart for two different variables (on the x-axis and y-axis, respectively) indicate whether a measurement should be viewed as in control or out of control in the T2 calculation; 95% of controlled variables fall within the small ellipse and 95% of uncontrolled variables fall within the large one.

GPC's probabilistic approach to factorial analysis can discriminate between variables by identifying whether a variation associated with a given measurement is exceptional or ordinary. GPC also provides important information on which variables are anomalous. A control chart for two different variables (on the x-axis and y-axis, respectively) highlights specific problems. These data enable engineers to determine whether a measurement should be viewed as out of control in the T2 calculation.

By carrying out the different GPC steps, engineers can completely automate the detection and identification of errors. Using this method, models can be created so that when errors are detected, corrective action can be performed manually after an alarm is sounded or automatically with the aid of the obtained models.

Implementing GPC on the Fab Floor

To determine GPC's process control capabilities, a test involving a chemical vapor deposition (CVD) tool was conducted at STMicroelectronics. A SilverBox from Si Automation provided equipment and sensor integration with the CVD tool and performed data collection. To obtain models for automatic fault detection and identification, the GPC Scan software module analyzed data used for automation and data collected during processing.

To implement automatic fault detection, classification, and identification (FDCI), information was first collected by the SilverBox through its equipment, peripheral, and sensor interfaces, as well as from its automation part. The SilverBox then prepared this information for the GPC methodology by transforming it into homogeneous data following Gaussian laws.

Automatic FDCI was obtained from this rescaled data by using the Silver-Box's GPC Guard. The GPC Guard used the results obtained by the GPC Scan and was responsible for fault detection by calculating the T2, for fault identification by classifying the error signatures, and for launching corrective action. The GPC Scan also prepared models for the GPC Guard. On historical data, it carried out fault detection by calculating the T2, fault classification by identifying the error signatures, and fault identification by iterative factorial analysis.

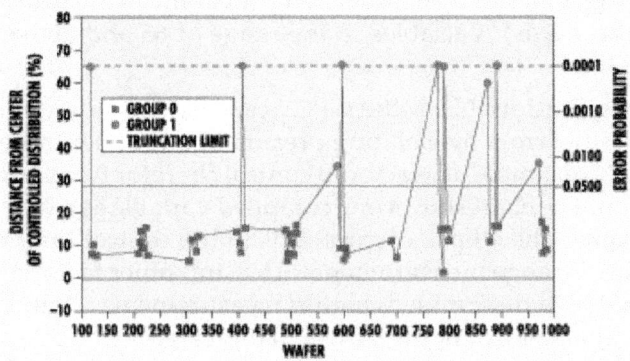

Fig. : Global control chart for error detection enabled engineers to monitor errors globally while minimizing undetected errors and false alerts.

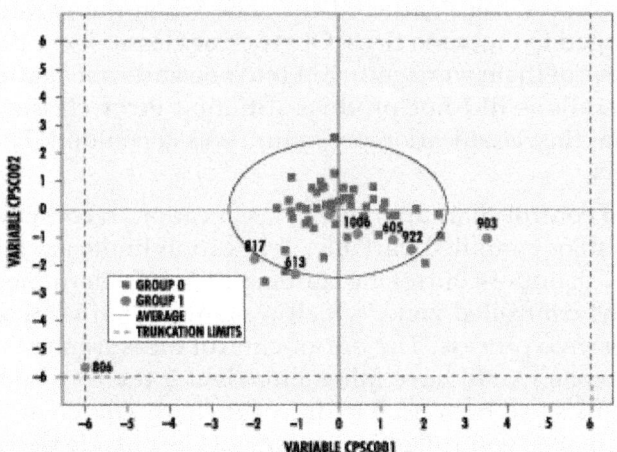

Fig. : Factorial control chart of two variables before error classification. This chart enabled engineers to determine whether measurements should be considered in control or out of control in the T2 calculation step. The ellipse corresponds to the control limit and truncation limits show out-of-range measurements.

Initial Univariate Analysis. The test was based on the GPC analysis of 11 variables collected during the process. Initially, 11 univariate SPC control charts of the monitored variables were prepared in which the variables were scaled on normal law for easier interpretation. Each chart displayed the absolute values of a given variable over a series of wafers and lots. The center of each graph showed the average of all the measurements for the given variable.

Fault Detection. A T2 control chart was then built for the 11 variables, which enabled the engineers to monitor errors globally while minimizing undetected errors and false alerts. The global control chart for error detection, displays the wafer numbers on the x-axis and the distance of the measurements from the center of the controlled distribution on the left-hand side of the y-axis. The control limit was set up to 5% (associated probabilities are shown on the right-hand side of the y-axis). Variables in the range of 65 and above are shown on the truncation line.

Fault Classification. Next, the GPC Scan performed an iterative factorial analysis to classify errors by isolating presumed out-of-control observations at each iteration. An example of a factorial control chart for two variables before error classification. The chart shows one compiled variable as a function of another compiled variable. The ellipse corresponds to the control limit and truncation limits show out-of-range measurements. This and other factorial control charts enabled engineers to determine whether measurements should be considered in control or out of control in the T2 calculation step.

The iterative factorial analysis step led to the error classification step. The iterative process allowed engineers to classify errors to obtain a stable classification. At first, each iteration resulted in a new error class. But during the fifth iteration, two classes were gathered into one when the iteration determined that they belonged to the same class. Other error classes were then determined, after which three of them were gathered into one and split again into two. Since subsequent iterations did not produce different error classes, the engineers determined that the classification procedure was concluded. The result was six classes of errors.

A factorial control chart after error classification. A compiled variable as a function of another compiled variable. Truncation limits are also shown. This time, however, all out-of-control measurements were determined. Most of them lay far from the controlled zone, which was much thinner than at the beginning of the iterative process. The out-of-control measurements that lay inside the controlled zone would have fallen outside it if the corresponding relevant compiled variables had been used to draw the chart. This chart is important because it shows that no controlled measurements lie outside the controlled zone.

The error classification process led to the creation of specific-error control charts for each type of error. These charts were used to focus on a given anomaly. From the useless initial univariate control charts for individual variables, GPC Scan built control charts by error type. Apart from focusing on a given anomaly, these control charts differentiated between errors of the same type according to their intensity and variability.

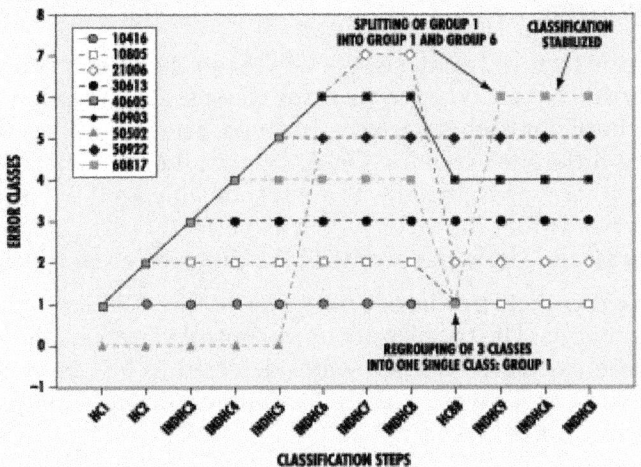

Fig. : Chart showing the classification step, which allowed engineers to obtain a stable classification.

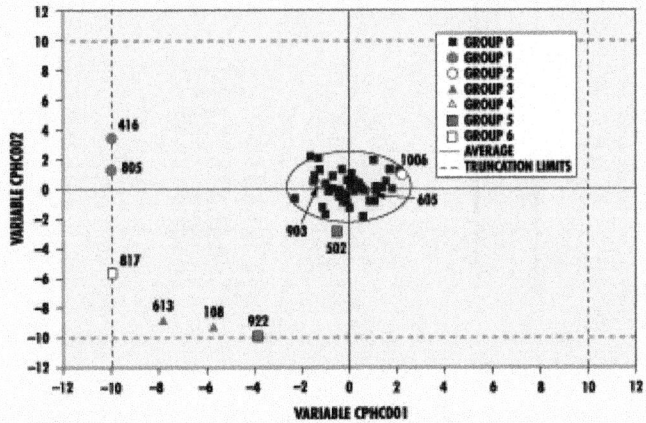

Fig. : Factorial control chart after error classification in which all out-of-control measurements, most of which lay far from the controlled zone, were determined. The chart shows error class INDHCA.

Specific-error control charts for the first three detected error types in this test. These charts show the distance of the measurements from the center of the controlled distribution. In each chart, the measurements correspond to the linear track of the specific error type; the results represent the distance of the measurements from the center of the distribution of the linear track in question. In each chart, blue crosses correspond to controlled measurements, red disks to the concerned error, and red circles to other error types. The x-axis shows the wafer number while the left-hand y-axis shows the absolute value of the distance from the center and the right-hand the associated probabilities of errors.

Fault Identification

Based on the factorial analysis, the GPC Scan then determined the origin of the detected errors for each of the six error classes. That step led to the creation of sensitivity charts for each detected error type. The sensitivity of each detected anomaly to the different variables. On the y-axis, 0 means that the variable had no impact—in other words, that it was not responsible for the type of error shown in the chart; 1 means that the error type was perfectly correlated to the variable. These charts typically show the profile of the different error types.

Now that a complete profile of errors existed, the GPC software was ready to interpret error origins. The profiles of out-of-control variables, shows corresponding values of the variables for each wafer detected to be out of control. Wafers with the same error types were gathered together. Variables that fall between the red lines were in control.

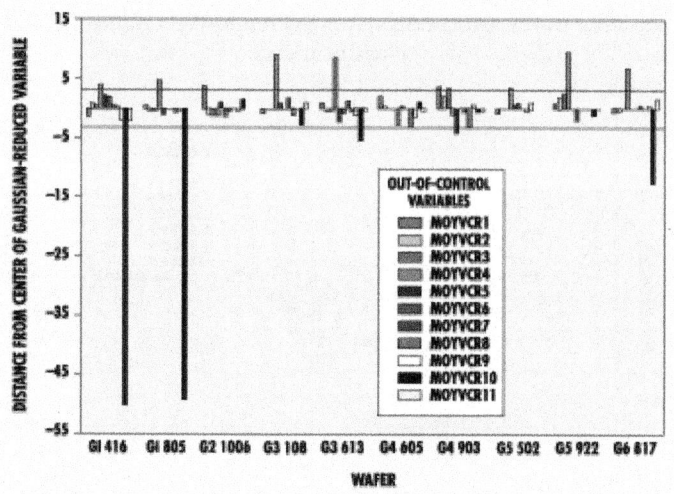

Fig. : Profiles of out-of-control variables showing corresponding values of the variables for each wafer detected to be out of control. Variables that fall between the red lines were in control. (MOYVCR = centered and reduced mean of variable.)

Decision Matrix Construction

In addition to factorial analyses, the sensitivity charts made it possible to create process control models, or decision matrices, which associated actions—whether alarms or automatic corrections—with identified errors. Each type of error had a signature, which was its linear track on the T2 graph and which corresponded to its sensibility chart. A decision matrix associates out-of-control action plans with each signature. Depending on the value of the measurements, these plans could be anything from alarms to no-action commands to a complete set of automated actions. Decision matrices allow engineers to control or correct errors automatically.

Applying GPC to Real-Time Control

After the GPC procedure, from the creation of the global T2 control chart to the construction of the decision matrix, had been performed off-line by the GPC Scan, it could then be applied to process tool control in real time. The GPC Guard software module uses the obtained process control models and decision matrices for run-to-run or wafer-to-wafer control. This reaction delay was chosen for the first versions of GPC Guard, even though GPC can theoretically be applied in real time. But compared with the huge delays for obtaining results from other available methodologies (days sometimes), run-to-run and wafer-to-wafer control is very close to real-time control.

Like any APC solution, GPC confronts the problem of implementation in the real world. Although many standards established by SEMI for equipment manufacturers and by Sematech for IC manufacturers regulate fab life, many fab areas remain beyond the scope of these standards, especially those that connect the SEMI world to the Sematech world. Consequently, the semiconductor industry has experienced several APC solutions, none of which has been sufficiently generic to offer an efficient solution for a broad range of applications.

Problems associated with implementing APC are manifold. How should relevant data be acquired and different data matched? On what platform should data acquisition and matching take place and how should the results be used? What type of control system has the least impact on manufacturing and product yields?

The GPC approach can offer a solution to many APC-related problems. While the test was performed in a pilot project for a CVD application, the results demonstrated that GPC could identify all out-of-control variables as the causes of equipment or process deviations. The use of the GPC classification tool permitted engineers to focus on relevant problems and ignore the signatures of second-order ones. For example, the signature of a first-wafer effect in dry etch, which was detected through GPC, could be automatically ignored in-line.

The flexible configuration possibilities of the data treatment phase of the GPC test adapted well to the customer's needs. Although the test was performed on dry etch equipment, the customer could quickly configure the GPC software to suit other types of tools.

This rapid reconfiguration capacity also allowed the customer to adapt GPC to the changing needs of the dry etch equipment. GPC was first applied to the critical phase of the dry etch process. After it proved its ability to detect and analyze all types of known errors in that phase, automatic corrective actions were integrated into the model, decreasing the need for human intervention. GPC also detected new error types, leading to its extension to all other process phases. For example, GPC detected and automatically corrected an error in the precritical phase caused by a problem in loading a particular gas. An error caused by an unloading problem in the last phase of the process also was detected and automatically corrected.

Although GPC allows real-time fault identification in an industrial context, automatic fault detection is not enough in the semiconductor industry; corrective action also must be taken. After identifying faults, GPC can automatically correct them. Actions taken to correct one fault can be used as a model to correct other faults of the same type. Most importantly, the GPC approach leads to the creation of corrective models based on the results of fault identification.

SAP GRC PROCESS CONTROL

SAP GRC Process Control (Process Control) is a solution for internal controls management that enables members of audit and internal controls teams to gain better visibility into key business processes and ensure a high level of reliability in financial statement reporting.

Process Control uses a controls-based approach to managing risk associated with business processes and to comply with Sec.404 of the Sarbanes-Oxley Act of 2002. It provides the necessary capabilities to fully document the control environment, evaluate the controls, certify the state of controls, and report and analyze control information. The solution extends value to key user segments such as audit managers, compliance managers and business process owners.

The four major capabilities of the solution are:

1. Control documentation
2. Control evaluation
3. Certification
4. Reporting and analysis

Control Documentation

Process Control supports a flexible environment for setting up corporate master data, which includes the following categories:

- **Organization structure**: A multi-level hierarchy showing all organizations within the scope of internal controls management.
- **Process catalog**: A catalog of key business processes, their underlying subprocesses and associated controls, created in hierarchical form.
- **Account groups**: A list of the key financial account groups and their related financial assertions.
- **Control objectives and risks**: A comprehensive, hierarchical catalog of control objectives and associated risks relevant to compliance.
- **Entity-level controls**: A catalog of central management controls, which may be classified by groups conforming to the five COSO components.

The elements in the central master data catalogs are mapped so that each subprocess is linked to one or more account groups, control objectives and risks. The subprocess, in turn, can be mapped to one or more organizations, making it local to that organization. At the same time, the parent process and child controls of the subprocess are also assigned to that organization. The controls can either be locally copied to the organization, which allows local maintenance, or referenced from the central catalog, which provides centralized maintenance. The end result is a fully documented master data repository comprising inter-related central and local objects, which provides the context for evaluation and reporting.

Control Evaluation

In Process Control, there are four main types of evaluation activity:

- **Assessments**: Process Control provides survey functionality, which enables users to assess the design of subprocesses and controls or perform self-assessments on a periodic basis. If the assessment results in a deficiency, the solution supports issue creation, remediation activities and subsequent reassessment.

- **Effectiveness Testing**: Process Control provides two methods for testing the effectiveness of controls – manual testing using test plans, and automated testing using system rules and programs.

- In the manual testing method, auditors or other testers use test plans from a repository to manually perform periodic tests of controls and follow-up remediation actions if required.

- In the automated testing method, controls are automatically tested on a predetermined schedule using system-driven rules.

- **Automated control monitoring**: Process Control also supports monitoring controls, which allow the tracking of noncompliance business events in an ERP system through a combination of automated programs and exception reports. The system also supports the ability to run ad hoc data queries for quick data retrieval, and to convert them into new controls.

- **Planning and Scheduling**: Central to the evaluation activities is the ability to create an evaluation plan, which includes specification of the test time frame, organizations involved, and type of evaluation.

- In Process Control, all evaluation activities are governed by a set of robust workflows and notifications. Based on evaluation schedules, users are automatically notified through workflow tasks which cover test performance, issue management, remediation and retesting.

Certification

To support regulatory requirements, Process Control provides the process of sign-off, which is the formal process of attestation by organization owners and officers of the company on the state of internal controls. Certification is achieved using a combination of sign-off surveys and workflows, and provides historic period reporting for audit requirements.

Reporting and Analysis

Process Control provides detailed and flexible reporting features that enable users to gain full visibility into the control environment and make data-driven decisions. It supports a catalog of 30 standard reports, which cover areas such as master data setup, evaluation activities, audit logs and authorization analysis. All reports come with selection criteria, export and print functionality, and drilldown capability for ease of use.

Process Control also provides Business Intelligence (BI) reporting to enable specialist users to view aggregate metrics, trends and patterns, and to provide decision support.

Process Control – An End-to-End Internal Controls Solution

With its combination of time-dependent data setup, robust workflows, detailed object-level security and analytic reports, Process Control meets the complex requirements of internal controls teams. Its configurability enables you to adapt easily to changes in your business and realize faster time-to-value. You can also improved visibility into compliance processes, reduce the overall cost of compliance, and gain a high degree of confidence in the quality of financial reporting.

Chapter 2

MODERN CONTROL IS BASED ON PROCESS DYNAMIC BEHAVIOR

MOTIVATION AND TERMINOLOGY OF AUTOMATIC PROCESS CONTROL

Automatic control systems enable us to operate our processes in a safe and profitable manner. Consider, as on this site, processes with streams comprised of gases, liquids, powders, slurries and melts. Control systems achieve this "safe and profitable" objective by continually measuring *process variables* such as temperature, pressure, level, flow and concentration – and taking actions such as opening valves, slowing down pumps and turning up heaters – all so that the measured process variables are maintained at operator specified set point values.

Safety First

The overriding motivation for automatic control is safety, which encompasses the safety of people, the environment and equipment.

The safety of plant personnel and people in the community are the highest priority in any plant operation. The design of a process and associated control system must always make human safety the primary objective.

The tradeoff between safety of the environment and safety of equipment is considered on a case by case basis. At the extremes, the control system of a multi-billion dollar nuclear power facility will permit the entire plant to become ruined rather than allow significant radiation to be leaked to the environment.

On the other hand, the control system of a coal-fired power plant may permit a large cloud of smoke to be released to the environment rather than allowing damage to occur to, say, a single pump or compressor worth a few thousand dollars.

The Profit Motive

When people, the environment and plant equipment are properly protected, our control objectives can focus on the profit motive. Automatic control systems offer strong benefits in this regard.

Plant-level control objectives motivated by profit include:

1. meeting final product specifications
2. minimizing waste production
3. minimizing environmental impact
4. minimizing energy use
5. maximizing overall production rate

It can be most profitable to operate as close as possible to these minimum or maximum objectives. For example, our customers often set our product specifications, and it is essential that we meet them if failing to do so means losing a sale.

Suppose we are making a film or sheet product. It takes more raw material to make a product thicker than the minimum our customers will accept on delivery. Consequently, the closer we can operate to the minimum permitted thickness constraint without going under, the less material we use and the greater our profit.

Or perhaps we sell a product that tends to be contaminated with an impurity and our customers have set a maximum acceptable value for this contaminant. It takes more processing effort (more money) to remove impurities, so the closer we can operate to the maximum permitted impurity constraint without going over, the greater the profit.

Whether it is a product specification, energy usage, production rate, or other objective, approaching these targets ultimately translates into operating the individual process units within the plant as close as possible to predetermined set point values for temperature, pressure, level, flow, concentration and the other measured process variables.

Controllers Reduce Variability

A poorly controlled process can exhibit large variability in a measured process variable (*e.g.*, temperature, pressure, level, flow, concentration) over time.

Suppose, as in this example, the measured process variable (PV) must not exceed a maximum value. And as is often the case, the closer we can run to this operating constraint, the greater our profit.

To ensure our operating constraint limit is not exceeded, the operator-specified set point (SP), that is, the point where we want the control system to maintain our PV, must be set far from the constraint to ensure it is never violated. Note in the plot that SP is set at 50% when our PV is poorly controlled.

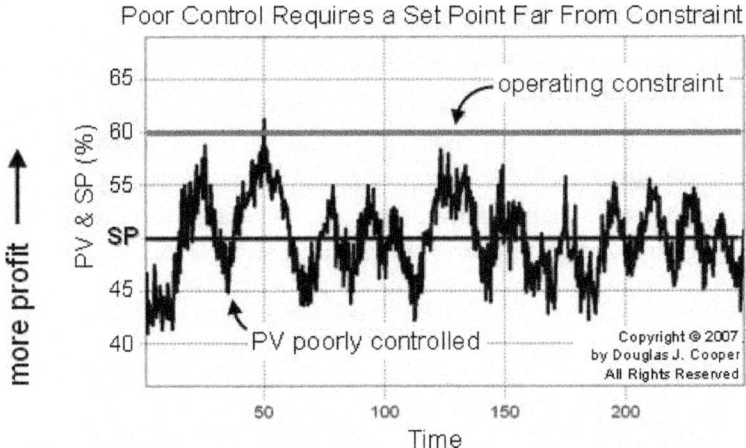

The same process with improved control. There is significantly less variability in the measured PV, and as a result, the SP can be moved closer to the operating constraint.

With the SP in the plot below moved to 55%, the average PV is maintained closer to the specification limit while still remaining below the maximum allowed value. The result is increased profitability of our operation.

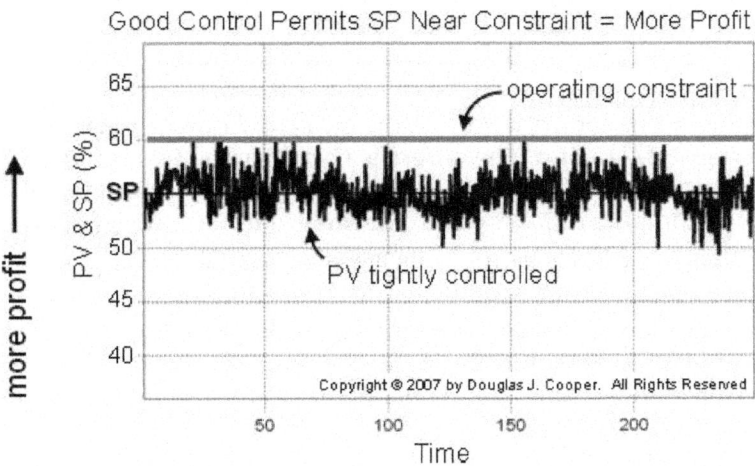

Terminology of Control

This is a simplistic example because a home furnace is either on or off. Most control challenges have a *final control element* (FCE), such as a valve, pump or compressor, that can receive and respond to a complete range of *controller output* (CO) signals between full on and full off. This would include, for example, a valve that can be open 37% or a pump that can be running at 73%.

For our home heating process, the *control objective* is to keep the *measured process variable* (PV) at the *set point value* (SP) in spite of *unmeasured disturbances* (D).

For our home heating system:

PV = process variable is house temperature

CO = controller output signal from thermostat to furnace valve

SP = set point is the desired temperature set on the thermostat by the home owner

D = heat loss disturbances from doors, walls and windows; changing outdoor temperature; sunrise and sunset; rain.

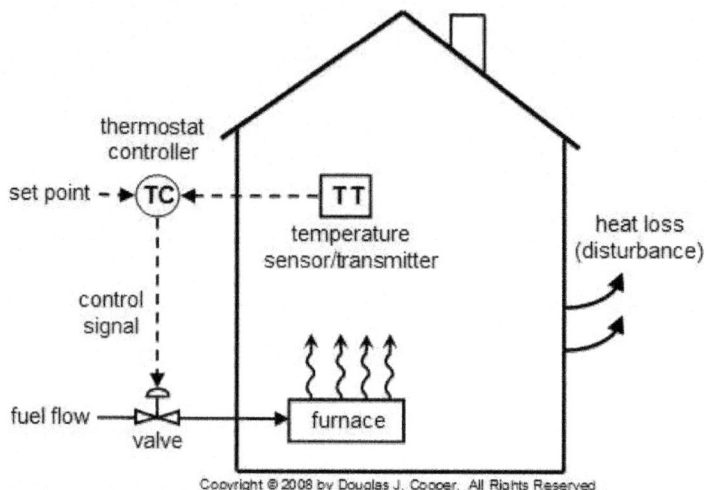

To achieve this control objective, the measured process variable is compared to the thermostat set point. The difference between the two is the *controller error*, which is used in a *control algorithm* such as a PID (proportional-integral-derivative) controller to compute a CO signal to the final control element (FCE).

The change in the controller output (CO) signal causes a response in the final control element (fuel flow valve), which subsequently causes a change in the *manipulated process variable* (flow of fuel to the furnace). If the manipulated process variable is moved in the right direction and by the right amount, the measured process variable will be maintained at set point, thus satisfying the control objective.

This example, like all in process control, involves a measurement, computation and action:

- is the measured temp colder than set point (SP – PV > 0)? Then open the valve.

- is the measured temp hotter than set point (SP – PV < 0)? Then close the valve.

Note that computing the necessary controller action is based on *controller error*, or the difference between the set point and the measured process variable, *i.e.*

$$e(t) = SP - PV \text{ (error = set point – measured process variable)}$$

In a home heating process, control is an on/off or open/close decision. It is a straightforward decision to make. The price of such simplicity, however, is that the capability to tightly regulate our measured PV is rather limited.

One situation not addressed above is the action to take when

$$PV = SP \ (i.e., \ e(t) = 0).$$

And in industrial practice, we are concerned with *variable position* final control elements, so the challenge elevates to computing:

- the direction to move the valve, pump, compressor, heating element…
- how far to move it at this moment
- how long to wait before moving it again
- whether there should be a delay between measurement and action

This Site

This site offers information and discussion on proven methods and practices for PID (proportional-integral-derivative) control and related architectures such as cascade, feed forward, Smith predictors, multivariable decoupling, and similar traditional and advanced classical strategies.

Applications focus on processes with streams comprised of gases, liquids, powders, slurries and melts. As stated above, final control elements for these applications tend to be valves, variable speed pumps and compressors, and cooling and heating elements.

Industries that operate such processes include chemical, bio-pharma, oil and gas, paints and coatings, food and beverages, cement and coal, polymers and plastics, metals and materials, pulp and paper, personal care products, and more.

AUTOMATIC CONTROL SYSTEM

An automatic control system is a preset closed-loop control system that requires no operatoraction. This assumes the process remains in the normal range for the control system. Anautomatic control system has two process variables associated with it: a controlled variable anda manipulated variable.

A *controlled variable* is the process variable that is maintained at a specified value or within aspecified range. In the previous example, the storage tank level is the controlled variable.

A *manipulated variable* is the process variable that is acted on by the control system to maintain the controlled variable at the specified value or within the specified range. In the previous example, the flow rate of the water supplied to the tank is the manipulated variable.

Functions of Automatic Control

In any automatic control system, the four basic functions that occur are:

- Measurement

- Comparison
- Computation
- Correction

In the water tank level control system in the example above, the level transmitter measures the level within the tank. The level transmitter sends a signal representing the tank level to the level control device, where it is compared to a desired tank level. The level control device then computes how far to open the supply valve to correct any difference between actual and desired tank levels.

Elements of Automatic Control

The three functional elements needed to perform the functions of an automatic control system are:

1. A measurement element
2. An error detection element
3. A final control element.

THE COMPONENTS OF A CONTROL LOOP

Components of a Control Loop

A controller seeks to maintain the measured process variable (PV) at set point (SP) in spite of unmeasured disturbances (D). The major components of a control system include a sensor, a controller and a final control element. To design and implement a controller, we must:

1. have identified a process variable we seek to regulate, be able to measure it (or something directly related to it) with a sensor, and be able to transmit that measurement as an electrical signal back to our controller, and
2. have a final control element (FCE) that can receive the controller output (CO) signal, react in some fashion to impact the process (*e.g.*, a valve moves), and as a result cause the process variable to respond in a consistent and predictable fashion.

Home Temperature Control

The home heating control system described in this chapter can be organized as a traditional control loop block diagram. Block diagrams help us visualize the components of a loop and see how the pieces are connected.

A home heating system is simple on/off control with many of the components contained in a small box mounted on our wall. Nevertheless, we introduce the idea of control loop diagrams by presenting a home heating system in the same way we would a more sophisticated commercial control application.

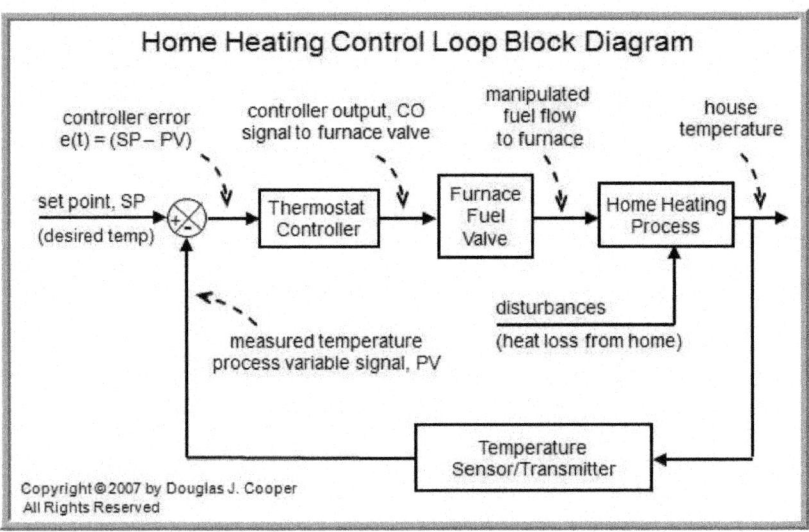

Starting from the far right in the diagram above, our process variable of interest is house temperature. A sensor, such as a thermistor in a modern digital thermostat, measures temperature and transmits a signal to the controller.

The measured temperature PV signal is subtracted from set point to compute controller error,

$$e(t) = SP - PV.$$

The action of the controller is based on this error, e(t).

In our home heating system, the controller output (CO) signal is limited to open/close for the fuel flow solenoid valve (our FCE). So in this example, if

$$e(t) = SP - PV > 0,$$

the controller signals to open the valve. If

$$e(t) = SP - PV < 0,$$

it signals to close the valve. As an aside, note that there also must be a safety interlock to ensure that the furnace burner switches on and off as the fuel flow valve opens and closes.

As the energy output of the furnace rises or falls, the temperature of our house increases or decreases and a feedback loop is complete. The important elements of a home heating control system can be organized like any commercial application:

- Control Objective: *maintain house temperature at SP in spite of disturbances*
- Process Variable: *house temperature*
- Measurement Sensor: *thermistor; or bimetallic strip coil on analog models*
- Measured Process Variable (PV) Signal: *signal transmitted from the thermistor*
- Set Point (SP): *desired house temperature*

- Controller Output (CO): *signal to fuel valve actuator and furnace burner*
- Final Control Element (FCE): *solenoid valve for fuel flow to furnace*
- Manipulated Variable: *fuel flow rate to furnace*
- Disturbances (D): *heat loss from doors, walls and windows; changing outdoor temperature; sunrise and sunset; rain…*

A General Control Loop and Intermediate Value Control

The home heating control loop above can be generalized into a block diagram pertinent to all feedback control loops.

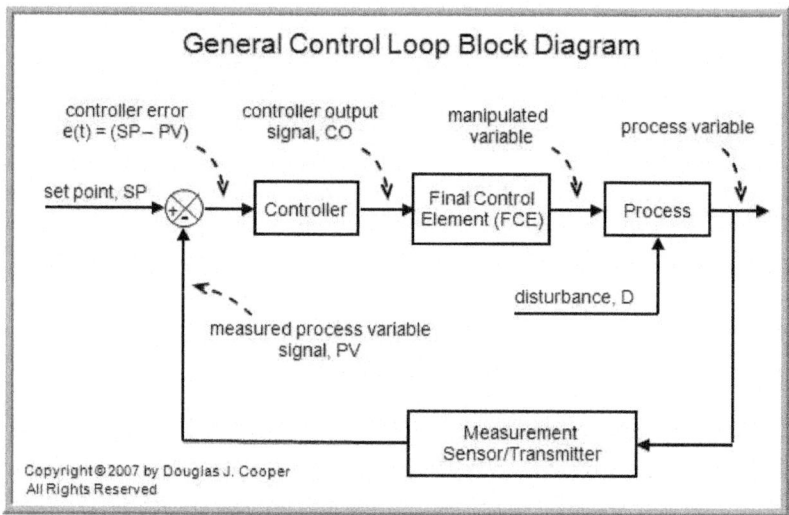

Both diagrams above show a closed loop system based on negative feedback. That is, the controller takes actions that counteract or oppose any drift in the measured PV signal from set point.

While the home heating system is on/off, our focus going forward shifts to intermediate value control loops. An intermediate value controller can generate a full range of CO signals anywhere between full on/off or open/closed. The PI algorithm and PID algorithm are examples of popular intermediate value controllers.

To implement intermediate value control, we require a sensor that can measure a full range of our process variable, and a final control element that can receive and assume a full range of intermediate positions between full on/off or open/ closed. This might include, for example, a process valve, variable speed pump or compressor, or heating or cooling element.

Note from the loop diagram that the process variable becomes our official PV only after it has been measured by a sensor and transmitted as an electrical signal to the controller. In industrial applications. these are most often implemented as 4-20 milliamps signals, though commercial instruments are available that have been calibrated in a host of amperage and voltage units.

With the loop closed as shown in the diagrams, we are said to be in automatic mode and the controller is making all adjustments to the FCE. If we were to open the loop and switch to manual mode, then we would be able to issue CO commands through buttons or a keyboard directly to the FCE. Hence:

- open loop = manual mode
- closed loop = automatic mode

Cruise Control and Measuring Our PV

Cruise control in a car is a reasonably common intermediate value control system. For those who are unfamiliar with cruise control, here is how it works.

We first enable the control system with a button on the car instrument panel. Once on the open road and at our desired cruising speed, we press a second button that switches the controller from manual mode (where car speed is adjusted by our foot) to automatic mode (where car speed is adjusted by the controller).

The speed of the car at the moment we close the loop and switch from manual to automatic becomes the set point. The controller then continually computes and transmits corrective actions to the gas pedal (throttle) to maintain measured speed at set point.

It is often cheaper and easier to measure and control a variable directly related to the process variable of interest. This idea is central to control system design and maintenance. And this is why the loop diagrams above distinguish between our "process variable" and our "measured PV signal."

Cruise control serves to illustrate this idea. Actual car speed is challenging to measure. But transmission rotational speed can be measured reliably and inexpensively. The transmission connects the engine to the wheels, so as it spins faster or slower, the car speed directly increases or decreases.

Thus, we attach a small magnet to the rotating output shaft of the car transmission and a magnetic field detector (loops of wire and a simple circuit) to the body of the car above the magnet. With each rotation, the magnet passes by the detector and the event is registered by the circuitry as a "click." As the drive shaft spins faster or slower, the click rate and car speed increase or decrease proportionally.

So a cruise control system really adjusts fuel flow rate to maintain click rate at the set point value. With this knowledge, we can organize cruise control into the essential design elements:

- Control Objective: *maintain car speed at SP in spite of disturbances*
- Process Variable: *car speed*
- Measurement Sensor: *magnet and coil to clock drive shaft rotation*
- Measured Process Variable (PV) Signal: *"click rate" signal from the magnet and coil*
- Set Point (SP): *desired car speed, recast in the controller as a desired click rate*

- Controller Output (CO): *signal to actuator that adjusts gas pedal (throttle)*
- Final Control Element (FCE): *gas pedal position*
- Manipulated Variable: *fuel flow rate*
- Disturbances (D): *hills, wind, curves, passing trucks…*

The traditional block diagram for cruise control is thus:

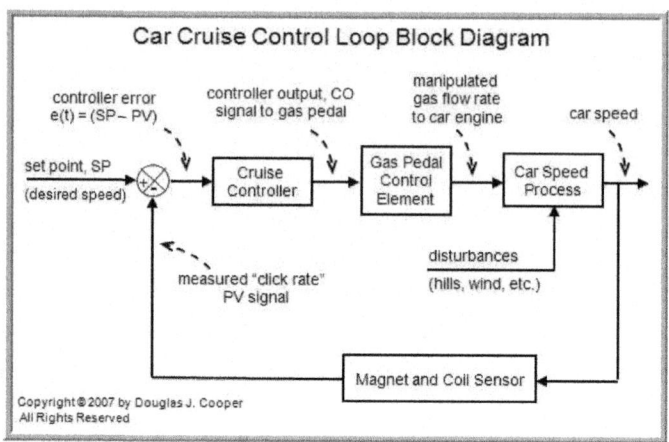

Instruments Should be Fast, Cheap and Easy

The above magnet and coil "click rate = car speed" example introduces the idea that when purchasing an instrument for process control, there are wider considerations that can make a loop faster, easier and cheaper to implement and maintain. Here is a "best practice" checklist to use when considering an instrument purchase:

1. Low cost
2. Easy to install and wire
3. Compatible with existing instrument interface
4. Low maintenance
5. Rugged and robust
6. Reliable and long lasting
7. Sufficiently accurate and precise
8. Fast to respond (small time constant and dead time)
9. Consistent with similar instrumentation already in the plant.

FEEDBACK LOOP CONTROL SYSTEMS

What is a **feedback loop control system** anyway? And **why do we care**? What makes them **unstable**?

It is appropriate to review some general concepts of feedback loop control systems. This page explains what they are and why they are so great, and introduces terminology for subsequent pages on feedback loop stability.

Feedback Loop Control System

There are four elements in any feedback loop control system.

1. **Sensor** of the position to be controlled
2. **Reference input** that specifies the value the controlled variable should have
3. **Comparator** that compares the actual sensed position, or feedback signal, with the desired position or reference signal. The output of the comparator is usually called an error signal, whose polarity determines which way a correction needs to be made.
4. **Control mechanism** which is activated by the error signal and results in a correction of the position. This is often called an actuator.

In our levitator, the sensor is the optical device that measures the position (or lack) of the suspended object. The reference input is establish by another optical device to measure the ambient light. The comparator is an electrical device that subtracts and amplifies the two inputs. The control mechanism is the electromagnetic lifting coil.

Since the four elements just mentioned are all essential to closed loop systems, it follows that any scheme to control something that lacks one or more of these items is *not* a feedback control system. Thus, it is easy to examine many legislative programs and obvious why so many of them fail. It also follows, from looking at things in a general way, that nothing can be controlled by feedback unless it can be measured.

Benefits of Feedback

The desired position of the suspended object is the only intentional input to the system. But several other factors such as weight and gravity, power supplies, and air currents can affect the position. Such inputs, being unwanted, are often called *disturbances*. Since they are subject to nonlinear effects and unknown change with time, they are responsible for the impossibility of merely balancing the coil strength with the weight of the object. The main reason for feedback control is to measure and compensate for the effect of disturbances.

In other types of systems, feedback allows the apparent response speed of a component such as a motor can be increased by overdriving it when rapid response is needed. Still another reason to use feedback is to provide a stiff output, which means an output that is not susceptible to being changed by disturbances.

And in other instances, it is desirable to have the output exactly proportional to the input, but the amplifiers and other components may not be perfectly linear. The use of feedback can greatly reduce nonlinearities in all other system components except the sensor used to provide the feedback signal. Finally, when systems are being mass-produced with inexpensive components that may have a considerable variation in values, feedback can greatly reduce the effect of differences between one unit and another.

Problems of Feedback

If all these benefits sound almost too good to be true, it is time for a reality check. Actually, they are true enough, but there is always the Dark side of the Force. There are two main costs:

1. There is an increase in system complexity, which may increase component count. Sometimes this may be offset by the possibility of using cheaper components.
2. Feedback introduces a stability problem, and this is much more serious. This problem is sufficiently troublesome that 90% of the pages on books about feedback are devoted to it.

By **stability problem** we mean a tendency to over control, or overshoot, when the input or a disturbance is felt. Alternatively, when looking at the frequency response, the gain may rise near the upper end of the pass band, which is usually undesireable. In an extreme case the gain can become high enough to cause oscillation, that is, a sustained cyclic response without any input. This effect generally renders the system useless or even destructive.

Causes of Instability

The stability problem is inevitable. It results from the fact that the feedback, which is connected so as to be negative at low frequencies, usually becomes positive at high frequencies. Good stability is usually possible provided the loop gain is low enough. The main reason the feedback ultimately becomes positive as the frequency increases is that both the control system and the load it is driving contain components that can store energy. *Capacitance* and *inductance* are electrical energy storage elements, and *mass* and *springs* and raising an object against *gravity* are mechanical energy storage elements.

Since the drive to physical devices is not infinite, the response must dwindle toward zero as the frequency approaches infinity, with an associated phase shift approaching 90°. Several phase shifts can add up so that the total around the loop equals 360°, which is positive feedback. Only 180° of additional shift from the energy storage elements is needed to cause positive feedback, since the connection at the comparator introduces 180° to make the feedback negative at low frequencies.

Historical Perspective

Historically, the stability problem was first clearly recognized when centrifugal fly-ball governors were applied to early steam engines shortly after their invention around the middle of the eighteenth century. It was approximately another century before the first mathematical analysis of this problem was carried out by the eminent scientist James

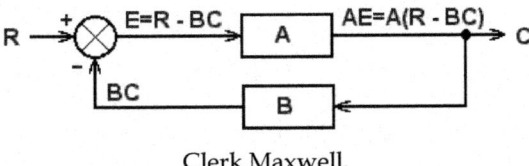

Clerk Maxwell.

It was not until well into the twentieth century that Nyquist, Bode, and many others laid the foundations of modern control theory.

FEEDBACK LOOP EQUATIONS

How does a **feedback loop control system** really work? Since it appears that every signal depends on every other stage in the loop, how do you compute the overall gain?

Loop Equations

This diagram shows the basic model for any feedback control system. It shows the four elements in an abstract manner.

Signal flow is clockwise around the loop. Arrows indicating direction are shown, although they are usually used only at the summing junction, or comparator, which is the circle with the X in it. The inputs (two in this case but there can be any number) have + or - signs to indicate whether each input is added or subtracted. With two inputs and the polarities shown, the summing junction is simply subtracting one signal from the other, in effect performing the comparison that is one of the functions needed for every feedback loop.

The input is labeled **R** for reference which, in this design, is the ambient light measured by a photodetector in units of volts. The output is **C** for controlled variable, which is the position measured in millimeters from the center position of the photodetector. The output of the summing junction is E for error signal. Photodetector **B** converts position into voltage, and the letter B also represents the sensitivity of the detector in units of volts/mm. Block **A** represents all the stages that process the error signal and drive the lifting coil.

From the equations in the diagram above, we eliminate E, since it is an internal parameter. Solving for the overall gain of the system, we get:

$$\frac{C}{R} = \frac{A}{1+AB}$$

$$\frac{C}{R} = \frac{1}{B} \times \frac{1}{\left(1+\dfrac{1}{AB}\right)}$$

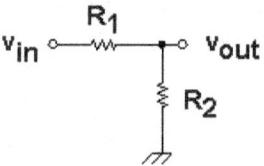

The second form, with $1/B$ factored out, looks a bit more complicated but is actually more convenient for most purposes, since C/R will be almost exactly $1/B$ for all useful feedback loops. The reason is that AB will usually be much greater than one, making the other factor in the second form almost unity.

Loop Gain

The overall gain, **C/R**, is called the **closed loop gain** since it is the gain from input to output with the loop closed and operating. It is the only gain of any final interest. This gain represents how much the input (reference) signal is amplified at the output.

The real loop gain is the product of all the gains around the loop, **AB**, and is referred to as the **open loop gain**. This gain could be measured (theoretically!) by opening the loop anywhere, inserting a small test signal, measuring the signal that appears on the other end of the break, and calculating the ratio.

Despite the simplicity of this equation, it completely describes the behavior of all feedback loop control systems in the world. The **transfer functions** A and B are arbitrary. These two blocks represent all the signal processing in the forward and reverse directions, and may be fantastically complicated. They may (and usually do!) have frequency-dependent elements and even nonlinear parts. This equation covers them all. The designer's big challenge is to characterize their particular circuit design into these two transfer functions.

Transfer Functions

The characteristics of loop components can be described either by mathematical expressions, called **transfer functions**, or by graphs. Transfer function is a fancy name for gain.

In the simplest situation the gain of a network or component is just a number that the input is multiplied by to give the output. For example, a two-resistor voltage divider network. Since a voltage divider attentuates a signal, instead of amplifying it, the gain is less than unity, which means that if it is given in decibels, it is negative. This transfer function is

$$V_{out}/V_{in} = R_2/ (R_1 + R_2).$$

If several components are connected in series, the individual gains are multiplied together to give the overall gain. For example, if two such resistor networks are cascaded together, with buffering to prevent loading effects, the overall gain (attentuation) would be the product of the individual network gains.

For a voltage divider, multiplying the input by the gain will give the output, regardless of the nature of the input. Whether the input is a dc value, sine wave, square wave, or a transient, the output is always the same fraction of the input at every instant. The reason for this simplicity in the case of a voltage divider is that no energy can be stored, so there is no time dependency between the input and output.

When energy storage elements are present, the output at any instant depends on the current value of the input, and also to some degree on previous values. Further, the way previous values affect the output depends on the waveform of the input. For example, the position of the object being lifted depends on its position an instant earlier, along with its previous speed and forces of gravity versus the lifting coil.

FEEDBACK LOOP AND BODE PLOT

How does a **Bode plot** graphically represent a **transfer function**? How does a **common RC network** look in a Bode plot?

Bode Plot

A graphical approach is usually the easiest way to analyze and design feedback loops. So we will review how to represent the transfer function graphically. There are several ways to do so, but the method suggested by H. W. Bode in the 1930s is particularly useful.

Bode's method consists of plotting two curves, the **log of gain**, and phase, as functions of the **log of frequency**.

Usually the gain in decibels, abbreviated dB, and the phase are plotted linearly along the y axis on graph paper that has several cycles of a log scale on the x axis. Each cycle represents a factor of ten in frequency. This special paper is known as semilog graph paper, and it or a computer program with log-log graphing are essential for making Bode plots.

Definition of Decibel

The decibel is a logarithmic measure of a voltage ratio, or gain. It is defined as

$$dB = 20\log (v_{out} / v_{in})$$

Or by the equivalent exponential form as

$$v_{out} / v_{in} = 10^{dB/20}$$

The calculation can be done mentally with the aid of a small table of values. Memorization is practical because reciprocal values of gain convert to the same value of decibel, except for sign.

Voltage ratio	dB
1/100	-40
1/10	-20
1/2	-6
1/SQRT(2)	-3
1	0
SQRT(2)	3
2	6
3.16	10
5	14
10	20
100	40

Cascading Networks

Because of the properties of logarithms, when networks are cascaded so their gains multiply, the overall gain in decibels is obtained by adding the decibels of the networks.

Therefore, three cascaded networks with gains of 2, 2, and 10 would have a total gain of $2 \times 2 \times 10 = 40$, or example, and the gains in decibels would combine as $6 + 6 + 20 = 32dB$. The same idea makes it easy to convert to and from decibels by breaking down a total into its components.

Another example going the other way is 34 dB, which is $14 + 14 + 6$ dB, so the gain is $5 \times 5 \times 2 = 50$.

Phase Lag Network

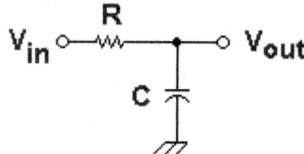

Whew. Having covered all this background, we are ready to make a Bode plot. Let's start with a "phase lag network" as shown in this schematic, also known as a low-pass filter.

The sole parameter characterizing this network is its time constant **T**, and we will arbitrarily take this to be 1 ms for this exercise. The break frequency is then

$$f = 1/(2 \text{ pi } T) = 160 \text{ Hz.}$$

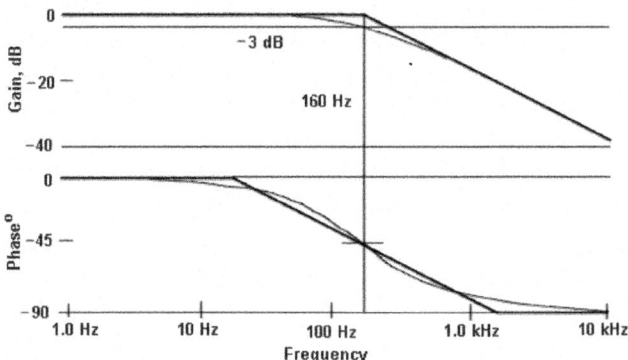

At low frequencies the gain is flat and unity, or 0 dB. At high frequencies the gain rolls off inversely with frequency, decreasing by a factor of 2 (or 6 dB) for every frequency doubling. This is an increase of one octave wherever it occurs.

Alternatively, the roll-off rate can be expressed as 20 dB per decade (a factor of 10 in frequency), which results in a straight line on semilog graph paper with a slope of -6 dB per/octave. It intersects the low-frequency curve at the break frequency.

The straight line segments show an asymptotic representation of the lag characteristic, which is not quite exact near the break frequency. There the gain is actually 1/SQRT(2) = 0.707, or -3 dB. Calculating more points enables us to draw the curve as accurately as desired, but the single 3 dB down point and the two asymptotes suffice to get the picture. Even that extra graphing is seldom done, however, since the process entails extra effort. The asymptotic form is generally more useful, since it shows the break frequency explicitly.

Although the x axis is a log frequency scale, the values of frequency are indicated directly for convenience. So, as far as the numbers are concerned, it is a frequency scale, but the markings are not spaced uniformly.

Looking at the phase plot, we see it is 0 degrees well below cutoff, -90 degrees well above, and -45 degrees at the break frequency. We see the transition is more gradual than that of the gain plot. A good approximation to the curve is a straight-line asymptote: 0 degrees at one-tenth of the break frequency and -90 degrees at ten times the break. On a Bode plot the line will be exact at the break frequency, showing a phase of -45 degrees.

For this lag network and for many others that constitute a subset known as *minimum phase networks*(MPNs), the phase characteristic contains no information in addition to that carried by the gain plot. Therefore, the phase curve is often not plotted at all.

PHASE LEAD NETWORK

The magnetic levitator is **not stable with only position information**. (Believe me, I've been there!) The problem is this: suppose the ball is a little higher than the reference point. The circuit reduces the coil strength to allow the ball to go down. But! It doesn't turn on the coil again until the ball is past the reference spot. The ball has picked up some speed, and now it's too late for the coil to overcome both the speed and weight of the ball at the additional distance. Remember that a coil's strength drops off rapidly over distance. So the ball drops out. We fix this problem by detecting the ball's speed even before it moves past the reference point, and adjusting the coil strength to anticipate the new position.

This circuit uses a capacitor to get both **speed and position** information. It is connected in an arrangement commonly known as a "phase lead network". Let's look at the Bode plot of these networks, and then study our levitator's network more closely.

Bode Plot of Phase Lead Network

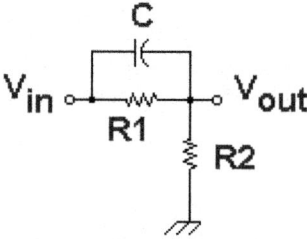

A general purpose phase advance schematic is shown. This network becomes a voltage divider at very low frequencies. It has unity gain at very high frequencies. In between, the transition is very much like the phase lag seen on the previous page.

The time constant of the lower break frequency is R_1C, and the ratio of break frequencies is the reciprocal of the gain (attentuation), resulting in the Bode plots shown below.

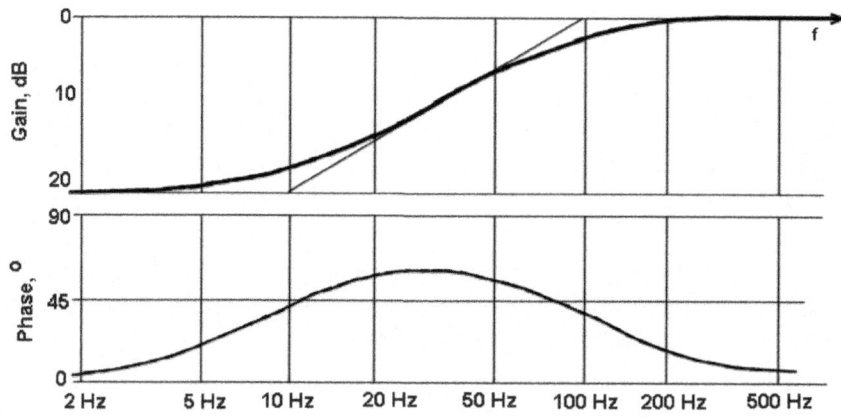

The sample Bode plot above is only meant to illustrate how a phase lead network can modify the gain and phase characteristics of a transfer function.

Properly chosen values can move the two break points to a wide variety of possible frequencies.

The Levitator's Phase Lead Network

First, a little disclaimer, and a **plea for help**. I don't know why, but adding this circuit made my levitator stable. Do you know why? Can you tell me how to analyze the transfer functions of the remaining parts of the levitator?

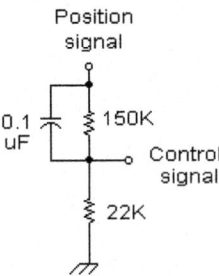

Look at the output from pin 6, slightly redrawn at left. Let's ignore the capacitor for a moment. The 150K and 22K resistors form a divider circuit. It reduces the voltage by the ratio of the two resistors. The "gain" will multiply the position signal by:

$$\text{Gain} = 22K / (22K + 150K) = 0.128$$

In other words it **reduces** the signal by a factor of **eight**. How come? Didn't we just amplify it by a factor of 9? Are we crazy or what?

The idea here is the signal through the capacitor bypasses the 150K resistor. This signal is the **speed**, ie, the **derivative of the position** signal. This speed signal goes through at full strength, and only the position signal is reduced. This results in the proper ratio of speed-to-position to stabilize the ball under the coil.

The lower breakpoint frequency is $f_1 = 1/(2\text{ pi }R_1C) = 10.6$ Hz.

The upper breakpoint frequency is $f_2 = f_1/\text{Gain} = 82.9$ Hz.

If you still find the suspended object is unstable, and knocks around a couple times a second, then you probably want to change f_1 to an even lower frequency. For example, **suppose you want to cut f_1 in half**. Do this by:

1. Swap out the 150K resistor, and put in a 75K resistor, *i.e.* about half the original value.

2. This also lowered the upper breakpoint by four, because

    (a) the attenuation was cut in half and

    (b) the lower breakpoint was cut in half.

So swap out the 22K resistor, and put in a 5.5K resistor, *i.e.* about one fourth the original value.

DESIGNING THE OUTPUT AMPLIFIER

This circuit amplifies the control signal in preparation for the power output transistor. Why do we need this stage at all? Because we reduced the whole signal by one-ninth in the speed-plus-position circuit.

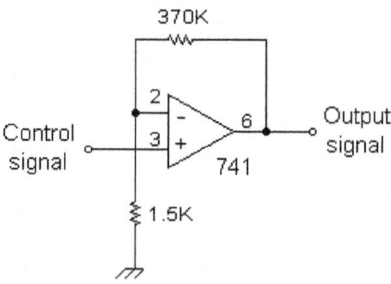

1. This op-amp amplifies the control signal to provide plenty of "punch" to the power transistor and coil.
2. And it isolates the rather large load of the power transistor from the capacitor's time constant and from the resistor divider bridge.

This 741 op-amp is wired as a standard non-inverting amplifier. The gain is computed from the feedback and input resistors:

$$\text{Gain} = (R_f + R_i) / R_i = (370K + 1.5K) / 1.5K = 247$$

That's a lot of gain! Do we really need that much? I'm not sure, but it worked for me! <grin> Actually, with this much gain it acts more like a binary on-off switch than a linear amplifier. It will pretty much ensure the coil is either fully "on" or completely "off". Which is a good idea for reducing power dissipation in the final output transistor. So you probably don't need a very big heat sink for your power output transistor. You might experiment with lower values of gain to see what happens.

DESIGNING THE LEVITATOR COIL DRIVER

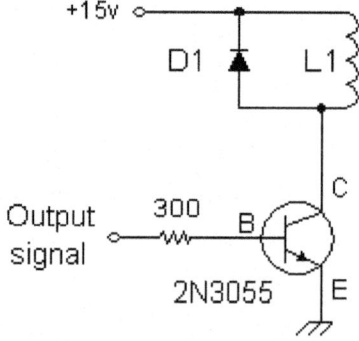

This circuit controls the current in the electromagnetic levitation coil L1. It is driven by a 741 op-amp, which is spec'd to source up to 40 mA of current. For any 2N3055 transistor with a beta = 50 or higher, it is sufficient for loads up to 2 amperes with no additional circuitry.

The base resistor provides some protection. It limits the output current demand on the op-amp. What is the actual limit? At worst case, the op-amp is at 15v and the base terminal is at 0.7v. So the maximum base current will be

$$I_b = (15 - 0.7) / 300 = 48 \text{ mA}.$$

Diode D1 protects the transistor. It is arranged to block the +12v from pin C. Normally D1 is non-conducting and all the current (if any) will flow through the coil L1 and the transistor. But at the instant the transistor turns off, the current circulating in the coil has to be safely dissipated somewhere. So the diode D1 turns on for a millisecond and allows the current to safely be discharged as heat in the coil.

How did I really choose the base resistor? I tried gradually increasing values of resistors while measuring the coil current. At some value, the coil current began to be reduced, so I went back to the next smaller resistor.

The actual maximum value of base resistor you can use depends on the maximum collector current required, and the gain (beta) of the particualr power transistor in the circuit. The value is not really critical, as long as the power transistor can run enough current through the coil. Indeed, you could probably leave out this resistor and connect the base directly to the op-amp output pin. You can experiment here, or just choose a safe value around 220 or 300 ohms.

Designing a PWM Driver

The switched controller that I used is nice, but a **PWM** (pulse width modulation) circuit is better. It provides cool-running operation *and* linear operation. It can be implemented with a small number of components.

I did not build this, but am including the concept for my next levitator.

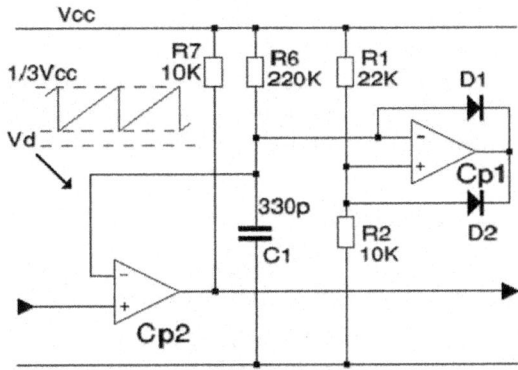

- At power-up, C1 is discharged and starts to charge via R6.
- The + input to Cp1 is held at approximately 1/3 Vcc by R1 and R2.

- C1 charges up to this level and the comparator switches on, pulling its output low.

- C1 now discharges through D1 and D2 holds the junction of R1 and R2 at one Vd above zero volts.

- When C1 is sufficiently discharged the current through D1 is limited by R6 whilst the current through D2 is limited by R2, so will be 10 times as high as that through D1.

- Because of this current ratio, it can be guaranteed that there will be less voltage dropped across D1 than across D2, so the negative input will always fall below the positive input and the oscillator will always reset.

The circuit has several advantages :

- It has a defined lower limit (trough) on the waveform - so needs no level-shifting

- Peak of waveform is easily adjusted without affecting the trough.

- It has a low component count.

- It has very good linearity.

- For even better linearity, you can replace resistor R6 with a current source.

To offset these advantages, the oscillator gives a sawtooth rather than a triangular waveform but there are very few occasions when this has any practical significance.

PROCESS DATA, DYNAMIC MODELING AND A RECIPE FOR PROFITABLE CONTROL

It is best practice to *follow a formal procedure or "recipe" when designing and tuning a PID* (proportional-integral-derivative) *controller. A recipe-based approach is the fastest method for moving a controller into operation. And perhaps most important, the performance of the controller will be superior to a controller tuned using a guess-and-test or trial-and-error method.*

Additionally, a recipe-based approach overcomes many of the concerns that make control projects challenging in a commercial operating environment. Specifically, the recipe-based method causes less disruption to the production schedule, wastes less raw material and utilities, requires less personnel time, and generates less off-spec product.

The recipe for success is short:

1. Establish the design level of operation (DLO), defined as the expected values for set point and major disturbances during normal operation

2. Bump the process and collect controller output (CO) to process variable (PV) dynamic process data around this design level

3. Approximate the process data behavior with a first order plus dead time (FOPDT) dynamic model

4. Use the model parameters from step 3 in rules and correlations to complete the controller design and tuning.

We explore each step of this recipe in detail in other articles on this site. For now, we introduce some initial thoughts about steps 2 and 4.

Step 2: Bumping Our Process and Collecting CO to PV Data

From a controller's view, a complete control loop goes from wire out to wire in as shown below. Whenever we mention controller output (CO) or process variable (PV) data anywhere on this site, we are specifically referring to the data signals exiting and entering our controller at the wire termination interface.

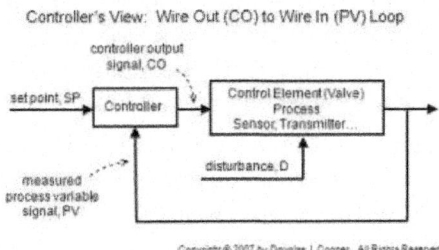

To generate CO to PV data, we bump our process. That is, we step or pulse the CO and record PV data as the process responds. Here are three basic rules we follow in all of our examples:

Start with the Process at Steady State and Record Everything

The point of bumping the CO is to learn about the cause and effect relationship between it and the PV. With the plant initially at steady state, we are starting with a clean slate. The dynamic behavior of the process is then clearly isolated as the PV responds. It is important that we start capturing data before we make the initial CO bump and then sample and record quickly as the PV responds.

Make Sure the PV Response Dominates the Process Noise

When performing a bump test, it is important that the CO moves far enough and fast enough to force a response that clearly dominates any noise or random error in the measured PV signal. If the CO to PV cause and effect response is clear enough to see by eye on a data plot, we can be confident that modern software can model it.

The Disturbances Should be Quiet During the Bump Test

We desire that the dynamic test data contain PV response data that has been clearly, and in the ideal world exclusively, forced by changes in the CO.

Data that has been corrupted by unmeasured disturbances is of little value for controller design and tuning. The model will then incorrectly describe the CO to

PV cause and effect relationship. And as a result, the controller will not perform correctly. If we are concerned that a disturbance event has corrupted test data, it is conservative to rerun the test.

Step 4: Using Model Parameters For Design and Tuning

The final step of the recipe states that once we have obtained model parameters that approximate the dynamic behavior of our process, we can complete the design and tuning of our PID controller.

We look ahead at this last step because this is where the payoff of the recipe-based approach is clear. To establish the merit, we assume for now that we have determined the design level of operation for our process (step 1), we have collected a proper data set rich in dynamic process information around this design level (step 2), and we have approximated the behavior revealed in the process data with a first order plus dead time (FOPDT) dynamic model (step 3).

Thankfully, we do not need to know what a FOPDT model is or even what it looks like. But we do need to know about the three model parameters that result when we fit this approximating model to process data.

The FOPDT (first order plus dead time) model parameters, listed below, tell us important information about the measured process variable (PV) behavior whenever there is a change in the controller output (CO) signal:

- process gain, Kp (tells the direction and how far PV will travel)
- process time constant, Tp (tells how fast PV moves after it begins its response)
- process dead time, θp (tells how much delay before PV first begins to respond)

Aside: we do not need to understand differential equations to appreciate the articles on this site. But for those interested, we note that the first order plus dead time (FOPDT) dynamic model has the form:

$$Tp\frac{dPV(t)}{dt} + PV(t) = Kp \times CO(t - \theta p)$$

Where:

PV(t) = measured process variable as a function of time

CO(t - θp) = controller output signal as a function of time and shifted by θp

θp = process dead time

t = time

The other variables are as listed above this box. It is a first order differential equation because it has one derivative with one time constant, Tp. It is called a first order plus dead time equation because it also directly accounts for a delay or dead time, θ, in the CO(t) to PV(t) behavior.

We study what these three model parameters are and how to compute them in other articles, but here is why process gain, Kp, process time constant, Tp, and process dead time, θp, are all important:

Tuning

These three model parameters can be plugged into proven correlations to directly compute P-Only, PI, PID, and PID with CO Filter tuning values. No more trial and error. No more tweaking our way to acceptable control.

Controller Action

Before implementing our controller, we must input the proper direction our controller should move to correct for growing errors. Some vendors use the term "reverse acting" and "direct acting." Others use terms like "up-up" and "up-down" (as CO goes up, then PV goes up or down). This specification is determined solely by the sign of the process gain, Kp.

Loop Sample Time, T

Process time constant, Tp, is the clock of a process. The size of Tp indicates the maximum desirable loop sample time. Best practice is to set loop sample time, T, at 10 times per time constant or faster ($T \leq 0.1Tp$). Sampling faster will not necessarily provide better performance, but it is a safer direction to move if we have any doubts. Sampling too slowly will have a negative impact on controller performance. Sampling slower than five times per time constant will lead to degraded performance.

Dead Time Problems

As dead time grows larger than the process time constant ($\theta p > Tp$), the control loop can benefit greatly from a model based dead time compensator such as a Smith predictor. The only way we know if $\theta p > Tp$ is if we have followed the recipe and computed the parameters of a FOPDT model.

Model Based Control

If we choose to employ a Smith predictor, a dynamic feed forward element, a multivariable decoupler, or any other model based controller, we need a dynamic model of the process to enter into the control computer. The FOPDT model from step 2 of the recipe is often appropriate for this task.

Fundamental to Success

With tuning values, loop specifications, performance diagnostics and advanced control all dependent on knowledge of a dynamic model, we begin to see that process gain, Kp; process time constant, Tp; and process dead time, θp; are parameters of fundamental importance to success in process control.

OPTIMIZATION-MODELING PROCESS

Optimization problems are ubiquitous in the mathematical modeling of real world systems and cover a very broad range of applications. These applications arise in all branches of Economics, Finance, Chemistry, Materials Science, Astronomy, Physics, Structural and Molecular Biology, Engineering, Computer Science, and Medicine.

Optimization modeling requires appropriate time. The general procedure that can be used in the process cycle of modeling is to:

(1) describe the problem,

(2) prescribe a solution, and

(3) control the problem by assessing/updating the optimal solution continuously, while changing the parameters and structure of the problem.

Clearly, there are always feedback loops among these general steps.

Mathematical Formulation of the Problem: As soon as you detect a problem, think about and understand it in order to adequately describe the problem in writing. Develop a mathematical model or framework to re-present reality in order to devise/use an optimization solution algorithm. The problem formulation must be validated before it is offered a solution. A good mathematical formulation for optimization must be both inclusive (*i.e.*, it includes what belongs to the problem) and exclusive (*i.e.*, shaved-off what does not belong to the problem).

Find an Optimal Solution: This is an identification of a solution algorithm and its implementation stage. The only good plan is an implemented plan, which stays implemented!

Managerial Interpretations of the Optimal Solution: Once you recognize the algorithm and determine the appropriate module of software to apply, utilize software to obtain the optimal strategy. Next, the solution will be presented to the decision-maker in the same style and language used by the decision-maker. This means providing managerial interpretations of the strategic solution in layman's terms, not just handing the decision-maker a computer printout.

Post-Solution Analysis: These activities include updating the optimal solution in order to control the problem. In this ever-changing world, it is crucial to periodically update the optimal solution to any given optimization problem. A model that was valid may lose validity due to changing conditions, thus becoming an inaccurate representation of reality and adversely affecting the ability of the decision-maker to make good decisions. The optimization model you create should be able to cope with changes.

The Importance of Feedback and Control: It is necessary to place heavy emphasis on the importance of thinking about the feedback and control aspects of an optimization problem. It would be a mistake to discuss the context of the optimization-modeling process and ignore the fact that one can never expect to find a never-changing, immutable solution to a decision problem. The very nature of the optimal strategy's environment is changing, and therefore feedback and control are an important part of the optimization-modeling process.

The above process is depicted as the Systems Analysis, Design, and Control stages in the following flow chart, including the validation and verification activities:

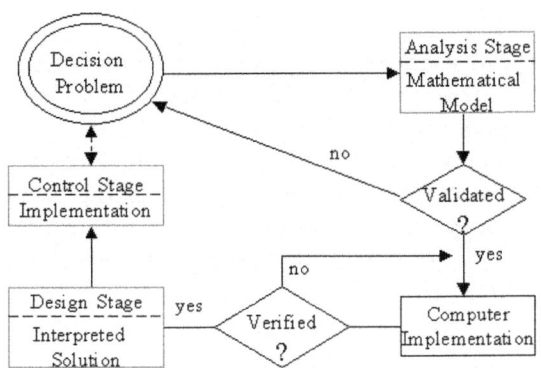

Describe the Problem, Prescribe a Solution, Update the Solution

Ingredients of Optimization Problems and Their Classification

The essence of all businesslike decisions, whether made for a firm, or an individual, is finding a course of action that leaves you with the largest profit.

Mankind has long sought, or professed to seek, better ways to carry out the daily tasks of life. Throughout human history, man has first searched for more effective sources of food and then later searched for materials, power, and mastery of the physical environment. However, relatively late in human history general questions began to quantitatively formulate first in words, and later developing into symbolic notations.

One pervasive aspect of these general questions was to seek the "best" or "optimum". Most of the time managers seek merely to obtain some improvement in the level of performance, or a "goal-seeking" problem. It should be emphasized that these words do not usually have precise meanings.

Efforts have been made to describe complex human and social situations. To have meaning, the problem should be written down in a mathematical expression containing one or more variables, in which the value of variables are to be determined. The question then asked, is what values should these variables have to ensure the mathematical expression has the greatest possible numerical value (maximization) or the least possible numerical value (minimization). This process of maximizing or minimizing is referred to as optimization.

Optimization, also called mathematical programming, helps find the answer that yields the best result--the one that attains the highest profit, output, or happiness, or the one that achieves the lowest cost, waste, or discomfort. Often these problems involve making the most efficient use of resources--including money, time, machinery, staff, inventory, and more. Optimization problems are often classified as linear or nonlinear, depending on whether the relationship in the problem is linear with respect to the variables. There are a variety of software packages to solve optimization problems. For example, LINDO or your WinQSB solve linear program models and LINGO and What's Best! solve nonlinear and linear problems.

Mathematical Programming, solves the problem of determining the optimal allocations of limited resources required to meet a given objective. The objective must represent the goal of the decision-maker. For example, the resources may correspond to people, materials, money, or land. Out of all permissible allocations of the resources, it is desired to find the one or ones that maximize or minimize some numerical quantity such as profit or cost. Optimization models are also called Prescriptive or Normative models since they seek to find the best possible strategy for decision-maker.

There are many optimization algorithms available. However, some methods are only appropriate for certain types of problems. It is important to be able to recognize the characteristics of a problem and identify an appropriate solution technique. Within each class of problems, there are different minimization methods, which vary in computational requirements, convergence properties, and so on. Optimization problems are classified according to the mathematical characteristics of the objective function, the constraints, and the controllable decision variables.

Optimization problems are made up of three basic ingredients:

- An objective function that we want to minimize or maximize. That is, the quantity you want to maximize or minimize is called the objective function. Most optimization problems have a single objective function, if they do not, they can often be reformulated so that they do. The two interesting exceptions to this rule are:

The goal seeking problem: In most business applications the manager wishes to achieve a specific goal, while satisfying the constraints of the model. The user does not particularly want to optimize anything so there is no reason to define an objective function. This type of problem is usually called a feasibility problem.

Multiple objective functions: Often, the user would actually like to optimize many different objectives at once. Usually, the different objectives are not compatible. The variables that optimize one objective may be far from optimal for the others. In practice, problems with multiple objectives are reformulated as single-objective problems by either forming a weighted combination of the different objectives or else by placing some objectives as "desirable" constraints.

- The controllable inputs are the set of decision variables which affect the value of the objective function. In the manufacturing problem, the variables might include the allocation of different available resources, or the labor spent on each activity. Decision variables are essential. If there are no variables, we cannot define the objective function and the problem constraints.

- The uncontrollable inputs are called parameters. The input values may be fixed numbers associated with the particular problem. We call these values parameters of the model. Often you will have several "cases" or variations of the same problem to solve, and the parameter values will change in each problem variation.

- Constraints are relations between decision variables and the parameters. A set of constraints allows some of the decision variables to take on certain values, and exclude others. For the manufacturing problem, it does not make sense to spend a negative amount of time on any activity, so we constrain all the "time" variables to be non-negative. Constraints are not always essential. In fact, the field of unconstrained optimization is a large and important one for which a lot of algorithms and software are available. In practice, answers that make good sense about the underlying physical or economic problem, cannot often be obtained without putting constraints on the decision variables.

Feasible and Optimal Solutions: A solution value for decision variables, where all of the constraints are satisfied, is called a feasible solution. Most solution algorithms proceed by first finding a feasible solution, then seeking to improve upon it, and finally changing the decision variables to move from one feasible solution to another feasible solution. This process is repeated until the objective function has reached its maximum or minimum. This result is called an optimal solution.

The basic goal of the optimization process is to find values of the variables that minimize or maximize the objective function while satisfying the constraints. This result is called an optimal solution.

There are well over 4000 solution algorithms for different kinds of optimization problems. The widely used solution algorithms are those developed for the following mathematical programs: convex programs, separable programs, quadratic programs and the geometric programs.

Linear Program

Linear programming deals with a class of optimization problems, where both the objective function to be optimized and all the constraints, are linear in terms of the decision variables.

A short history of Linear Programming:

1. In 1762, Lagrange solved tractable optimization problems with simple equality constraints.

2. In 1820, Gauss solved linear system of equations by what is now call Causssian elimination. In 1866 Wilhelm Jordan refinmened the method

to finding least squared errors as a measure of goodness-of-fit. Now it is referred to as the Gauss-Jordan Method.

3. In 1945, Digital computer emerged.

4. In 1947, Dantzig invented the Simplex Methods.

5. In 1968, Fiacco and McCormick introduced the Interior Point Method.

6. In 1984, Karmarkar applied the Interior Method to solve Linear Programs adding his innovative analysis.

Linear programming has proven to be an extremely powerful tool, both in modeling real-world problems and as a widely applicable mathematical theory. However, many interesting optimization problems are nonlinear. The study of such problems involves a diverse blend of linear algebra, multivariate calculus, numerical analysis, and computing techniques.

Important areas include the design of computational algorithms (including interior point techniques for linear programming), the geometry and analysis of convex sets and functions, and the study of specially structured problems such as quadratic programming. Nonlinear optimization provides fundamental insights into mathematical analysis and is widely used in a variety of fields such as engineering design, regression analysis, inventory control, geophysical exploration, and economics.

Quadratic Program

Quadratic Program (QP) comprises an area of optimization whose broad range of applicability is second only to linear programs. A wide variety of applications fall naturally into the form of QP. The kinetic energy of a projectile is a quadratic function of its velocity. The least-square regression with side constraints has been modeled as a QP.

Certain problems in production planning, location analysis, econometrics, activation analysis in chemical mixtures problem, and in financial portfolio management and selection are often treated as QP. There are numerous solution algorithms available for the case under the restricted additional condition, where the objective function is convex.

Constraint Satisfaction

Many industrial decision problems involving continuous constraints can be modeled as continuous constraint satisfaction and optimization problems. Constraint Satisfaction problems are large in size and in most cases involve transcendental functions. They are widely used in chemical processes and cost restrictions modeling and optimization.

Convex Program

Convex Program (CP) covers a broad class of optimization problems. When the objective function is convex and the feasible region is a convex set, both of these assumptions are enough to ensure that local minimum is a global minimum.

Data Envelopment Analysis

The Data Envelopment Analysis (DEA) is a performance metric that is grounded in the frontier analysis methods from the economics and finance literature. Frontier efficiency (output/input) analysis methods identify best practice performance frontier, which refers to the maximal outputs that can be obtained from a given set of inputs with respect to a sample of decision making units using a comparable process to transform inputs to outputs.

The strength of DEA relies partly on the fact that it is a non-parametric approach, which does not require specification of any functional form of relationships between the inputs and the outputs. DEA output reduces multiple performance measures to a single one to use linear programming techniques. The weighting of performance measures reacts to the decision-maker's utility.

Dynamic Programming

Dynamic programming (DP) is essentially bottom-up recursion where you store the answers in a table starting from the base case(s) and building up to larger and larger parameters using the recursive rule(s). You would use this technique instead of recursion when you need to calculate the solutions to all the sub-problems and the recursive solution would solve some of the sub-problems repeatedly. While generally DP is capable of solving many diverse problems, it may require huge computer storage in most cases.

Separable Program

Separable Program (SP) includes a special case of convex programs, where the objective function and the constraints are separable functions, *i.e.*, each term involves just a single variable.

Geometric Program

Geometric Program (GP) belongs to Nonconvex programming, and has many applications in particular in engineering design problems.

Fractional Program

In this class of problems, the objective function is in the form of a fraction (*i.e.*, ratio of two functions). Fractional Program (FP) arises, for example, when maximizing the ratio of profit capital to capital expended, or as a performance measure wastage ratio.

Heuristic Optimization

A heuristic is something "providing aid in the direction of the solution of a problem but otherwise unjustified or incapable of justification." So heuristic arguments are used to show what we might later attempt to prove, or what we might expect to find in a computer run. They are, at best, educated guesses.

Several heuristic tools have evolved in the last decade that facilitate solving optimization problems that were previously difficult or impossible to solve. These tools include evolutionary computation, simulated annealing, tabu search, particle swarm, *etc.*

Common approaches include, but are not limited to:

1. comparing solution quality to optimum on benchmark problems with known optima, average difference from optimum, frequency with which the heuristic finds the optimum.

2. comparing solution quality to a best known bound for benchmark problems whose optimal solutions cannot be determined.

3. comparing your heuristic to published heuristics for the same problem type, difference in solution quality for a given run time and, if relevant, memory limit.

4. profiling average solution quality as a function of run time, for instance, plotting mean and either min and max or 5th and 95th percentiles of solution value as a function of time -- this assumes that one has multiple benchmark problem instances that are comparable.

Global Optimization

The aim of Global Optimization (GO) is to find the best solution of decision models, in presence of the multiple local solutions. While *constrained optimization* is dealing with finding the optimum of the objective function subject to constraints on its decision variables, in contrast, *unconstrained optimization* seeks the global maximum or minimum of a function over its entire domain space, without any restrictions on decision variables.

Nonconvex Program

A Nonconvex Program (NC) encompasses all nonlinear programming problems that do not satisfy the convexity assumptions. However, even if you are successful at finding a local minimum, there is no assurance that it will also be a global minimum. Therefore, there is no algorithm that will guarantee finding an optimal solution for all such problem.

Nonsmooth Program

Nonsmooth Programs (NSP) contain functions for which the first derivative does not exist. NSP are arising in several important applications of science and engineering, including contact phenomena in statics and dynamics or delamination effects in composites. These applications require the consideration of nonsmoothness and nonconvexity.

Metaheuristics

Most metaheuristics have been created for solving discrete combinatorial optimization problems. Practical applications in engineering, however, usually require techniques, which handle continuous variables, or miscellaneous continuous and discrete variables. As a consequence, a large research effort has focused on fitting several well-known metaheuristics, like Simulated Annealing (SA), Tabu Search (TS), Genetic Algorithms (GA), Ant Colony Optimization (ACO), to the continuous cases. The general metaheuristics aim at transforming discrete domains of application into continuous ones, by means of:

1. Methodological developments aimed at adapting some metaheuristics, especially SA, TS, GA, ACO, GRASP, variable neighborhood search, guided local search, scatter search, to continuous or discrete/continuous variable problems.

2. Theoretical and experimental studies on metaheuristics adapted to continuous optimization, *e.g.,* convergence analysis, performance evaluation methodology, test-case generators, constraint handling, *etc.*

3. Software implementations and algorithms for metaheuristics adapted to continuous optimization.

4. Real applications of discrete metaheuristics adapted to continuous optimization.

5. Performance comparisons of discrete metaheuristics (adapted to continuous optimization) with that of competitive approaches, *e.g.,* Particle Swarm Optimization (PSO), Estimation of Distribution Algorithms (EDA), Evolutionary Strategies (ES), specifically created for continuous optimization.

Multilevel Optimization

In many decision processes there is a hierarchy of decision makers and decisions are taken at different levels in thishierarchy. Multilevel Optimization focuses on the whole hierarchy structure. The field of multilevel optimization has become a well known and important research field. Hierarchical structures can be found in scientific disciplines such as environment, ecology, biology, chemical engineering, mechanics, classification theory, databases, network design, transportation, supply chain, game theory and economics. Moreover, new applications are constantly being introduced.

Multiobjective Program

Multiobjective Program (MP) known also as Goal Program, is where a single objective characteristic of an optimization problem is replaced by several goals. In solving MP, one may represent some of the goals as constraints to be satisfied, while the other objectives can be weighted to make a composite single objective function.

Multiple objective optimization differs from the single objective case in several ways:

- The usual meaning of the optimum makes no sense in the multiple objective case because the solution optimizing all objectives simultaneously is, in general, impractical; instead, a search is launched for a feasible solution yielding the best compromise among objectives on a set of, so called, efficient solutions;

- The identification of a best compromise solution requires taking into account the preferences expressed by the decision-maker;

- The multiple objectives encountered in real-life problems are often mathematical functions of contrasting forms.

- A key element of a goal programming model is the achievement function; that is, the function that measures the degree of minimisation of the unwanted deviation variables of the goals considered in the model.

A Business Application: In credit card portfolio management, predicting the cardholder's spending behavior is a key to reduce the risk of bankruptcy. Given a set of attributes for major aspects of credit cardholders and predefined classes for spending behaviors, one might construct a classification model by using multiple criteria linear programming to discover behavior patterns of credit cardholders.

Non-Binary Constraints Program

Over the years, the constraint programming community has paid considerable attention to modeling and solving problems by using binary constraints. Only recently has non-binary constraints captured attention, due to growing number of real-life applications. A non-binary constraint is a constraint that is defined on k variables, where k is normally greater than two.

A non-binary constraint can be seen as a more global constraint. Modeling a problem as a non-binary constraint has two main advantages: It facilitates the expression of the problem; and it enables more powerful constraint propagation as more global information becomes available.

Success in timetabling, scheduling, and routing, has proven that the use of non-binary constraints is a promising direction of research. In fact, a growing number of OR/MS/DS workers feel that this topic is crucial to making constraint technology a realistic way to model and solve real-life problems.

Bilevel Optimization

Most of the mathematical programming models deal with decision-making with a single objective function. The bilevel programming on the other hand is developed for applications in decentralized planning systems in which the first level is termed as the leader and the second level pertains to the objective of the follower.

In the bilevel programming problem, each decision maker tries to optimize its own objective function without considering the objective of the other party, but the decision of each party affects the objective value of the other party as well as the decision space.

Bilevel programming problems are hierarchical optimization problems where the constraints of one problem are defined in part by a second parametric optimization problem. If the second problem has a unique optimal solution for all parameter values, this problem is equivalent to usual optimization problem having an implicitly defined objective function. However, when the problem has non-unique optimal solutions, the optimistic (or weak) and the pessimistic (or strong) approaches are being applied.

Combinatorial Optimization

Combinatorial generally means that the state space is discrete (*e.g.*, symbols, not necessarily numbers). This space could be finite or denumerable sets. For example, a discrete problem is combinatorial. Problems where the state space is totally ordered can often be solved by mapping them to the integers and applying "numerical" methods. If the state space is unordered or only partially ordered, these methods fail. This means that the **heuristics methods** becomes necessary, such as simulated annealing.

Combinatorial optimization is the study of packing, covering, and partitioning, which are applications of integer programs. They are the principle mathematical topics in the interface between combinatorics and optimization. These problems deal with the classification of integer programming problems according to the complexity of known algorithms, and the design of good algorithms for solving special subclasses. In particular, problems of network flows, matching, and their matroid generalizations are studied. This subject is one of the unifying elements of combinatorics, optimization, operations research, and computer science.

Evolutionary Techniques

Nature is a robust optimizer. By analyzing nature's optimization mechanism we may find acceptable solution techniques to intractable problems. Two concepts that have most promise are simulated annealing and the genetic techniques. Scheduling and timetabling are amongst the most successful applications of evolutionary techniques.

Genetic Algorithms (GAs) have become a highly effective tool for solving hard optimization problems. However, its theoretical foundation is still rather fragmented.

Particle Swarm Optimization

Particle Swarm Optimization (PSO) is a stochastic, population-based optimization algorithm. Instead of competition/selection, like say in Evolutionary

Computation, PSO makes use of cooperation, according to a paradigm sometimes called "swarm intelligence". Such systems are typically made up of a population of simple interacting agents without any centralized control, and inspired by cases that can be found in nature, such as ant colonies, bird flocking, animal herding, bacteria molding, fish schooling, *etc.*

There are many variants of PSO including constrained, multiobjective, and discrete or combinatorial versions, and applications have been developed using PSO in many fields.

Swarm Intelligence

Biologists studied the behavior of social insects for a long time. After millions of years of evolution all these species have developed incredible solutions for a wide range of problems. The intelligent solutions to problems naturally emerge from the self-organization and indirect communication of these individuals. Indirect interactions occur between two individuals when one of them modifies the environment and the other responds to the new environment at a later time.

Swarm Intelligence is an innovative distributed intelligent paradigm for solving optimization problems that originally took its inspiration from the biological examples by swarming, flocking and herding phenomena in vertebrates. Data Mining is an analytic process designed to explore large amounts of data in search of consistent patterns and/or systematic relationships between variables, and then to validate the findings by applying the detected patterns to new subsets of data.

Online Optimization

Whether costs are to be reduced, profits to be maximized, or scarce resources to be used wisely, optimization methods are available to guide decision-making. In online optimization, the main issue is incomplete data and the scientific challenge: how well can an online algorithm perform? Can one guarantee solution quality, even without knowing all data in advance? In real-time optimization there is an additional requirement: decisions have to be computed very fast in relation to the time frame we are considering.

Linear Programming

Linear programming is often a favorite topic for both professors and students. The ability to introduce LP using a graphical approach, the relative ease of the solution method, the widespread availability of LP software packages, and the wide range of applications make LP accessible even to students with relatively weak mathematical backgrounds. Additionally, LP provides an excellent opportunity to introduce the idea of "what-if" analysis, due to the powerful tools for post-optimality analysis developed for the LP model.

Linear Programming (LP) is a mathematical procedure for determining optimal allocation of scarce resources. LP is a procedure that has found practical application in almost all facets of business, from advertising to production planning. Transportation, distribution, and aggregate production planning problems are the most typical objects of LP analysis. In the petroleum industry, for example a data processing manager at a large oil company recently estimated that from 5 to 10 percent of the firm's computer time was devoted to the processing of LP and LP-like models.

Linear programming deals with a class of programming problems where both the objective function to be optimized is linear and all relations among the variables corresponding to resources are linear. This problem was first formulated and solved in the late 1940's. Rarely has a new mathematical technique found such a wide range of practical business, commerce, and industrial applications and simultaneously received so thorough a theoretical development, in such a short period of time.

Today, this theory is being successfully applied to problems of capital budgeting, design of diets, conservation of resources, games of strategy, economic growth prediction, and transportation systems. In very recent times, linear programming theory has also helped resolve and unify many outstanding applications.

It is important for the reader to appreciate, at the outset, that the "programming" in Linear Programming is of a different flavor than the "programming" in Computer Programming. In the former case, it means to plan and organize as in "Get with the program!", it programs you by its solution. While in the latter case, it means to write codes for performing calculations. Training in one kind of programming has very little direct relevance to the other. In fact, the term "linear programming" was coined before the word "programming" became closely associated with computer software. This confusion is sometimes avoided by using the term linear optimization as a synonym for linear programming.

Any LP problem consists of an objective function and a set of constraints. In most cases, constraints come from the environment in which you work to achieve your objective. When you want to achieve the desirable objective, you will realize that the environment is setting some constraints (*i.e.*, the difficulties, restrictions) in fulfilling your desire or objective. This is why religions such as Buddhism, among others, prescribe living an abstemious life. No desire, no pain. Can you take this advice with respect to your business objective?

What is a function: A function is a thing that does something. For example, a coffee grinding machine is a function that transform the coffee beans into powder. The (objective) function maps and translates the input domain (called the feasible region) into output range, with the two end-values called the maximum and the minimum values.

When you formulate a decision-making problem as a linear program, you must check the following conditions:

1. The objective function must be linear. That is, check if all variables have power of 1 and they are added or subtracted (not divided or multiplied)

2. The objective must be either maximization or minimization of a linear function. The objective must represent the goal of the decision-maker

3. The constraints must also be linear. Moreover, the constraint must be of the following forms ($\leq$, $\geq$, or =, that is, the LP-constraints are always *closed*).

For example, the following problem is not an LP: Max X, subject to X < 1. This very simple problem has no solution.

As always, one must be careful in categorizing an optimization problem as an LP problem. Here is a question for you. Is the following problem an LP problem?

Max X2

subject to:

X1 + X2 $\leq$ 0

$X1^2 - 4 \leq 0$

Although the second constraint looks "as if" it is a nonlinear constraint, this constraint can equivalently be written as:

X1 $\geq$ -2, and X2 $\leq$ 2.

Therefore, the above problem is indeed an LP problem.

For most LP problems one can think of two important classes of objects: The first is limited resources such as land, plant capacity, or sales force size; the second, is activities such as "produce low carbon steel", "produce stainless steel", and "produce high carbon steel". Each activity consumes or possibly contributes additional amounts of the resources. There must be an objective function, *i.e.* a way to tell bad from good, from an even better decision.

The problem is to determine the best combination of activity levels, which do not use more resources than are actually available. Many managers are faced with this task everyday. Fortunately, when a well-formulated model is input, linear programming software helps to determine the best combination.

The Simplex method is a widely used solution algorithm for solving linear programs. An algorithm is a series of steps that will accomplish a certain task.

LP Problem Formulation Process and Its Applications

To formulate an LP problem, I recommend using the following guidelines after reading the problem statement carefully a few times.

Any linear program consists of four parts: a set of decision variables, the parameters, the objective function, and a set of constraints. In formulating a given decision problem in mathematical form, you should practice understanding the problem (*i.e.*, formulating a mental model) by carefully reading and re-reading the problem statement. While trying to understand the problem, ask yourself the following general questions:

1. What are the decision variables? That is, what are controllable inputs? Define the decision variables precisely, using descriptive names. Remem-

ber that the controllable inputs are also known as controllable activities, decision variables, and decision activities.

2. What are the parameters? That is, what are the uncontrollable inputs? These are usually the given constant numerical values. Define the parameters precisely, using descriptive names.

3. What is the objective? What is the objective function? Also, what does the owner of the problem want? How the objective is related to his decision variables? Is it a maximization or minimization problem? The objective represents the goal of the decision-maker.

4. What are the constraints? That is, what requirements must be met? Should I use inequality or equality type of constraint? What are the connections among variables? Write them out in words before putting them in mathematical form.

Learn that the feasible region has nothing or little to do with the objective function (min or max). These two parts in any LP formulation come mostly from two distinct and different sources. The objective function is set up to fulfill the decision-maker's desire (objective), whereas the constraints which shape the feasible region usually comes from the decision-maker's environment putting some restrictions/conditions on achieving his/her objective.

The following is a very simple illustrative problem. However, the way we approach the problem is the same for a wide variety of decision-making problems, and the size and complexity may differ. The first example is a product-mix problem.

SAMPLE TIME IMPACTS CONTROLLER PERFORMANCE

There are two sample times, T, used in process controller design and tuning.

One is the control loop sample time that specifies how often the controller samples the measured process variable (PV) and then computes and transmits a new controller output (CO) signal.

The other is the rate at which CO and PV data are sampled and recorded during a bump test of our process. Bump test data is used to design and tune our controller prior to implementation.

In both cases, sampling too slow will have a negative impact on controller performance. Sampling faster will not necessarily provide better performance, but it is a safer direction to move if we have any doubts.

Fast and slow are relative terms defined by the process time constant, T_p. Best practice for both control loop sample time and bump test data collection are the same:

Best Practice:

Sample time should be 10 times per process time constant or faster ($T \leq 0.1T_p$).

We explore this "best practice" rule in a detailed study here.

Yet perhaps we can gain an appreciation for how sample time impacts controller design and tuning with this thought experiment:

Suppose you see me standing on your left. You close your eyes for a time, open them, and now I am standing on your right. Do you know how long I have been at my new spot? Did I just arrive or have I been there for a while? What path did I take to get there? Did I move around in front or in back of you? Maybe I even jumped over you?

Now suppose your challenge is to keep your hands at your side until I pass by, and just as I do, you are to reach out and touch me. What are your chances with your eyes closed (and loud music is playing so you cannot hear me)?

Now lets say you are permitted to blink open your eyes briefly once per minute. Do you think you will have a better chance of touching me? How about blinking once every ten seconds? Clearly, as you start blinking say, two or three times a second, the task of touching me becomes easy. That's because you are sampling fast enough to see my "process" behavior fully and completely.

Based on this thought experiment, sampling too slow is problematic and sampling faster is generally better.

Keep in mind the "T ≤ 0.1Tp" rule as we study PID control. This applies both to sampling during data collection, and the "measure and act" loop sample time when we implement our controller.

REAL-TIME CONTROL IN TEST CELL APPLICATIONS

Real-time test cells encompass a wide array of applications, ranging from simple dynamometers to complex multi-axis servo-hydraulic simulators. The goal of all these test systems is to apply a load or strain on the device under test (DUT) to validate its performance. The results indicate characteristics of the DUT such as efficiency, durability, and operating limits.

What is Real-Time Control?

The primary requirement in any real-time control application is determinism. Determinism is a guarantee that a specified event will occur within a fixed period of time. This can include control algorithm calculations, alarm monitoring, signal I/O, or any other system function. For example, a deterministic system is required to ensure that an alarm is triggered when a mechanical system goes out of its safe operating range. For true deterministic control, most major test control systems utilize a real-time operating system (RTOS) to provide determinism and stability.

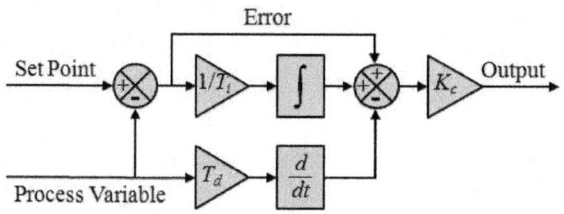

The term "real-time control" generally refers to closed loop PID control. "Closed loop" indicates a feedback control mechanism where a desired response is fed back into the control algorithm and compared against the desired setpoint.

Requirements of a Real-Time Test System

In addition to real-time control, most modern control systems provide digital data acquisition, stimulus generation, safety monitoring, and data logging. These tasks must all execute deterministically in order to validate the performance of the DUT and to ensure that tests are executed safely.

Data Acquisition

Data acquisition provides sensor data for many tasks such as closed loop control, alarm monitoring, data logging, and pass/fail analysis. The importance of these tasks requires that the correct data acquisition modules be chosen based on application requirements. Important parameters for data acquisition parameters include:

1. **Sampling and Generation Rate:** Data must be sampled or generated at a required rate plus or minus a known acceptable tolerance. Without this determinism, control loop calculations may run based on expired data causing tests to fail or compromising system safety.

2. **Sampling Resolution:** Data acquisition modules usually range in resolution from 8 to 24 bits. It is important to choose a resolution that adequately represents the system under test. For example, if we are testing a motor that rotates 0 to 10,000 RPM, using 12-bit acquisition would provide an ideal signal resolution of 10,000 / 4096 or 2.44 RPM while 16-bit acquisition would provide 10,000 / 65536 or 0.15 RPM. The requirements of the application will determine how precisely the system must be measured.

3. **Response Time:** Analog-to-digital converters (ADCs) require time to sample signals and provide data to the control system. This time varies based on the type of data acquisition module and can significantly impact the control system performance. Some modules can sample in parallel while others may multiplex multiple signals with one ADC. Single converter modules often cost less but introduce additional delays between samples. There are also modules that implement delta-sigma converters which introduce a signal delay as they fill their internal pipeline. They may acquire at a high rate with high accuracy, but the signal coming out of the converter is delayed many milliseconds before providing data to control loops and alarms.

Timing and Synchronization

Coupled tightly with a test system's data acquisition is the requirement for synchronization and coordination between various tasks within the control system.

These tasks include data acquisition, multi-axis control, and safety operations; events within these tasks must occur at the same time or in relation to one another.

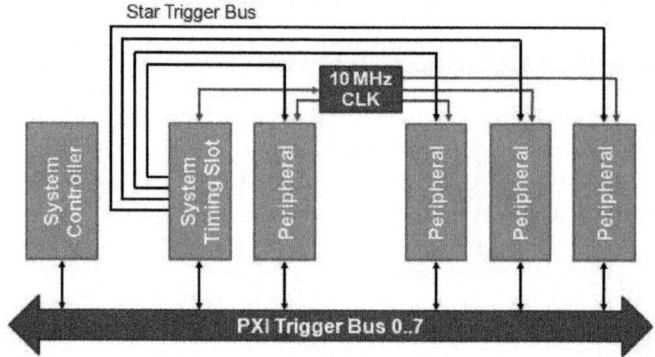

Data acquisition tasks must be synchronized to ensure sampling across all devices to prevent samples from different devices from either drifting (*e.g.* where 1 kHz on one device is not exactly equal to 1 kHz on another) or lagging (*e.g.* where samples from one device are sampled at a later time than another).

Multi-axis control processes require significant coordination to ensure synchronized motion commands. For example, if you are controlling two hydraulic actuators to apply a load, you must be able to deterministically specify the load for each actuator based on an absolute time. If the actuator commands are out of synchronization, the actuators will produce erroneous loads and possible destroy themselves and the DUT.

Safety Monitoring

In addition to synchronized data acquisition and coordinated axis control, a test cell control system must also include safety mechanisms to prevent equipment damage and ensure personnel safety. Safety systems typically incorporate three levels of protection:

- Physical protection such as barriers, guarding and direct electrical disconnects.
- Real-time software monitoring such as PLCs or real-time controllers that monitor the system and respond deterministically. Typically, these systems respond by opening disconnects such as safety relays.
- Supervisory software monitoring such as a Windows PC. This level of safety applies to non-critical tasks or warnings that, if ignored, will eventually lead to a real-time controller safety action.

A test cell control system must respond deterministically to safety critical channels that go outside of their operating limits. The control system must have a process dedicated to safety monitoring and alarm processing.

PROCESS CONTROLLERS

Basic Process Controllers with Proportional, Integrating and Derivative Functions

Basic Controller

The Basic Controller for an application can be visualized as

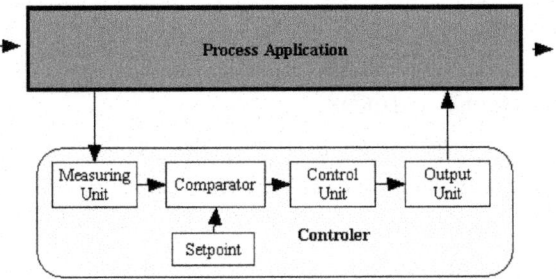

www.engineeringtoolbox.com

The controller consists of :

- a measuring unit with an appropriate instrument to measure the state of process, a temperature transmitter, pressure transmitter or similar.
- a input set point device to set the desired value.
- a comparator for comparing the measured value with the set point, calculating the difference or error between the two.
- a control unit to calculate the output magnitude and direction to compensate the deviation from the desired value.
- a output unit converting the output from the controller to physical action, a control valve, a motor or similar.

Controller Principles

The Control Units are in general build on the control principles

- proportional controller

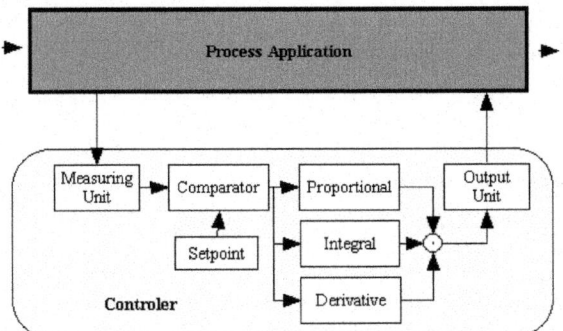

www.engineeringtoolbox.com

- integral controller
- derivative controller

Proportional Controller (P-Controller)

One of the most used controllers is the Proportional Controller (P-Controller) who produce an output action that is proportional to the deviation between the set point and the measured process value.

$$O_P = -k_P \, Er$$

where

O_P = *output proportional controller*

k_P = *proportional gain or action factor of the controller*

Er = *error or deviation between the set point value and the measured value*

The gain or action factor - k_P

- influence on the output with a magnitude of k_P
- determines how fast the system responds. If the value is too large the system will be in danger to oscillate and/or become unstable. If the value is too small the system error or deviation from set point will be very large.
- can be regarded linear only for very small variations.

The gain k_P can be expressed as

$$k_P = 100 / P$$

where

P = *proportional band*

The proportional band P, express the value necessary for *100%* controller output. If $P = 0$, the gain or action factor k_P would be infinity - the control action would be *ON/OFF*.

Note! A proportional controller will have the effect of reducing the rise time and will reduce, but never eliminate, the steady-state error.

Integral Controller (I-Controller)

With integral action, the controller output is proportional to the amount of time the error is present. Integral action eliminates offset.

$$O_I = -k_I \, \Sigma(Er \, dt)$$

where

O_I = *output integrating controller*

k_I = *integrating gain or action factor of the controller*

dt = *time sample*

The integral controller produce an output proportional to the summarized deviation between the set point and measured value and integrating gain or action factor.

Integral controllers tend to respond slowly at first, but over a long period of time they tend to eliminate errors.

The integral controller eliminates the steady-state error, but may make the transient response worse. The controller may be unstable.

The integral regulator may also cause problems during shutdowns and start up as a result of the integral saturation or wind up effect. An integrating regulator with over time deviation (typical during plant shut downs) will summarize the output to +/- 100%. During start up the output is set to 100%m which may be catastrophic.

Derivative Controller (D-Controller)

With derivative action, the controller output is proportional to the rate of change of the measurement or error. The controller output is calculated by the rate of change of the deviation or error with time.

$$O_D = - k_D \, dEr / dt$$

where

O_D = *output derivative controller*

k_D = *derivative gain or action factor of the controller*

dEr = *deviation change over time sample dt*

dt = *time sample*

The derivative or differential controller is never used alone. With sudden changes in the system the derivative controller will compensate the output fast. The long term effects the controller allow huge steady state errors.

A derivative controller will in general have the effect of increasing the stability of the system, reducing the overshoot, and improving the transient response.

Proportional, Integral, Derivative Controller (PID-Controller)

The functions of the individual proportional, integral and derivative controllers complements each other. If they are combined its possible to make a system that responds quickly to changes (derivative), tracks required positions (proportional), and reduces steady state errors (integral).

Note that these correlations may not be exactly accurate, because P, I and D are dependent of each other. Changing one of these variables can change the effect of the other two.

Controller Response	Rise Time	Overshoot	Settling Time	Steady State Error
P	Decrease	Increase	Small Change	Decrease
I	Decrease	Increase	Increase	Eliminate
D	Small Change	Decrease	Decrease	Small Change

STEP TEST DATA FROM THE HEAT EXCHANGER PROCESS

The first order plus dead time (FOPDT) dynamic model and discussed how this model, when used to approximate the controller output (CO) to process variable (PV) behavior of proper data from our process, yields the all-important model parameters:

- process gain, Kp (*tells the direction and how far PV will travel*)
- process time constant, Tp (*tells how fast PV moves after it begins its response*)
- process dead time, θp (*tells how much delay before PV first begins to respond*)

The previous articles also mentioned that these FOPDT model parameters can be used to determine PID tuning values, proper sample time, whether the controller should be direct or reverse acting, whether dead time is large enough to cause concern, and more.

A Hands-On Study

There is an old saying (a Google search shows a host of attributed authors) that goes something like this: *I hear and I forget, I see and I remember, I do and I understand.*

Since our goal is to understand, this means we must "do." To that end, we take a hands-on approach in this case study that will help us appreciate what each FOPDT model parameter is telling us about our process and empower us to act accordingly as we explore best practices for controller design and tuning.

To proceed, we require a process we can manipulate freely. We start with a heat exchanger because they are common to a great many industries.

Heat Exchanger Process

The heat exchanger we will study is really a process simulation from commercial software. The simulation is developed from first-principles theory, so its response behavior is realistic. The benefit of a simulation is that we can manipulate process variables whenever and however we desire without risk to people or profit.

The heat exchanger is shown below in manual mode (also called open loop). Its behavior is that of a counter-current, shell and tube, hot liquid cooler.

The measured process variable is the hot liquid temperature exiting the exchanger on the tube side. To regulate this hot exit temperature, the controller moves a valve to manipulate the flow rate of a cooling liquid entering on the shell side.

The hot tube side and cool shell side liquids do not mix. Rather, the cooling liquid surrounds the hot tubes and pulls off heat energy as it passes through the exchanger. As the flow rate of cooling liquid around the tubes increases (as the valve opens), more heat is removed and the temperature of the exiting hot liquid decreases.

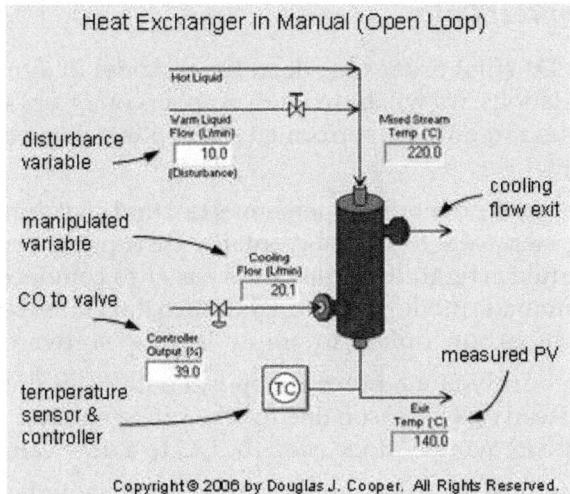

A side stream of warm liquid combines with the hot liquid entering the exchanger and acts as a disturbance to our process in this case study. As the warm stream flow rate increases, the mixed stream temperature decreases (and vice versa).

The heat exchanger in automatic mode (also called closed loop) using the standard nomenclature for this site.

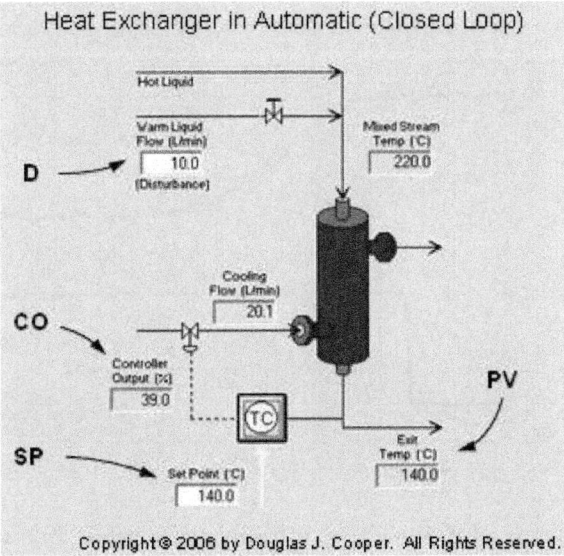

The measured process variable (PV) is the hot liquid temperature exiting the exchanger. The controller output (CO) signal moves a valve to manipulate the flow rate of cooling liquid on the shell side to maintain the PV at set point (SP). The warm liquid flow acts as a disturbance (D) to the process.

Generating Step Test Data

To fit a FOPDT (first order plus dead time) model to dynamic process data using hand calculations, we will be reading numbers off of a plot. Such a graphical analysis technique can only be performed on step test data collected in manual mode (open loop).

Practitioner's Note: operations personnel can find switching to manual mode and performing step tests to be unacceptably disruptive, especially when the production schedule is tight. It is sometimes easier to convince them to perform a closed loop (automatic mode) pulse test, but such data must be analyzed by software and this reduces our "doing" to simply "seeing" software provide answers.

To generate our dynamic process step test data, wait until the CO and PV appear to be as steady as is reasonable for the process under study. Then, after confirming we are in manual mode, step the CO to a new value.

The CO step must be large enough and sudden enough to cause the PV to move in a clear response that dominates all noise in the measurement signal. Data collection must begin before the CO step is implemented and continue until the PV reaches a new steady state.

The plot below shows dynamic step test data from the heat exchanger. Note that the PV signal (the upper trace in the plot) includes a small amount of random measurement noise. This is added in the simulation to create a more realistic process behavior.

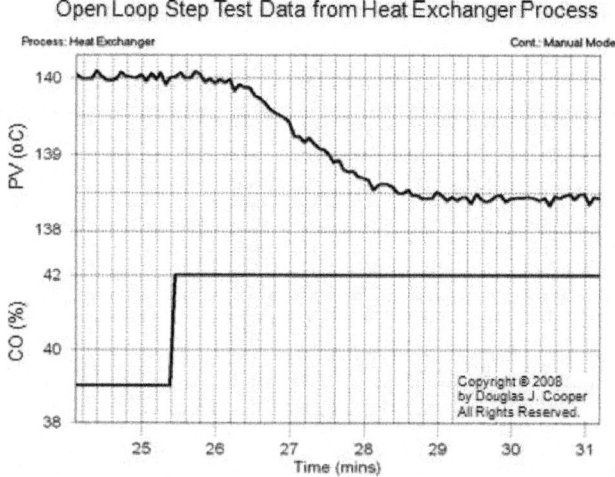

As shown in the plot, the CO is initially constant at 39% while the exit temperature PV is steady at about 140 °C. The CO is then stepped from 39% up to 42%.

The step increase in CO causes the valve to open, increasing the flow rate of cooling liquid into the shell side of the exchanger. The additional cooling liquid causes the measured PV (exit temperature on the tube side) to decrease from its initial steady state value of 140 °C down to a new value of about 138.4 °C.

We will refer back to the this dynamic process test data in future articles as we work through the details of computing process gain, Kp; process time constant, Tp; and process dead time, θp.

Practitioner's Note: the heat exchanger graphic shows that this process has one disturbance variable, D. It is a side stream of warm liquid that mixes with the hot liquid on the tube side. When generating the step test data above, disturbance D is held constant. Yet real processes can have many disturbances, and by their very nature, disturbances are often beyond our ability to monitor, let alone control. While quiet disturbances are something we can guarantee in a simulation, we may not be so lucky in the plant. Yet to accurately model the dynamics of a process, it is essential that the influential disturbances remain quiet when generating dynamic process test data. Whether the disturbances have remained quiet during a dynamic test is something you must "know" about your process. Otherwise, you should not be adjusting any controller settings.

To appreciate this sentiment, recognize that you "know" when your car is acting up. You can sense when it shows a slight but clearly different behavior that needs attention. Someone planning to adjust the controllers in an industrial operation should have this same level of familiarity with their process.

HEAT EXCHANGER PERFORMANCE TESTING

ISO 17025 Heat Exchanger Testing Services By Graftel

Graftel has over 20 years of support experience proving both equipment and on-site services to the nuclear power industry.

Heat exchanger performance testing, requires the onsite installation of instrumentation and data collection of information.

This information is processed using quality software to calculate parameters used to satisfy acceptance criterion which satisfy plant technical specification requirements.

Graftel's approach to heat exchanger testing and analysis is consistent with the methodology presented in the EPRI document *TR-107397, "Service Water Heat Exchanger Testing Guidelines."*

The testing services provided support the site's heat exchanger testing program while keeping it consistent with EPRI guidelines utilizing the four step method outlined below.

- Test plan development. A testing plan is developed that includes the below items:
  - o Establish the test objectives, acceptance criterion, analytical methodology, test conditions and overall target test uncertainty
  - o Specification of measurement system and test methodology
  - o Test uncertainty analysis
- Test procedure development. The test procedure is developed in accordance with utility specific procedure requirements. The test procedure will provide the step-by-step process for ensuring the requirements of the test plan are carried out in a manner that will ensure the test is conclusive.
- The procedure will include instructions for proper installation of required sensors so that their performance stays within criteria specified by the test plan.
- Test implementation. This includes PCs loaded with data collection software, all required instrumentation and cabling, instrument calibration, on-site support installing instrumentation; obtaining and re-cording all test data as well as providing recommendations for operations personnel in establishing optimum system configurations and heat loads.
- Performance evaluation. This evaluation includes the following key elements: -Statistical analysis of the test data reducing it to a set of input parameters for thermal performance analysis.
  - o Thermal performance analysis utilizing PROTO-HX™ or a utility-provided analytical pack-age.
  - o A detailed post-test uncertainty analysis that provides the 95 percent confidence range around the calculated heat exchanger performance parameter.
  - o A final test report containing the above information.

Graftel, LLC. maintains an Appendix B QA program implementing 10 CFR 50 Part 21. Our calibration, software and engineering services have been recently audited and recorded in the NUPIC database.

In addition, Graftel's is an accredited ISO 17025 instrument calibration laboratory, certified by LAB. Graftel's team of recently nuclear plant badged personnel are instrumentation, calibration and data collection experts.

Proto-Power's engineering services are subcontracted by Graftel. They provide heat exchanger expertise as well as the analysis software, (PROTO-HX) which is recognized as the industry standard. Their personnel have extensive experience in heat exchanger testing.

TEMPERATURE CONTROL IN A HEAT EXCHANGER

This example shows how to design feedback and feedforward compensators to regulate the temperature of a chemical reactor through a heat exchanger.

- Heat Exchanger Process
- Using Measured Data to Model The Heat Exchanger Dynamics
- Feedback Control
- Feedforward Control
- Combined Feedforward-Feedback Control
- Interactive Simulation

Heat Exchanger Process

A chemical reactor called "stirring tank" is depicted below. The top inlet delivers liquid to be mixed in the tank. The tank liquid must be maintained at a constant temperature by varying the amount of steam supplied to the heat exchanger (bottom pipe) via its control valve. Variations in the temperature of the inlet flow are the main source of disturbances in this process.

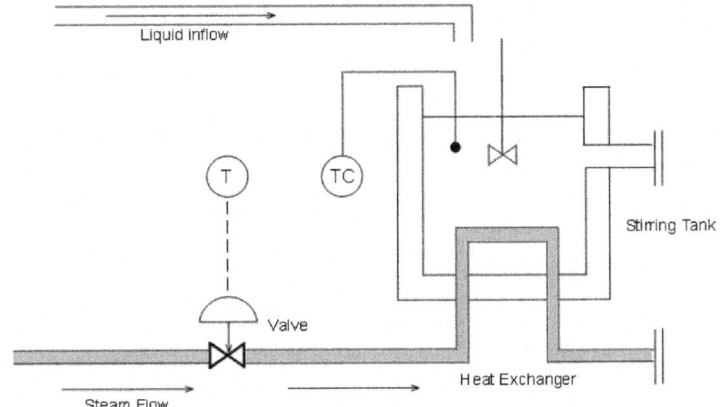

Fig. : Stirring Reactor with Heat Exchanger.

Using Measured Data to Model The Heat Exchanger Dynamics

To derive a first-order-plus-deadtime model of the heat exchanger characteristics, inject a step disturbance in valve voltage V and record the effect on the tank temperature T over time. The measured response in normalized units is shown below:

heatex_plotdata

title ('Measured response to step change in steam valve voltage');

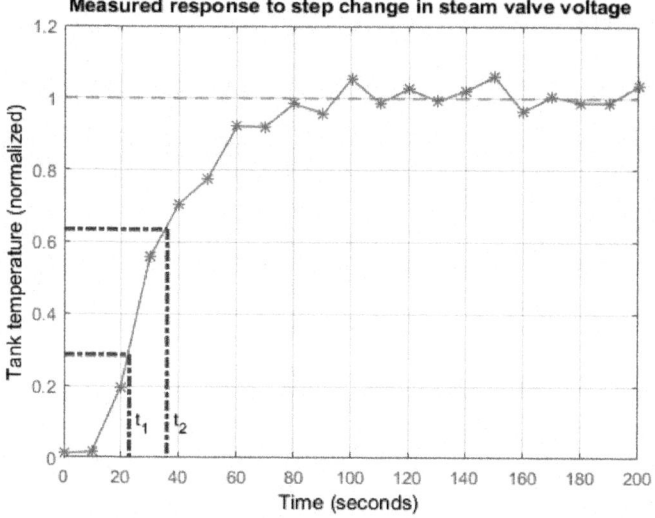

The values t1 and t2 are the times where the response attains 28.3% and 63.2% of its final value. You can use these values to estimate the time constant tau and dead time theta for the heat exchanger:

```
t1 = 21.8; t2 = 36.0;
tau = 3/2 * (t2 - t1)
theta = t2 - tau
tau = 21.3000
theta = 14.7000
```

Verify these calculations by comparing the first-order-plus-deadtime response with the measured response:

```
s = tf('s');
Gp = exp(-theta*s)/(1+tau*s)
Gp =

                  1
exp(-14.7*s) * ----------
               21.3 s + 1
```

```
Continuous-time transfer function.
hold on, step(Gp), hold off
title('Experimental vs. simulated response to step change');
```

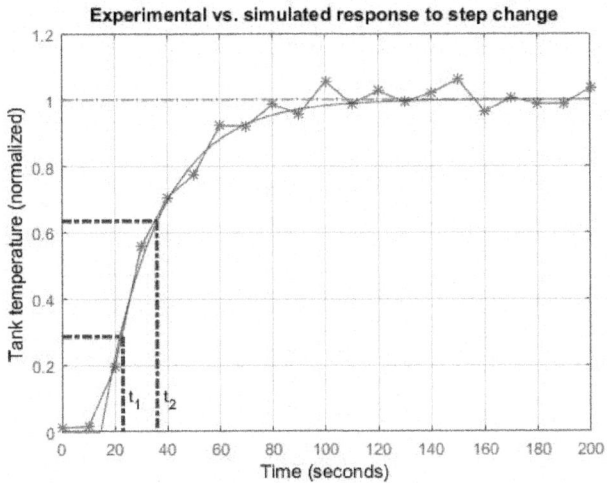

The model response and the experimental data are in good agreement. A similar bump test experiment could be conducted to estimate the first-order response to a step disturbance in inflow temperature. Equipped with models for the heat exchanger and inflow disturbance, we are ready to design the control algorithm.

Feedback Control

A block diagram representation of the open-loop process is shown below.

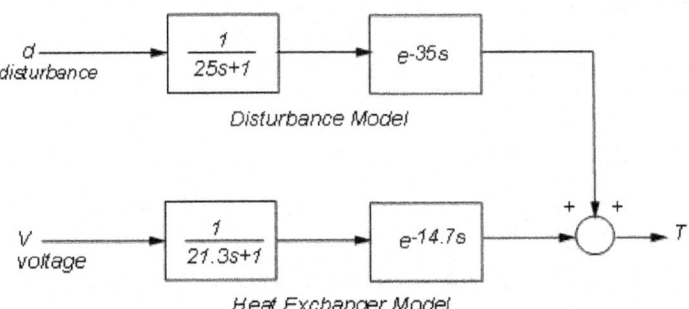

Fig. : Open-Loop Process.

The transfer function

$$G_p(s) = \frac{e^{-14.7s}}{21.3s + 1}$$

models how a change in the voltage V driving the steam valve opening affects the tank temperature T, while the transfer function

$$G_p(s) = \frac{e^{-35s}}{25s+1}$$

models how a change d in inflow temperature affects T. To regulate the tank temperature T around a given setpoint Tsp, we can use the following feedback architecture to control the valve opening (voltage V):

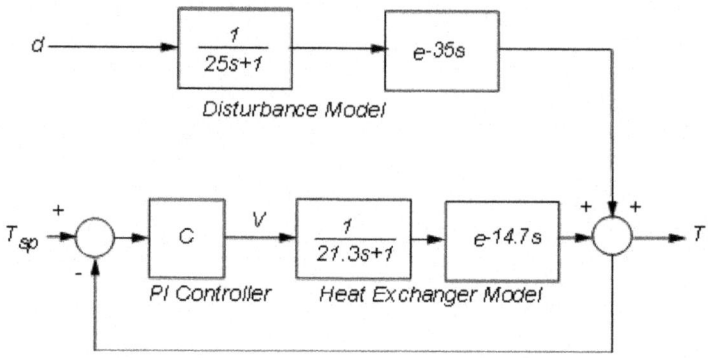

Fig. : Feedback Control.

In this configuration, the proportional-integral (PI) controller

$$C(s) = K_c \left(1 + \frac{1}{\tau_c s} \right)$$

calculates the voltage V based on the gap Tsp-T between the desired and measured temperatures. You can use the ITAE formulas to pick adequate values for the controller parameters:

$$K_c = 0.859(\theta/\tau)^{-0.977}, \quad \tau_c = (\theta/\tau)^{0.680}\,\tau/0.674$$

```
Kc = 0.859 * (theta / tau)^(-0.977)
tauc = (tau / 0.674) * (theta / tau)^0.680
C = Kc * (1 + 1/(tauc*s));
Kc = 1.2341
tauc = 24.5582
```

To see how well the ITAE controller performs, close the feedback loop and simulate the response to a set point change:

```
Tfb = feedback(ss(Gp*C),1);
step(Tfb), grid on
title('Response to step change in temperature setpoint T_{sp}')
ylabel('Tank temperature')
```

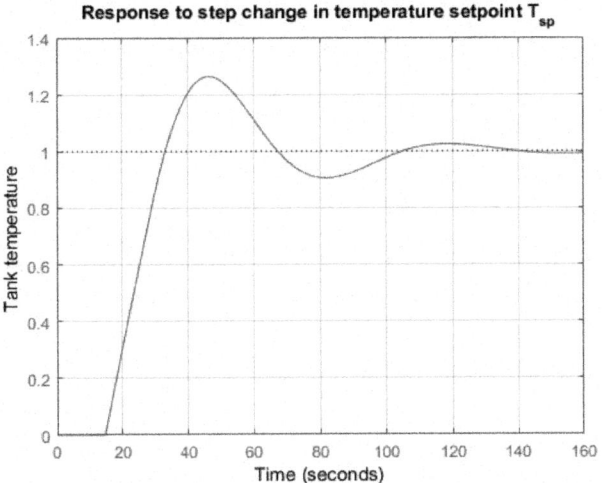

The response is fairly fast with some overshoot. Looking at the stability margins confirms that the gain margin is weak:

```
margin(Gp*C), grid
```

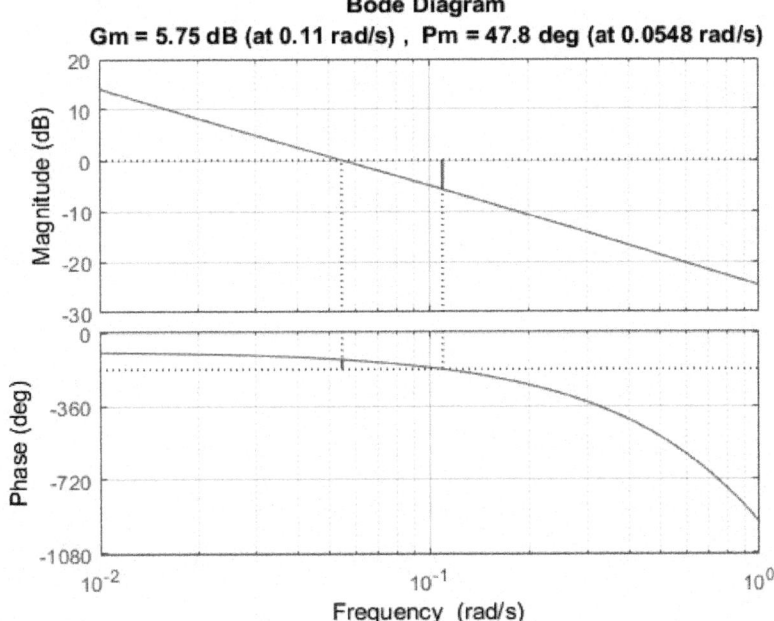

Reducing the proportional gain Kc strengthens stability at the expense of performance:

```
C1 = 0.9 * (1 + 1/(tauc*s)); % reduce Kc from 1.23 to 0.9
margin(Gp*C1), grid
```

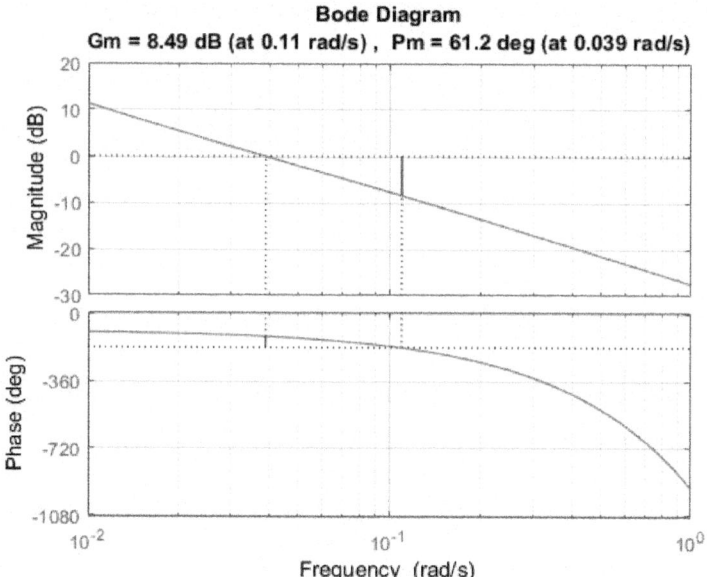

```
step(Tfb,'b', feedback(ss(Gp*C1),1),'r')
legend('Kc = 1.23','Kc = 0.9')
```

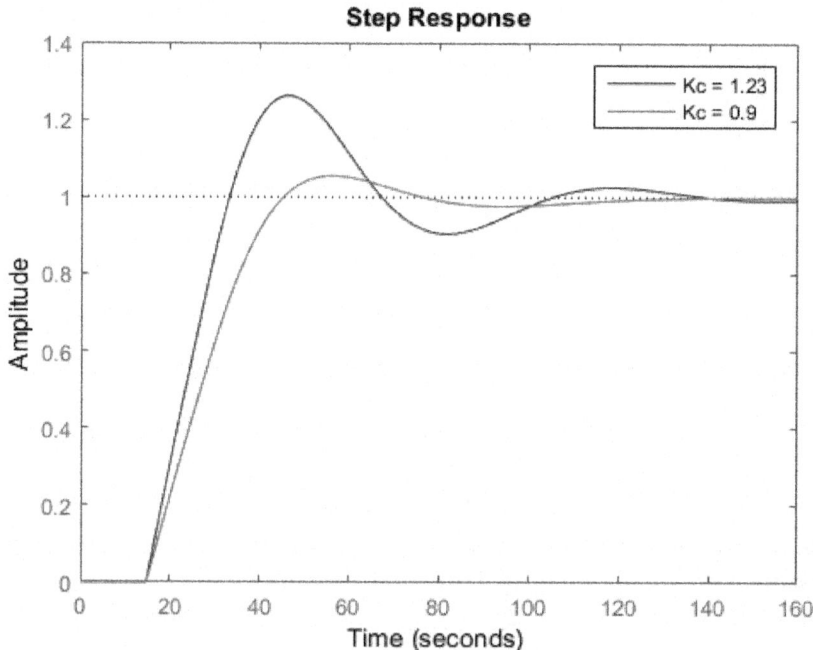

Feedforward Control

Recall that changes in inflow temperature are the main source of temperature fluctuations in the tank. To reject such disturbances, an alternative to feedback control is the feedforward architecture shown below:

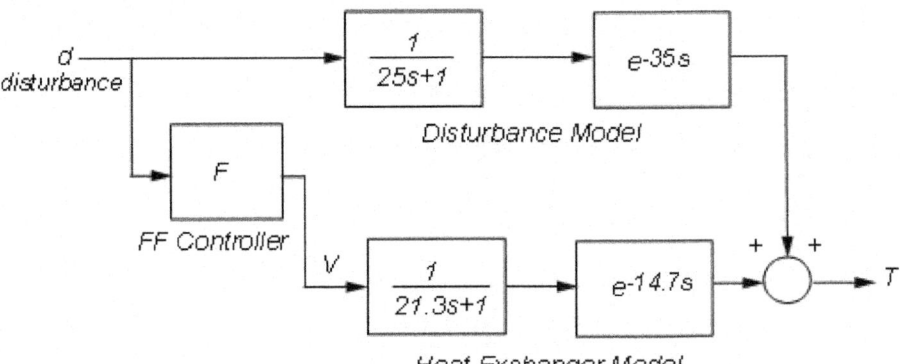

Fig. : Feedforward Control.

In this configuration, the feedforward controller F uses measurements of the inflow temperature to adjust the steam valve opening (voltage V). Feedforward control thus anticipates and preempts the effect of inflow temperature changes.

Straightforward calculation shows that the overall transfer from temperature disturbance d to tank temperature T is

$$T = (G_pF + G_d)d$$

Perfect disturbance rejection requires

$$G_pF + G_d = 0 \rightarrow F = -\frac{G_d}{G_p} = -\frac{21.3s+1}{25s+1}e^{-20.3s}$$

In reality, modeling inaccuracies prevent exact disturbance rejection, but feed forward control will help minimize temperature fluctuations due to inflow disturbances. To get a better sense of how the feedforward scheme would perform, increase the ideal feed forward delay by 5 seconds and simulate the response to a step change in inflow temperature:

```
Gd = exp(-35*s)/(25*s+1);
F = -(21.3*s+1)/(25*s+1) * exp(-25*s);
Tff = Gp * ss(F) + Gd; % d->T transfer with feedforward control
step(Tff), grid
title('Effect of a step disturbance in inflow temperature')
ylabel('Tank temperature')
```

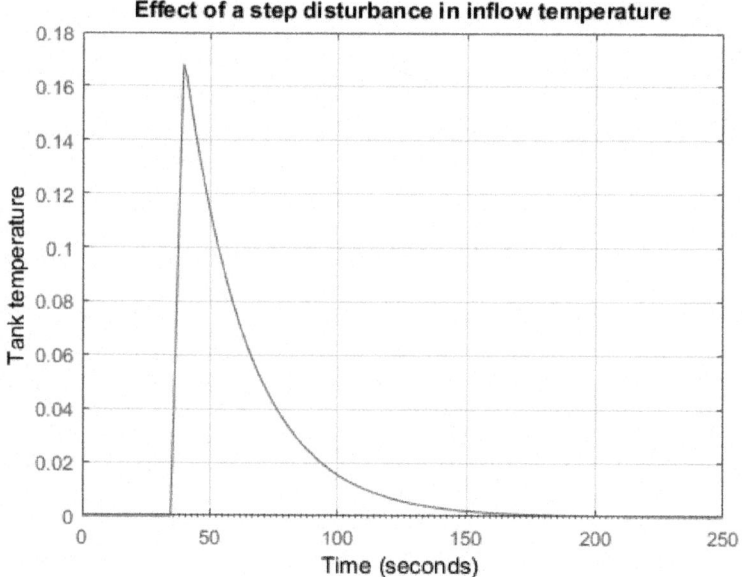

Combined Feed forward-Feedback Control

Feedback control is good for set point tracking in general, while feed forward control can help with rejection of measured disturbances. Next we look at the benefits of combining both schemes. The corresponding control architecture is shown below:

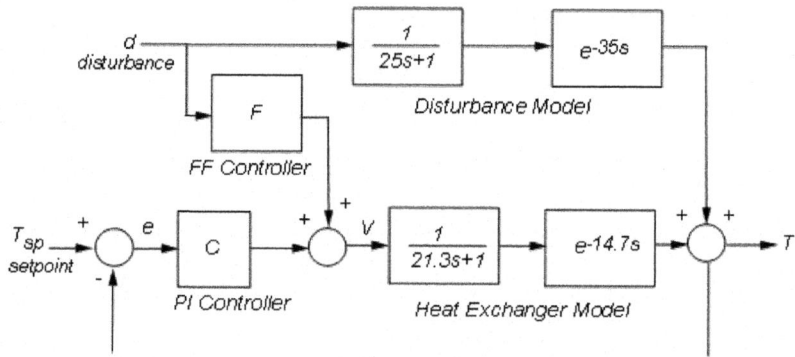

Fig. : Feed forward-Feedback Control.

Use connect to build the corresponding closed-loop model from Tsp,d to T. First name the input and output channels of each block, then let connect automatically wire the diagram:

```
Gd.u = 'd'; Gd.y = 'Td';
Gp.u = 'V'; Gp.y = 'Tp';
F.u = 'd'; F.y = 'Vf';
```

```
C.u = 'e'; C.y = 'Vc';
Sum1 = sumblk('e = Tsp - T');
Sum2 = sumblk('V = Vf + Vc');
Sum3 = sumblk('T = Tp + Td');
Tffb = connect(Gp,Gd,C,F,Sum1,Sum2,Sum3,{'Tsp','d'},'T');
```

To compare the closed-loop responses with and without feedforward control, calculate the corresponding closed-loop transfer function for the feedback-only configuration:

```
C.u = 'e'; C.y = 'V';
Tfb = connect(Gp,Gd,C,Sum1,Sum3,{'Tsp','d'},'T');
```

Now compare the two designs:

```
step(Tfb,'b',Tffb,'r--'), grid
```

```
title('Closed-loop response to setpoint and disturbance step change')
```

```
ylabel('Tank temperature')
```

```
legend('Feedback only','Feedforward + feedback')
```

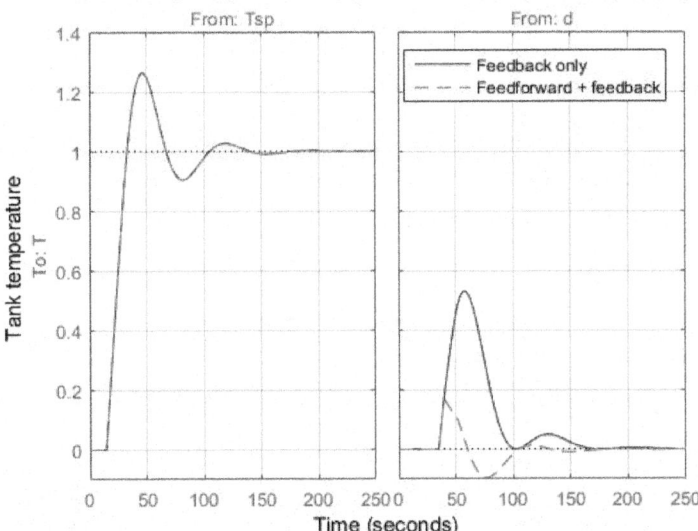

The two designs have identical performance for setpoint tracking, but the addition of feedforward control is clearly beneficial for disturbance rejection. This is also visible on the closed-loop Bode plot

```
bodemag(Tfb,'b',Tffb,'r--',{1e-3,1e1})
```

```
legend('Feedback only','Feedforward + feedback','Location','SouthEast')
```

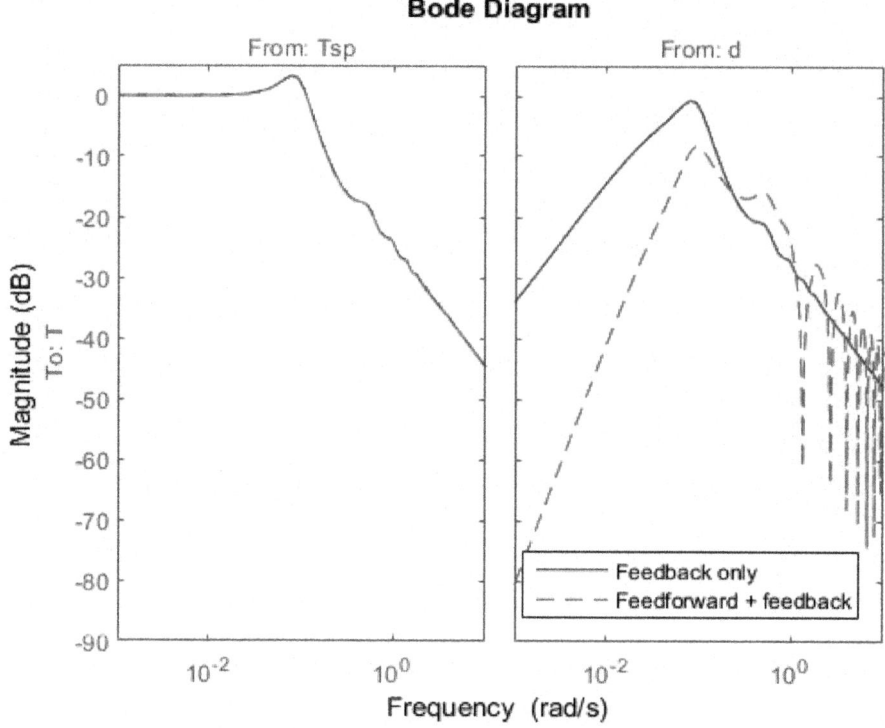

Interactive Simulation

To gain additional insight and interactively tune the feedforward and feedback gains, use the companion GUI and Simulink® model.

Open the Heat Exchanger model and GUI

heatex

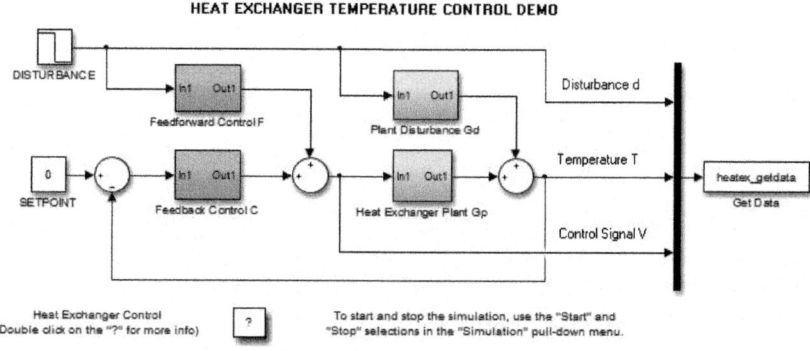

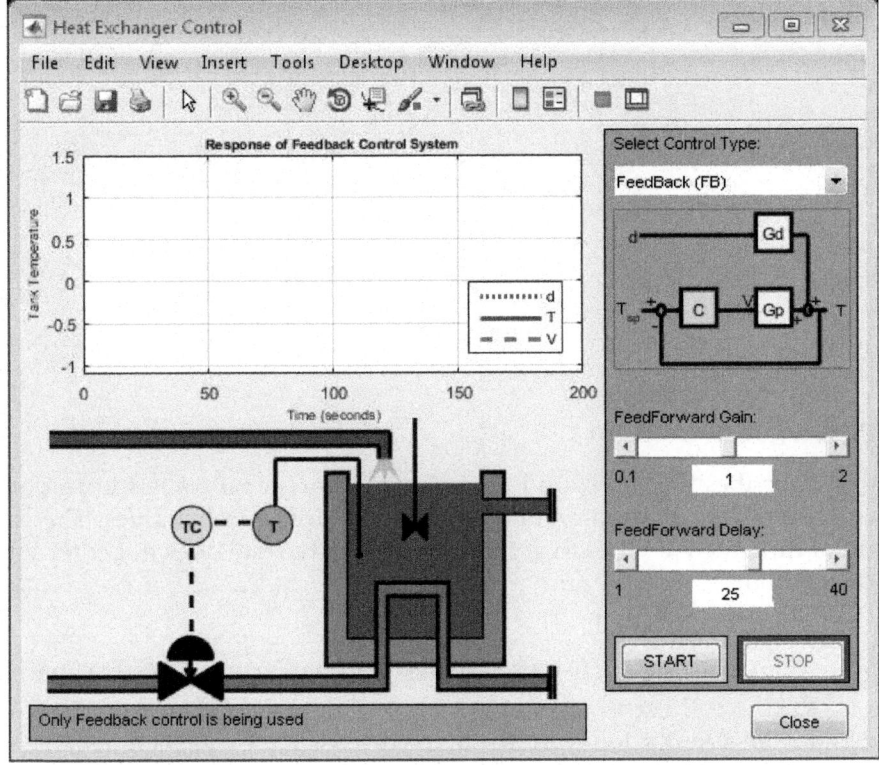

PROCESS GAIN IS THE "HOW FAR" VARIABLE

Step 3 of our controller design and tuning recipe is to approximate the often complex behavior contained in our dynamic process test data with a simple first order plus dead time (FOPDT) dynamic model.

We focus on process gain, Kp, and seek to understand what it is, how it is computed, and what it implies for controller design and tuning. Corresponding articles present details of the other two FOPDT model parameters: process time constant,_Tp; and process dead time, θp.

Heat Exchanger Step Test Data

We explore Kp by analyzing step test data from a heat exchanger. The heat exchanger is a realistic simulation where the measured process variable (PV) is the temperature of hot liquid exiting the exchanger. To regulate this PV, the controller output (CO) signal moves a valve to manipulate the flow rate of a cooling liquid into the exchanger.

The step test data below was generated by moving the process from one steady state to another. As shown, the CO was stepped from 39% up to 42%, causing the measured PV to decrease from 140 °C down to approximately 138.4 °C.

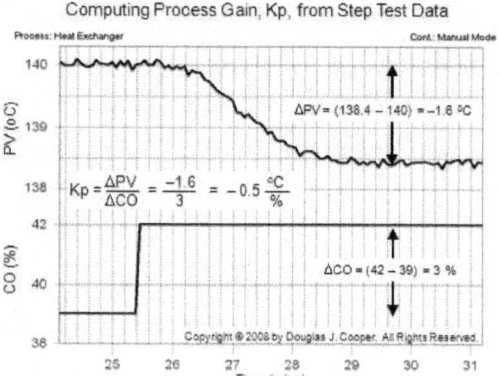

Computing Process Gain, Kp, from Step Test Data

Computing Process Gain

Kp describes the direction PV moves and how far it travels in response to a change in CO. It is based on the difference in steady state values. The path or length of time the PV takes to get to its new steady state does not enter into the Kp calculation.

Thus, Kp is computed:

> where ΔPV and ΔCO represent the total change from initial to final steady state.

Aside: the assumptions implicit in the discussion above include that:

> the process is properly instrumented as a CO to PV pair,

> major disturbances remained reasonably quiet during the test, and

> the process itself is self regulating. That is, it naturally seeks to run at a steady state if left uncontrolled and disturbances remain quiet. Most, but certainly not all, processes are self regulating. Certain configurations of something as simple as liquid level in a pumped tank can be non-self regulating.

Reading numbers off the above plot:

The CO was stepped from 39% up to 42%, so the $\Delta CO = 3\%$.

The PV was initially steady at 140 °C and moved down to a new steady state value of 138.4 °C. Since it decreased, the $\Delta PV = -1.6$ °C.

Using these ΔCO and ΔPV values in the Kp equation above, the process gain for the heat exchanger is computed:

$$Kp = \frac{\Delta PV}{\Delta CO} = \frac{138.4 - 140°C}{42 - 39\%} = -0.53\frac{°C}{\%}$$

Practitioner's Note: Real plant data is rarely as clean as that shown in the plot above and we should be cautious not to try and extract more information from our data than it actually contains. When used in tuning correlations, rounding the Kp value to -0.5 °C/% will provide virtually the same performance.

Kp Impacts Control

Process gain, Kp, is the "how far" variable because it describes how far the PV will travel for a given change in CO. It is sometimes called the sensitivity of the process.

If a process has a large Kp, then a small change in the CO will cause the PV to move a large amount. If a process has a small Kp, the same CO change will move the PV a small amount.

As a thought experiment, let's suppose a disturbance moves our measured PV away from set point (SP). If the process has a large Kp, then the PV is very sensitive to CO changes and the controller should make small CO moves to correct the error. Conversely, if the process has a small Kp, then the controller needs to make large CO actions to correct the same error.

This is the same as saying that a process with a large process gain, Kp, should have a controller with a small controller gain, Kc (and vice versa).

Looking ahead to the PI tuning correlations we will use in our case studies:

$$Kc = \frac{1}{Kp} \frac{Tp}{(\theta p + Tc)} \quad \text{and} \quad Ti = Tp$$

Where:

Kc = controller gain, a tuning parameter

Ti = reset time, a tuning parameter

Notice that in the Kc correlation, a large Kp in the denominator will yield a small Kc value (that is, Kc is inversely proportional to Kp). Thus, the tuning correlation tells us the same thing as our thought experiment above.

Sign of Kp Tells Direction

The sign of Kp tells us the direction the PV moves relative to the CO change. The negative value found above means that as the CO goes up, the PV goes down. We see this "up-down" relationship in the plot. For a process where a CO increase causes the PV to move up, the Kp would be positive and this would be an "up-up" process.

When implementing a controller, we need to know if our process is up-up or up-down. If we tell the controller the wrong relationship between CO actions and the direction of the PV responses, our mistake may prove costly. Rather than correcting for errors, the controller will quickly amplify them as it drives the CO signal, and thus the valve, pump or other final control element (FCE), to the maximum or minimum value.

Units of Kp

If we are computing Kp and want the results to be meaningful for control, then we must be analyzing wire out to wire in CO to PV data as used by the controller.

The heat exchanger data plot indicates that the data arriving on the PV wire into the controller has been scaled (or is being scaled in the controller) into units of temperature. And this means the the controller gain, Kc, needs to reflect the units of temperature as well.

This may be confusing, at least initially, since most commercial controllers do not require that units be entered. The good news is that, as long as our computations use the same "wire out to wire in" data as collected and displayed by our controller, the units will be consistent and we need not dwell on this issue.

Aside: for the Kc tuning correlation above, the units of time (dead time and time constants) cancel out. Hence, the controller gain, Kc, will have the reciprocal or inverse units of Kp. For the heat exchanger, this means Kc has units of %/°C. With modern computer control systems, scaling for unit conversions is becoming more common in the controller signal path. Sometimes a display has been scaled but the signal in the loop path has not. You must pay attention to this detail and make sure you are using the correct units in your computations. It has been suggested that the gain of some controllers do not have units since both the CO and PV are in units of %. Actually, the Kc will have units of "% of CO signal" divided by "% of PV signal," which mathematically do not cancel out.

Practitioner's Note: Step test data is practical in the sense that all model fitting computations can be performed by reading numbers off of a plot. However, when dealing with production processes, operations personnel tend to prefer quick "bumps" rather than complete step tests.Step tests move the plant from one steady state to another. This takes a long time, may have safety implications, and can create expensive off-spec product. Pulse and doublet tests are examples of quick bumps that returns our plant to desired operating conditions as soon as the process data shows a clear response to a controller output (CO) signal change.Getting our plant back to a safe, profitable operation as quickly as possible is a popular concept at all levels of operation and management. Using pulse tests requires the use of inexpensive commercial software to analyze the bump test results, however.

PROCESS TIME CONSTANT IS THE "HOW FAST" VARIABLE

Step 3 of our controller design and tuning recipe is to approximate the often complex behavior contained in our dynamic process test data with a simple first order plus dead time (FOPDT) dynamic model.

We focus on process time constant, Tp, and seek to understand what it is, how it is computed, and what it implies for controller design and tuning. Corresponding articles present details of the other two FOPDT model parameters: process gain, Kp; and process dead time, θp.

Heat Exchanger Step Test Data

We seek to understand Tp by analyzing step test data from a heat exchanger. The heat exchanger is a realistic simulation where the measured process variable

(PV) is the temperature of hot liquid exiting the exchanger. To regulate this PV, the controller output (CO) moves a valve to manipulate the flow rate of a cooling liquid into the exchanger.

The step test data below was generated by moving the process from one steady state to another. As shown, the CO was stepped from 39% up to 42%, causing the measured PV to decrease from 140 °C down to approximately 138.4 °C.

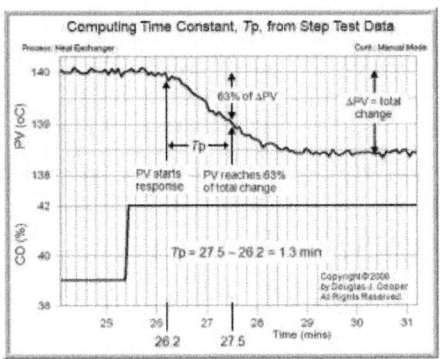

Time Constant in Words

In general terms, the time constant, Tp, describes how fast the PV moves in response to a change in the CO.

The time constant must be positive and it must have units of time. For controllers used on processes comprised of gases, liquids, powders, slurries and melts, Tp most often has units of minutes or seconds.

We can be more precise in our word definition if we restrict ourselves to step test data such as that shown in the plot above. Please recognize that while it is easier to describe Tp in words using step test data, it is a parameter that always describes "how fast" PV moves in response to any sort of CO change.

Step test data implies that the process is in manual mode (open loop) and initially at steady state. A step in the CO has forced a response in the PV, which moves from its original steady state value to a final steady state.

With these restrictions, we compute Tp in five steps:

1. Determine ΔPV, the total change that is going to occur in PV, computed as "final minus initial steady state"

2. Compute the value of the PV that is 63% of the total change that is going to occur, or "initial steady state PV + 0.63(ΔPV)"

3. Note the time when the PV passes through the 63% point of "initial steady state PV + 0.63(ΔPV)"

4. Subtract from it the time when the "PV starts a clear response" to the step change in the CO

5. The passage of time from step 4 minus step 3 is the process time constant, Tp.

Summarizing in one sentence, for step test data, Tp is the time that passes from when the PV shows its first response to the CO step, until when the PV reaches 63% of the total DPV change that is going to occur.

Computing Tp for the Heat Exchanger

Following the steps above for the heat exchanger step test data:

1. The PV was initially steady at 140 °C and moved down to a final steady state of 138.4 °C. The total change, ΔPV, is "final minus initial steady state" or:

$$\Delta PV = 138.4 - 140 = -1.6 \text{ °C}$$

2. The value of the PV that is 63% of this total change is "initial steady state PV + 0.63ΔPV" or:

$$\text{initial PV} + 0.63(\Delta PV) = 140 + 0.63(-1.6)$$
$$= 140 - 1.0$$
$$= 139 \text{ °C}$$

3. From the plot, the time when the PV passes through the "initial steady state PV + 0.63ΔPV" point of 139 °C is:

$$\text{Time to } 0.63(\Delta PV) = \text{Time to } 139 \text{ °C}$$
$$= 27.5 \text{ min}$$

4. From the plot, the time when the "PV starts a response" to the CO step is:

$$\text{Time PV response starts} = 26.2 \text{ min}$$

5. The time constant is "time to 63%(ΔPV)" minus "time PV response starts" or:

$$Tp = 27.5 - 26.2 = \textbf{1.3 min}$$

DEAD TIME VERSUS TIME CONSTANT

The dynamic response of self-regulating processes can be described reasonably accurately with a simple model consisting of process gain, dead time and lag (time constant). The process gain describes how much the process will respond to a change in controller output, while the dead time and time constant describes how quickly the process will respond.

Although the dead time and time constant both seem to describe the same thing, there are several fundamental differences between how dead time and time constant affects a control loop. The first difference is that dead time describes how long it takes before a process begins to respond to a change in controller output, and the time constant describes how fast the process responds once it has begun moving.

Measuring the Dead Time and Time Constant of a Process

Let's begin with the measurement of dead time and time constant of a self-regulating process. Typically, one will place the controller in manual control mode, wait for the process variable to settle down, and then make a step change of a few percent in the controller output. At first the process variable does nothing (dead time) and then it begins changing (time constant) until finally it settles out at a new level.

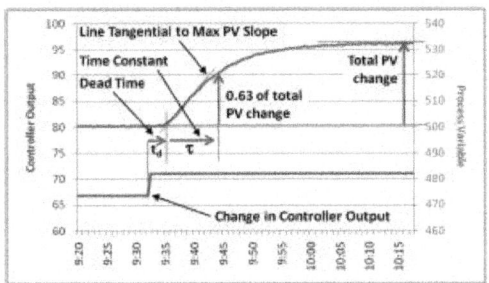

Fig. : Measuring Dead Time and Time Constant.

To measure the dead time and time constant, draw a horizontal line at the same level as the original process variable. We'll call this the baseline. Then find the maximum vertical slope of the process variable response curve. Draw a line tangential to the maximum slope all the way to cross the baseline. We'll call this crossing the *intersection*.

The process dead time is measured along the time axis as the time spanned between the step change in controller output and the intersection.

Next, measure the total change in process variable. Then find the point on the process response curve where the process variable has changed by 0.63 of the total change in process variable. We'll call this point P63.

The process time constant is measured along the time axis as the time spanned between the intersection (described previously) and P63.

Dead Time versus Time Constant

We can draw a chart with a continuum of dead time through time constant. Processes woth dynamics consisting of pure dead time will be on the left and pure lag (time constant) on the right. In the middle the process dead time will equal its time constant.

We'll find that flow loops and liquid pressure loops fall just about in the middle of the continuum, because their dead time and time constant are almost equal. Gas pressure and temperature loops will be located more toward the right – they are lag (time constant) dominant. Serpentine channels in water treatment plants and conveyors with downstream mass meters will appear on the left side – they are dead-time dominant.

Level loops should actually be treated differently, but can be approximated on the continuum by replacing the time constant with their residence time (time they will take to fill or empty out at full flow rate.) Most level loops will be located far to the right, having relatively short dead times.

The ratio of dead time to time constant affects the controller modes and tuning rules we use, the controllability of the process, and the minimum possible loop settling time.

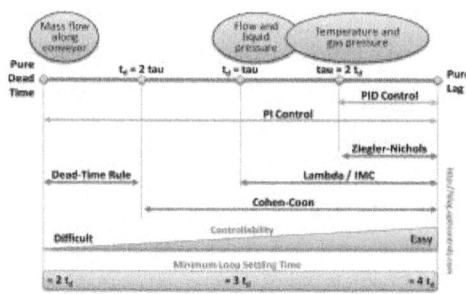

Fig. : A continuum from pure Dead Time to pure Lag.

Controller Modes

The derivative control mode works well where process variables continue to move in the same direction for some time, *i.e.* lag-dominant processes. Derivative control does not work well on processes where the process variable changes sporadically – typically processes with relatively short time constants, located in the middle and to the left on the continuum.

Applicability of Tuning Rules

Most tuning rules will work on lag-dominant processes. However, the Ziegler-Nichols rules have only a narrow range of applicability. Lambda / IMC tuning rules apply to a broader spectrum of processes, while Cohen-Coon has the widest coverage. The Dead-Time tuning rule, applies to processes on the left, as its name implies.

Controllability

Lag-dominant loops are easier to control than dead-time-dominant loops. Operators find that lag-dominant processes respond much more intuitively than dead-time-dominant processes and are easier to control in manual mode.

Loop Settling Time

When tuning a loop for the shortest possible settling time, one finds that there is a minimum limit on settling time. If you tune the controller any tighter, the loop will begin oscillating. The minimum settling time depends mostly on the amount of dead time in a control loop, and will be between two and four times the length of the dead time. The ratio of time constant to dead time determines where the minimum settling time falls between two and four times the process dead time.

DEAD TIME IS THE "HOW MUCH DELAY" VARIABLE

Step 3 of our controller design and tuning recipe is to approximate the often complex behavior contained in our dynamic process test data with a simple first order plus dead time (FOPDT) dynamic model.

We focus on process dead time, θp, and seek to understand what it is, how it is computed, and what it implies for controller design and tuning. Corresponding articles present details of the other two FOPDT model parameters: process gain, Kp; and process time constant, Tp.

Dead Time is the Killer of Control

Dead time is the delay from when a controller output (CO) signal is issued until when the measured process variable (PV) first begins to respond. The presence of dead time, θp, is never a good thing in a control loop.

Think about driving your car with a dead time between the steering wheel and the tires. Every time you turn the steering wheel, the tires do not respond for, say, two seconds. Yikes.

For any process, as θp becomes larger, the control challenge becomes greater and tight performance becomes more difficult to achieve.

Causes for Dead Time

Dead time can arise in a control loop for a number of reasons:

Control loops typically have "sample and hold" measurement instrumentation that introduces a minimum dead time of one sample time, T, into every loop. This is rarely an issue for tuning, but indicates that every loop has at least some dead time.

The time it takes for material to travel from one point to another can add dead time to a loop. If a property (*e.g.* a concentration or temperature) is changed at one end of a pipe and the sensor is located at the other end, the change will not be detected until the material has moved down the length of the pipe. The travel time is dead time. This is not a problem that occurs only in big plants with long pipes. A bench top process can have fluid creeping along a tube. The distance may only be an arm's length, but a low enough flow velocity can translate into a meaningful delay.

Sensors and analyzer can take precious time to yield their measurement results. For example, suppose a thermocouple is heavily shielded so it can survive in a harsh environment. The mass of the shield can add troublesome delay to the detection of temperature changes in the fluid being measured.

Higher order processes have an inflection point that can be reasonably approximated as dead time for the purpose of controller design and tuning. Note that modeling for tuning with the simple FOPDT form is different from modeling for simulation, where process complexities should be addressed with more sophisticated model forms.

Sometimes dead time issues can be addressed through a simple design change. It might be possible to locate a sensor closer to the action, or perhaps switch to a faster responding device. Other times, the dead time is a permanent feature of the control loop and can only be addressed through detuning or implementation of a dead time compensator (*e.g.* Smith predictor).

Heat Exchanger Test Data

We seek to understand Kp by analyzing step test data from a heat exchanger. The heat exchanger is a realistic simulation where the measured process variable (PV) is the temperature of hot liquid exiting the exchanger. To regulate this PV, the controller output (CO) moves a valve to manipulate the flow rate of a cooling liquid into the exchanger.

The step test data below was generated by moving the process from one steady state to another. In particular, CO was stepped from 39% up to 42%, causing the measured PV to decrease from 140 °C down to approximately 138.4 °C.

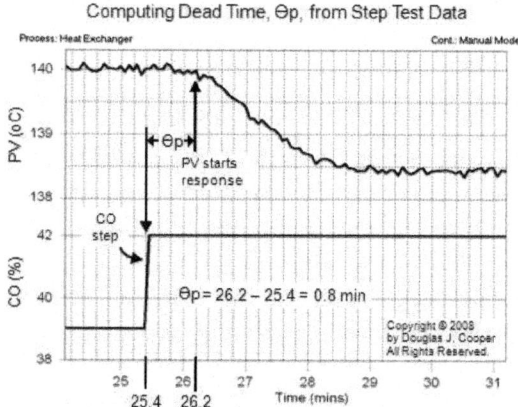

Computing Dead Time, Өp, from Step Test Data

Computing Dead Time

Estimating dead time, θp, from step test data is a three step procedure:

1. Locate the point in time when the "PV starts a clear response" to the step change in the CO. This is the same point we identified when we computed *T*p.

2. Locate the point in time when the CO was stepped from its original value to its new value.

3. Dead time, θp, is the difference in time of step 1 minus step 2.

Applying the three step procedure to the step test plot above:

1. As we had determined in the previous Tp, the PV starts a clear response to the CO step at 26.2 min.

2. Reading off the plot, the CO step occurred at 25.4 min, and thus,

3. **θp = 26.2 - 25.4 = 0.8 min**

We analyze step test data here to make the computation straightforward, but please recognize that dead time describes "how much delay" occurs from when any sort of CO change is made until when the PV first responds to that change.

Like a time constant, dead time has units of time and must always be positive. For the types of processes explored on this site (streams comprised of gasses, liquids, powders, slurries and melts), dead time is most often expressed in minutes or seconds.

During a dynamic analysis study, it is best practice to express Tp and θp in the same units (*e.g.* both in minutes or both in seconds). The tuning correlations and design rules assume consistent units. Control is challenging enough without adding computational error to our problems.

Implications for Control

- Dead time, θp, is large or small only in comparison to Tp, the clock of the process. Tight control becomes more challenging when $\theta p > Tp$. As dead time becomes much greater than Tp, a dead time compensator such as a Smith predictor offers benefit. A Smith predictor employs a dynamic process model (such as an FOPDT model) directly within the architecture of the controller. It requires additional engineering time to design, implement and maintain, so be sure the loop is important to safety or profitability before undertaking such a project.

- It is more conservative to overestimate dead time when the goal is tuning. Computing θp requires a judgment of when the "PV starts a clear response." If your judgment says that the "clear response" is sooner in time (maybe you choose 26.0 min for this case), then Tp increases, but dead time, θp, decreases by the same amount (and vice versa). We can see the impact this has by looking ahead to the PI tuning correlations:

$$Kc = \frac{1}{Kp} \frac{Tp}{(\theta p + Tc)} \quad \text{and} \quad Ti = Tp$$

Where:

Kc = controller gain, a tuning parameter

Ti = reset time, a tuning parameter

Since θp is in the denominator of the Kc correlation, as dead time gets larger, the controller gain gets smaller. A smaller Kc implies a less active controller. Overly aggressive controllers cause more trouble than sluggish controllers, at least in the first moments after being put into automatic. Hence, a larger dead time estimate is a more cautious or conservative estimate.

> Practitioner's Note on the "$\theta p,min = T$" Rule for Controller Tuning: Consider that all controllers measure, act, then wait until next sample time; measure, act, then wait until next sample time. This "measure, act, wait" procedure has a delay (or dead time) of one sample time, T, built naturally into its structure.

> Thus, the minimum dead time, θp, in any real control implementation is the loop sample time, T. Dead time can certainly be larger than T (and it usually is), but it cannot be smaller. Thus, if our model fit yields $\theta p < T$ (a dead time that is less than the controller sample time), we must recognize that this is an impossible outcome. Best practice in such a situation is to substitute $\theta p = T$ everywhere when using our controller tuning correlations and other design rules.

> This is the "$\theta p,min = T$" rule for controller tuning.

VALIDATING OUR HEAT EXCHANGER FOPDT MODEL

Over a series of articles, we generated step test data from a heat exchanger process simulation and then explored details of how to perform a graphical analysis of the plot data to compute values for a first order plus dead time (FOPDT) dynamic model.

The claim has been that the parameters from this FOPDT model will offer us a good approximation of the complex dynamic behavior of our process. And this, in turn, will provide us information critical for the design and tuning of a PID controller.

So appropriate questions at this point might be:

- How well does our FOPDT model fit the heat exchanger data?
- How does this help us design and tune our PID controller?

Comparing Graphical Model to Data

Here are the FOPDT model parameter estimates we computed from the graphical analysis of the step test data:

- Process gain (how far), $Kp = -0.53\ °C/\%$
- Time constant (how fast), $Tp = 1.3$ min
- Dead time (how much delay), $\theta p = 0.8$ min

Aside: the FOPDT dynamic model has the general controller output (CO) to process variable (PV) form:

$$Tp\frac{dPV(t)}{dt} + PV(t) = Kp \cdot CO(t - \theta p)$$

And this means we are claiming that the dynamic behavior of the heat exchanger can be reasonably approximated as:

$$1.3\frac{dPV(t)}{dt} + PV(t) = -0.53.CO(t-0.8)$$

With units: t [=] min, PV(t) [=] °C, CO(t – θp) [=] %

The FOPDT model prediction for PV in the plot below was generated by solving the above differential equation using the actual CO signal trace as shown in the plot.

The plot below compares step test data from the heat exchanger process to the FOPDT model using the parameters we computed in the previous articles.

Visual inspection reveals that the simple FOPDT model provides a very good approximation of the dynamic response behavior between the controller output (CO) signal and the measured process variable (PV) for this process.

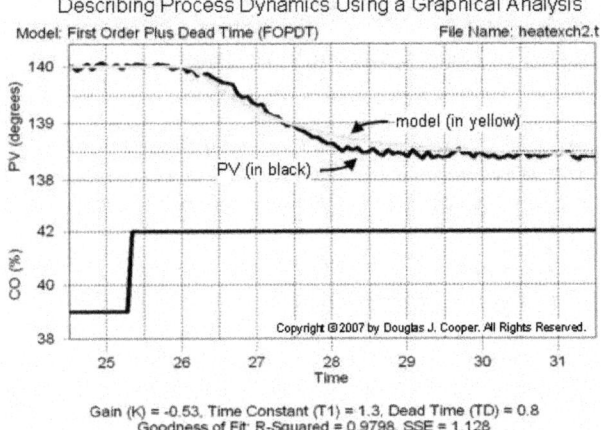

Gain (K) = -0.53, Time Constant (T1) = 1.3, Dead Time (TD) = 0.8
Goodness of Fit: R-Squared = 0.9798, SSE = 1.128

This validates that, for the heat exchanger process at this design level of operation, we have good knowledge of:

the direction the PV moves given a change in the CO

how far the PV ultimately travels for a given change in the CO

how fast the PV moves as it heads toward its new steady state

how much delay occurs between when the CO changes and the PV first begins to respond

Modeling Using Software

For comparison, we show a FOPDT model fit of the same heat exchanger data using a software modeling tool. The model parameters are somewhat different and the model appears to match the PV data a little better based on a visual inspection.

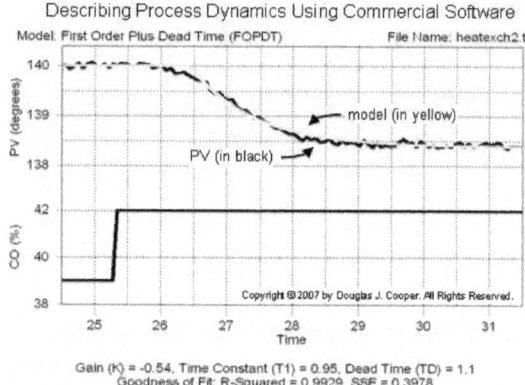

Describing Process Dynamics Using Commercial Software

Gain (K) = -0.54, Time Constant (T1) = 0.95, Dead Time (TD) = 1.1
Goodness of Fit: R-Squared = 0.9929, SSE = 0.3978

The benefits of commercial software for controller design and tuning. For now, it is enough to know that:

the option is available,

FOPDT modeling can occur with just a few mouse clicks instead of using graph paper and hand calculations, and

the results are more descriptive when compared to the graphical method.

With that said, graphical modeling is important to understand because it helps us isolate the role of each FOPDT model parameter, and in particular, appreciate what each says about the controller output (CO) to process variable (PV) relationship.

Using the Model in Controller Design

With a FOPDT model in hand, we can complete the controller design and tuning recipe to implement P-Only, PI, PID and PID with CO Filter control of the heat exchanger process.

The exciting result is that we achieve our desired controller performance using the information from this single step test. Our results are achieved quickly, there is no trial and error, we produce minimal off-spec product, and thus, there is no wasted time or expense.

The method of approximating complex behavior with a FOPDT model and then following a recipe for controller design and tuning has been proven on a wide variety of processes. In fact, it is an approach being used to improve profitability and safety in many plants today.

SIMPLE, MODEL-BASED PROCESS CONTROL

Nonlinear Control Techniques can be Used for more Efficient Process Operation

Often, nonlinearity is the primary problem for single-input-single-output (SISO) chemical process control. One solution is to design the process for PID control success - for linear responses, or with large inventories to reduce interaction and temper upsets. However, design is an act of balancing multiple objectives; other desirable issues, such as capital cost, flexibility, resource use, energy integration, and sustainability are usually sacrificed to accommodate process control. Practicable techniques for nonlinear control can ease design constraints and permit the operation of more competitive processes.

For nonlinear processes, gain scheduling is a conventional solution to accommodate nonlinearity. In gain scheduling, the controller coefficients are changed to reflect the operating region; the engineer uses process knowledge to create tuning values, which are either placed in a look-up table or expressed in equations.

However, if the engineer's process knowledge is expressed as a dynamic nonlinear model, it is almost as easy to implement a process model-based controller (PMBC) as it is gain scheduling. Several process model-based control approaches have found success in industrial applications. There are several commercial products for relatively simple nonlinear process model-based controllers, engineers have implemented versions in-house, and control integrators and service providers have been implementing model-based controllers for decades.

Process-model based control (PMBC) has several advantages over either classic PID or gain-scheduled PID, even with ratio and feedforward enhancements of advanced regulatory control (ARC). PMBC has a single-tuning parameter, has nonlinear compensation throughout the entire operating range, preserves process knowledge, and provides continuous monitoring of the process - for health, predictive maintenance, and constraint recognition, and economic optimization of setpoints.

PMBC can also be used within a horizon-predictive, constraint-handling framework. And, for multi-input-multi-output (MIMO) processes, PMBC can additionally decouple nonlinear interaction, balance deviations from setpoints when manipulated variable (MV) constraints are hit, and determine economic optimum MV values when there are extra degrees of freedom.

However, this chapter addresses only the multi-input-single-output (MISO), single step-ahead control approach. This approach can solve many problems and can be implemented by a process engineer.

The references describe PMBC applications on commercial-scale, pilot-scale, and lab-scale processes. The references cite SISO and MISO applications for control of fluid flow rate, heat exchanger temperature, distillation bottoms composition, plasma reactor pressure, and pH. MIMO applications include distillation dual-end composition control and fluidized bed gasification.

Modeling Perspectives

These, and similar process models, do not pretend to be rigorous; fortunately for control, the model perfection is not needed. Consider that the controller decides a control action that is only 75 percent perfect, leaving 25 percent error. In simplistic qualitative reasoning, at the next control action the controller will correct the 25 percent error by 75 percent. With a rule of thumb that there should be 30 control actions within a process settling time, the continual "steering" by the controller will have the process at the setpoint, with only a 0.25^{30} (vanishingly small) error. Balancing precision with sufficiency, a model that would represent a homework grade of "B-" or "C+" is fully adequate for control.

This robustness to model imprecision has several advantages. It lessens the burden on the control engineer to achieve model perfection. And it permits the controller to remain functional as a process continually changes because of fouling, reactivity, yield, filter/screen accumulation, or piping modifications, or such change in time.

Additionally, such simple models often represent the engineer's understanding of the process, and reflect methods already in use for process analysis, device sizing, and analytical trouble-shooting. The process engineer can develop such models, and imbedding them in controllers promulgates process knowledge.

These models do not exactly match the process. There is uncertainty in co-efficients (*e.g.*, heat transfer fouling, catalyst reactivity, tray efficiency, friction losses in pipes and fittings, concentration of non-key components, amount of non-condensable gases, *etc.*). Accordingly, when used for control, there needs to be a feedback correction.

THE GRAVITY DRAINED TANKS PROCESS

Self Regulating vs Integrating Process Behavior

This case study considers the control of liquid level in a gravity drained tanks process. Like the heat exchanger, the gravity drained tanks displays a typical self regulating process behavior. That is, the measured process variable (PV) naturally seeks a steady operating level if the controller output (CO) and major disturbances are held constant for a sufficient length of time.

It is important to recognize that not all processes, and especially not all liquid level processes, exhibit a self regulating behavior. Liquid level control of a pumped tank process, for example, displays a classical integrating (or on-self-regulating) behavior.

The control of integrating processes presents unique challenges that we will explore in later articles. For now, it is enough to recognize that controller design and tuning for integrating processes has special considerations.

We note that, like the heat exchanger and pumped tank process, the gravity drained tanks case study is a sophisticated simulation derived from first-principles theory and available incommercial software. Simulations let us study different ideas without risking safety or profit. Yet the rules and procedures we develop here are directly applicable to the broad world of real processes with streams comprised of liquids, gases, powders, slurries and melts.

Gravity Drained Tanks Process

The gravity drained tanks process, shown below in manual mode, is comprised of two tanks stacked one above the other. They are essentially two drums or barrels with holes punched in the bottom.

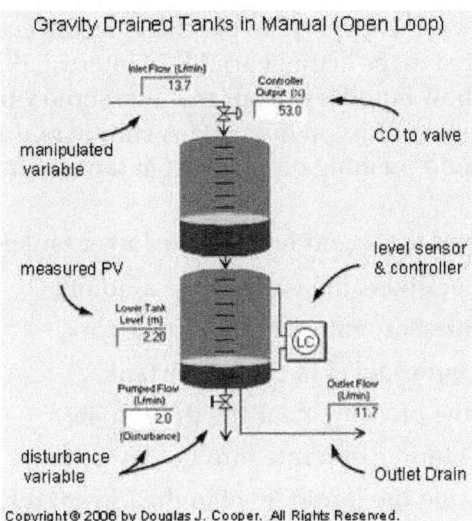

A variable position control valve manipulates the inlet flow rate feeding the upper tank. The liquid drains freely out through the hole in the bottom of the upper tank to feed the lower tank. From there, the liquid exits either by an outlet drain (another free-draining hole) or by a pumped flow stream.

The next graphic shows the process in automatic mode using our standard nomenclature.

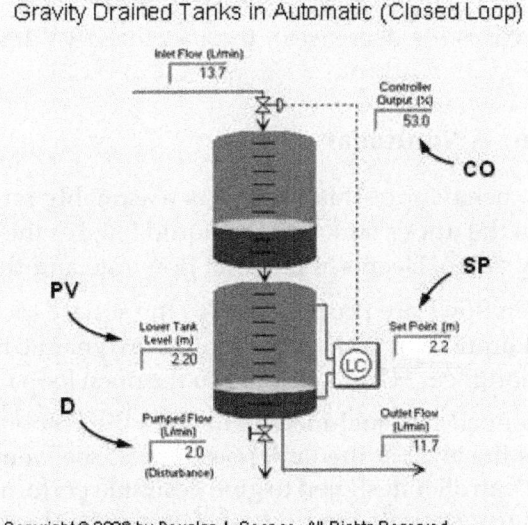

The measured process variable (PV) is liquid level in the lower tank. The controller output (CO) adjusts the valve to maintain the PV at set point (SP).

Aside: The graphic shows tap lines out of the top and bottom of the lower tank and entering the level sensor/controller. This configuration hints at the use of pressure drop as the level measurement method. We often choose to use sensors that are inexpensive to purchase, install and maintain to monitor parameters related to the actual variable of interest. While our trend plots in this case study show liquid level, our true measured variable is pressure drop across the liquid inventory in the tank. A simple multiplier block translates the weight of liquid pushing on the bottom tap into this level measurement display.

So, for example, if the liquid level in the lower tank is below set point:

- the controller opens the valve some amount,
- increasing the flow rate into the upper tank,
- raising the liquid level in the upper tank,
- increasing the pressure near the drain hole,
- raising the liquid drain rate into the lower tank,
- thus increasing the liquid level in the lower tank.

The Disturbance Stream

The pumped flow stream out of the lower tank acts as a disturbance to this process. The disturbance flow (D) is controlled independently, as if by another process (which is why it is a disturbance to our process).

Because the pumped flow rate, D, runs through a positive displacement pump, it is not affected by liquid level, though it drops to zero if the tank empties.

When D increases (or decreases), the measured PV level quickly falls (or rises) in response.

Process Behavior is Nonlinear

The dynamic behavior of this process is reasonably intuitive. Increase the inlet flow rate into the upper tank and the liquid level in the lower tank eventually rises to a new value. Decrease the inlet flow rate and the liquid level falls.

Gravity driven flows are proportional to the square root of the hydrostatic head, or height of liquid in a tank. As a result, the dynamic behavior of the process is modestly nonlinear. This is evident in the open loop response plot below.

The CO is stepped in equal increments, yet the response shape of the PV clearly changes as the level in the tank rises. The consequence of this nonlinear behavior is that a controller designed to give desirable performance at one operating level may not give desirable performance at another level.

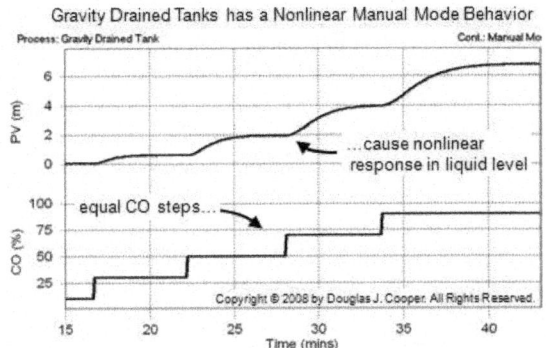

Gravity Drained Tanks has a Nonlinear Manual Mode Behavior

Modeling Dynamic Process Behavior

We next explore dynamic modeling of process behavior for the gravity drained tanks.

DYNAMIC "BUMP" TESTING OF THE GRAVITY DRAINED TANKS PROCESS

We introduced the gravity drained tanks process and established that it displays a self regulating behavior. We also learned that it exhibits a nonlinear behavior, though to a lesser degree than that of the heat exchanger.

Our control objective is to maintain liquid level in the lower tank at set point in spite of unplanned and unmeasured disturbances. The controller will achieve this by manipulating the inlet flow rate into the upper tank.

To proceed, we follow our controller design and tuning recipe:

1. Establish the design level of operation (DLO), defined as the expected values for set point and major disturbances during normal operation

2. Bump the process and collect controller output (CO) to process variable (PV) dynamic process data around this design level

3. Approximate the process data behavior with a first order plus dead time (FOPDT) dynamic model to obtain estimates for process gain, Kp (how far variable), process time constant, Tp (how fast variable), and the process dead time, θp (with how much delay variable).

4. Use the model parameters from step 3 in rules and correlations to complete the controller design and tuning.

Step 1: Design Level of Operation (DLO)

Nonlinear behavior is a common characteristic of processes with streams comprised of liquids, gases, powders, slurries and melts. Nonlinear behavior implies that Kp, Tp and/or θp changes as operating level changes.

Since we use Kp, Tp and θp values in correlations to complete the controller design and tuning, the fact that they change gives us pause. It implies that a controller tuned to provide a desired performance at one operating level will not give that same performance at another level. In fact, we demonstrated this on the heat exchanger.

The nonlinear nature of the gravity drained tanks process is evident in the manual mode (open loop) response plot below.

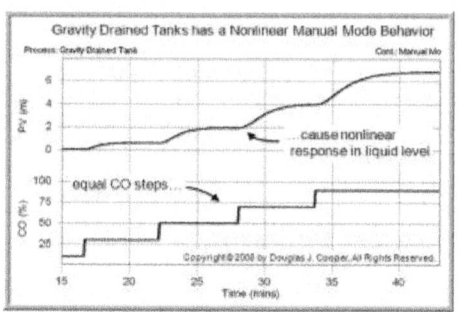

As shown, the CO is stepped in equal increments, yet the response (and thus the Kp, Tp, and/or θp) changes as the level in the tank rises.

We address this concern by specifying a design level of operation (DLO) as the first step of our controller design and tuning recipe. If we are careful about how and where we collect our test data, we heighten the probability that the recipe will yield a controller with our desired performance.

The DLO includes where we expect the set point, SP, and measured process variable, PV, to be during normal operation, and the range of values the SP and PV might assume so we can explore the nature of the process across that range.

For the gravity drained tanks, the PV is liquid level in the lower tank. For this control study, we choose:

Design PV and SP = 2.2 m with range of 2.0 to 2.4 m

The DLO also considers our major disturbances. We should know the normal or typical values for our major disturbances and be reasonably confident that they are quiet so we may proceed with a dynamic (bump) test.

The gravity drained tanks process has one major disturbance variable, the pumped flow disturbance, D. For this study, D is normally steady at about 2 L/min, but certain operations in the plant cause it to momentarily spike up to 5 L/min for brief periods.

Rejecting this disturbance is a major objective in our controller design. For this study, then:

Design D = 2 L/min with occasional spikes up to 5 L/min

Step 2: Collect Data at the DLO (Design Level of Operation)

The next step in our recipe is to collect dynamic process data as near as practical to our design level of operation. We do this with a **bump test**, where we step or pulse the CO and collect data as the PV responds.

It is important to wait until the CO, PV and D have settled out and are as near to constant values as is possible for our particular operation before we start a bump test. The point of bumping a process is to learn about the cause and effect relationship between the CO and PV.

With the process steady, we are starting with a clean slate and as the PV responds to the CO bumps, the dynamic cause and effect behavior is isolated and evident in the data. On a practical note, be sure the data capture routine is enabled before the initial bump so all relevant data is collected.

While closed loop testing is an option, here we consider two open loop (manual mode) methods: the step test and the doublet test.

For either method, the CO must be moved far enough and fast enough to force a response in the PV that dominates the measurement noise.

Also, our bump should move the PV both above and below the DLO during testing. With data from each side of the DLO, the FOPDT model will be able to average out the nonlinear effects.

- *Step Test*
- To collect data that will "average out" to our design level of operation, we start the test with the PV on one side of the DLO. Then, we step the CO so that the measurement moves across to settle on the other side of the DLO.

We acknowledge that it may be unrealistic to attempt such a precise step test in some production environments. But we should understand why we propose this ideal approach (answer: to average nonlinear process effects).

Recall that our DLO is a PV = 2.2 m and D = 2 L/min (though not shown on the plots, disturbance D remains constant at 2 L/min throughout the test).

Below, we set CO = 55% and the process steadies at a PV = 2.4 L.

We step the CO to 51% and the PV settles at abut 2.0 L/min. Thus, we have collected data "around" our design level of operation.

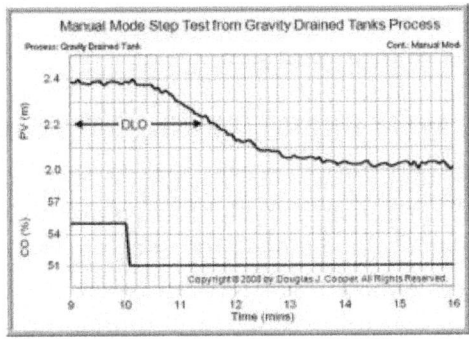

Note that we can start high and step the CO down (as above), or start low and step the CO up. Both methods produce dynamic data of equal value for our design and tuning recipe.

- *Doublet Test*
- A doublet test, is two CO pulses performed in rapid succession and in opposite direction.

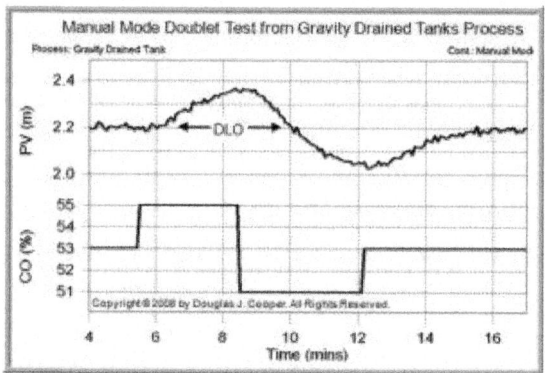

The second pulse is implemented as soon as the process has shown a clear response to the first pulse that dominates the noise in the PV. It is not necessary to wait for the process to respond to steady state for either pulse.

The doublet test offers important benefits. Specifically, it:

starts from and quickly returns to the design level of operation,

produces data both above and below the design level to "average out" the nonlinear effects, and

the PV always stays close to the DLO, thus minimizing off-spec production.

For these reasons, many industrial practitioners find the doublet to be the preferred method for generating open loop dynamic process data, though it does require that we use acommercial software tool for the model fitting task.

Modeling Process Dynamics

Next, we model the dynamics of the gravity drained tanks process and use the result in a series of control studies.

MODELING GRAVITY DRAINED TANKS DATA USING SOFTWARE

We have investigated a graphical analysis method for fitting a first order plus dead time (FOPDT) dynamic model to step test data for both the heat exchanger and the gravity drained tanks processes in previous articles.

Describing process behavior with an approximating FOPDT dynamic model is the third step of our controller design and tuning recipe. Thus, it is a critical step for quickly achieving desired controller performance while avoiding time consuming and expensive trial and error methods.

The reason we studied graphical modeling is because it is a useful way to isolate the role of each FOPDT model parameter, and in particular, appreciate what each says about a controller output (CO) to process variable (PV) relationship.

As we learned in these investigations, for a change in CO:

- Process gain, Kp, describes the direction and how far the PV moves,
- Time constant, Tp, describes how fast the PV responds,
- Dead time, θp, describes how much delay occurs before the PV first begins to move.

But in industrial practice, graphical modeling methods are very limiting for (at least) two reasons.

First, they restrict us to often-impractical step test data. With software, we can fit models to a broad range of dynamic data sets, including closed loop (automatic mode)set point response data.

And instead of using pencil, paper, calculator and ruler to analyze a step test, software can produce a reliable fit and present the results for our inspection almost as fast as we can click on the program icons.

Software Requires Electronic Data

One requirement for using a commercial software package for dynamic modeling and controller design is that the process data must be available in some sort of electronic form, ranging from a simple text file to an Excel spreadsheet format.

If the controller for the loop being modeled is software based (such as on a DCS), if the hardware is OPC enabled, if the controller is connected to a data historian, or if we have capable in-house tech support, then we should have access to process data in a file format.

Practitioner's note: If you believe process control is important to plant safety and profitability, yet your process data is not available in an electronic format, then perhaps your management is not convinced that process control is important to plant safety and profitability.

The file must contain the sampled CO signal paired with the corresponding PV measurement for the entire bump test. If the data was collected at a constant sample rate, T, then we also must know this value. Otherwise, the data file must match each CO and PV pair with a sample time stamp.

Model Fitting of Doublet Data

For our gravity drained tanks study, we have previously discussed that the PV is liquid level in the lower tank, that the set point, SP, is held constant during production, and that our main objective is rejecting disruptions from D, the pumped flow disturbance.

We also presented process data from a doublet test around our design level of operation (DLO), which for this study is:

- design PV and SP = 2.2 m with range of 2.0 to 2.4 m
- design D = 2 L/min with occasional spikes up to 5 L/min

A doublet test, is two CO pulses performed in rapid succession and in opposite direction. The second pulse is implemented as soon as the process has shown a response to the first pulse that clearly dominates the noise in the PV. It is not necessary to wait for the process to respond to steady state for either pulse.

The doublet test offers important benefits as a testing method, including that it starts from and quickly returns to the DLO. It also produces data both above and below the design level to "average out" the nonlinear effects. And the PV always stays close to the design level of operation, thus minimizing off-spec production.

Using the commercial software offered by Control Station, Inc, we read our data file into the software, select "First Order Plus Dead Time" from the model library, and then click "fit model."

The results of the automated model fit are displayed below:

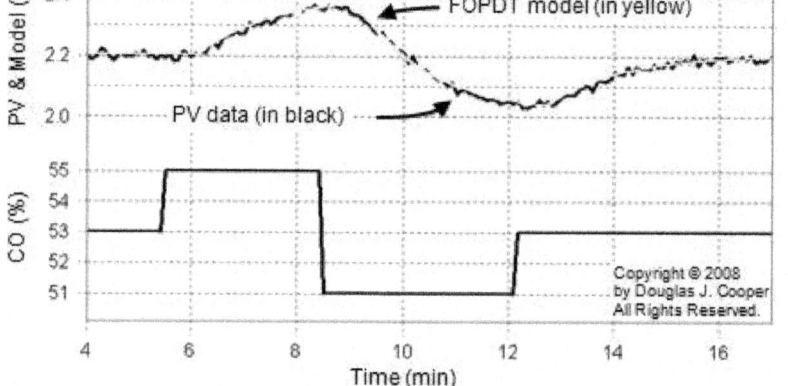

FOPDT Model Fit of Doublet Data

Model: First Order Plus Dead Time (FOPDT)   File Name: gdt doublet save.txt

Gain (K) = 0.095, Time Constant (T1) = 1.35, Dead Time (TD) = 0.57
Goodness of Fit: R-Squared = 0.983, SSE = 0.02217

The sampled data is the black trace in the above plot and the FOPDT model is displayed in yellow. The model parameters from the doublet test model fit are listed in the table below.

For comparison, the results from our previous step analysis are also listed:

	Step	Doublet
Process gain, Kp (m/%)	0.09	0.095
Time constant, Tp (min)	1.4	1.35
Dead time, θp (min)	0.5	0.57

The software model fit is consistent with the step test graphical analysis. The extra accuracy of the computer output, though displayed by the software, does not necessarily hold significance as process data rarely contains such precise dynamic information. For real processes, these numbers are essentially equal.

Model Fit Minimizes SSE

The model fitting software performs a systematic search for a combination of model parameters that minimizes the sum of squared errors (SSE), computed as:

$$SSE = \sum_{i=1}^{N} [\text{Measured PV}_i - \text{Model PV}_i]^2$$

The Measured PV is the actual data collected from our process. The Model PV is computed using the model parameters from the search routine and the actual CO data from the file. N is the total number of samples in the file.

In general, the smaller the SSE, the better the model describes the data.

Software Offers Benefits

When the external disturbances and noise in the PV signal are small, the doublet can be quite modest in size yet still yield data for a meaningful fit as shown below:

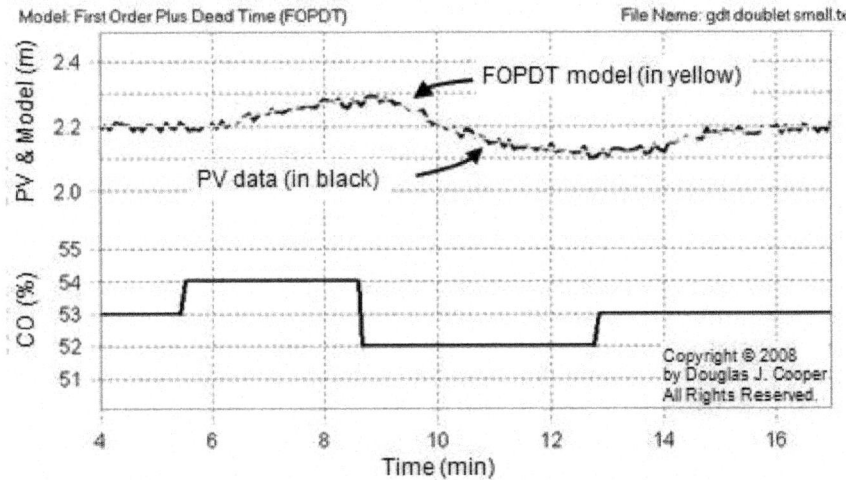

Doublet Can Be Small When Data Has Little Noise

Gain (K) = 0.093, Time Constant (T1) = 1.31, Dead Time (TD) = 0.65
Goodness of Fit: R-Squared = 0.9659, SSE = 0.01314

The software can also model data that contains significant noise in the PV signal, as long as the external disturbances are quiet.

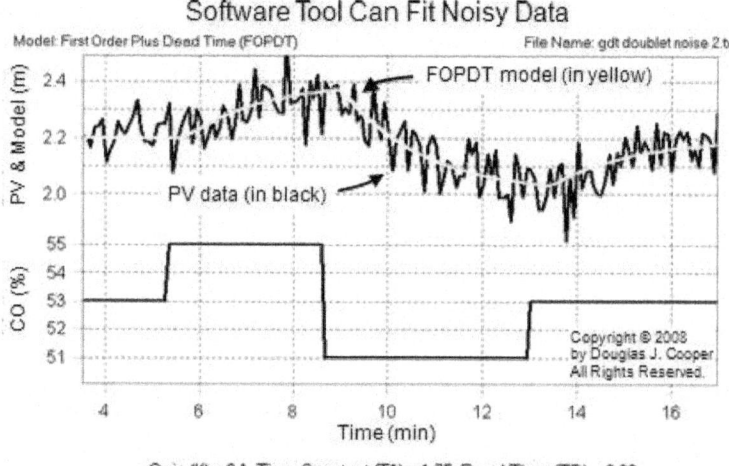

Software Tool Can Fit Noisy Data

Model: First Order Plus Dead Time (FOPDT)   File Name: gdt doublet noise 2.txt

Gain (K) = 0.1, Time Constant (T1) = 1.75, Dead Time (TD) = 0.38
Goodness of Fit: R-Squared = 0.6571, SSE = 0.8163

For comparison, the model parameters from all of the above fits are summarized below:

	Step	Doublet	Small	Noisy
Process gain, Kp (m/%)	0.09	0.095	0.093	0.10
Time constant, Tp (min)	1.4	1.35	1.31	1.75
Dead time, Өp (min)	0.5	0.57	0.65	0.38

From a controller design and tuning perspective, each set of model parameters are similar enough that each will yield a controller with virtually identical performance and capability.

Noise Band Guides Test

When generating dynamic process data, it is important that the CO change is large enough and fast enough to force a response in the measured PV that clearly dominates the higher-frequency signal noise and lower-frequency random process variations.

One way to quantify the amount of noise and random variation for a process is with a noise band.

While there are formal approaches to defining a noise band, a simple approach as illustrated below is to:

collect data for a period of time when the CO is held constant (*i.e.* the controller is in manual).

draw lines that bracket most of the data.

the separation in the brackets is the "noise band."

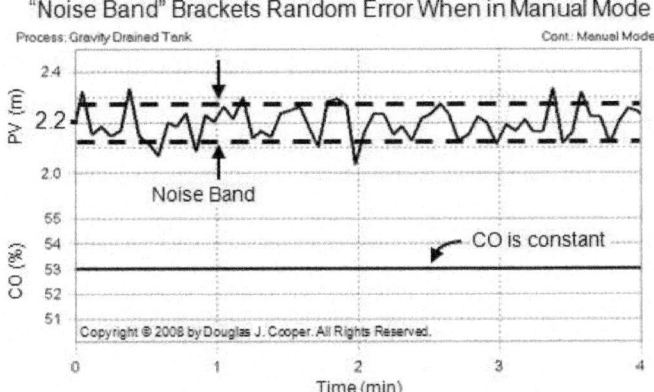

If the data is to be used for modeling, it is best practice to make changes in the CO that force the PV to move *at least* 5 times the noise band. In fact, some experts recommend that the PV moves 10 times the noise band to ensure a reliable result.

The noisy doublet example above did not meet this noise band rule, yet the fit was still reasonable. This is true in part because the process is a simulation and we could be certain that no process disturbances occurred to corrupt the data.

Advances in software also enable us to extract more information from a data set, which is why the noise band rule mentioned above has grown smaller in recent years.

Step 4: Controller Tuning and Testing

In later articles, we use the above model data and move quickly through the range of PID controllers. We focus on disturbance rejection and highlight the differences and similarities with the set point tracking studies we presented for the heat exchanger process.

DESIGN LEVEL OF OPERATION FOR THE JACKETED STIRRED REACTOR PROCESS

Like the heat exchanger and gravity drained tanks case studies, the jacketed stirred reactor is a self regulating processes. That is, the measured process variable (PV) naturally seeks a steady operating level if the controller output (CO) and major disturbance (D) are held constant for a sufficient length of time.

And like the heat exchanger and the gravity drained tanks, the jacketed stirred reactor process is actually a sophisticated simulation derived from first-principles theory and available in commercial software. Nevertheless, the methods and procedures we establish during these investigations are directly applicable to a broad range of industrial processes with streams comprised of liquids, gases, powders, slurries and melts.

The Jacketed Stirred Reactor

The process, shown below in manual mode, is often called a continuously stirred tank reactor (CSTR).

As labeled in the figure, a reactant feed stream enters the top of the vessel. A chemical reaction converts most of this feed into the desired product as the material passes through what is essentially a stirred tank. The stream exiting the bottom of the vessel includes the newly created product plus that portion of the feed that did not convert while in the vessel.

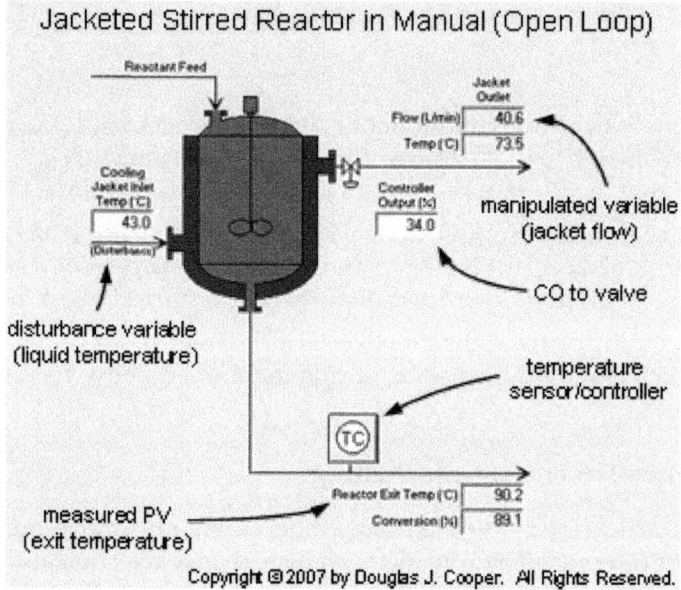

Copyright © 2007 by Douglas J. Cooper. All Rights Reserved.

This well mixed reactor has additional considerations we use later in the discussion:

The residence time, or overall flow rate of reactant feed plus product through the vessel, is constant.

The chemical reaction that occurs is exothermic, which means that heat energy is released as feed converts to product.

The Function of the Cooling Jacket

The chemical reaction releases heat and this energy causes the temperature of the material in the vessel to rise. As temperature rises, the conversion of feed to product proceeds faster, leading to the release of even more heat.

To stop the upward spiral of hotter temperatures increasing the rate of reaction that produces even more heat, the vessel is enclosed with a jacket (or outer shell). A cooling liquid flows through the jacket, collecting heat energy from the outer surface of the reactor vessel and carrying it away as the cooling liquid exits at the jacket outlet.

When the flow of cooling liquid through the jacket increases, more heat is removed. This lowers the reactor temperature, slowing the rate of the reaction, and thus decreasing the amount of feed converted to product during passage through the reactor.

When the flow of cooling liquid through the jacket decreases, some of the energy from the heat-producing reaction, rather than being carried away with the cooling liquid, accumulates in the vessel and drives the reactor temperature higher. The result is an increased conversion of reactant feed to product.

The flow rate of cooling liquid is adjusted with a valve on the cooling jacket outlet stream.

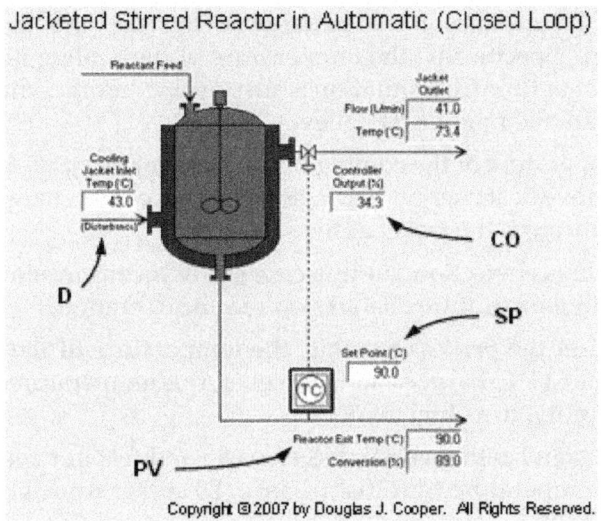

Control Conversion by Controlling Temperature

In this case study, we do not seek 100% conversion of reactant feed to product. Rather, our operating specification is a precise 89% conversion.

Because the reactor has a constant residence time, the amount of heat energy released inside the vessel is directly related to the percent of feed converted to product. By controlling the temperature in the reactor, we can maintain the percent conversion to the desired value.

The vessel is well mixed, so the bulk temperature inside the reactor is about the same as the temperature flowing out the exit stream. Thus, as shown in the process graphic above, we place a temperature sensor in the stream at the bottom of the vessel and our measured process variable (PV) becomes reactor exit temperature.

We can achieve our desired 89% conversion by maintaining the reactor exit stream temperature at 90 °C. From an operational view, this means the reactor exit temperature will have a fixed set point (SP) value of 90 °C.

During bump testing, operations personnel tell us that we may briefly move the reactor exit temperature up and down by 2 °C, but they strongly discourage anything more. Thus:

design PV and SP = 90 °C with approval for brief dynamic testing of ±2 °C

The Disturbance

Because we seek to hold conversion to a constant 89% at all times (which is achieved by holding reactor exit stream temperature at 90 °C), disturbance rejection becomes our main controller design concern.

The major disturbance in this jacketed stirred reactor is the result of an unfortunate design. Specifically, the temperature of the cooling liquid entering the jacket changes over time (this situation is surprisingly more common in industrial installations than one might first believe).

As the temperature of the cooling liquid entering the jacket changes, so does its ability to remove heat energy. Warm liquid removes less energy than cool liquid when flowing through the jacket at the same rate.

So "disturbance rejection" in this case study means minimizing the impact of cooling liquid temperature changes on reactor operation.

As labeled in the process graphic, the temperature of the cooling liquid is normally at about 43 °C. On occasion, however, this temperature can climb, sometimes rather rapidly, to as high as 50 °C.

We will design for the worst-case scenario and test our controller when the cooling liquid temperature (our disturbance, D) spikes from 43 °C up to 50 °C in a single step. Thus:

design D = 43 °C with spikes up to 50 °C

The Design Level of Operation (DLO)

The first step of the our four step design and tuning recipe is to establish the design level of operation. We have completed step 1 by establishing the DLO as:

- Design PV and SP = 90 °C with approval for brief dynamic testing of ±2 °C
- Design D = 43 °C with spikes up to 50 °C

MODELING THE DYNAMICS OF THE JACKETED STIRRED REACTOR WITH SOFTWARE

The control objective of the jacketed reactor case study is disturbance rejection. More specifically, we seek a controller design that will minimize the impact on reactor operation when the temperature of the liquid entering the cooling jacket changes.

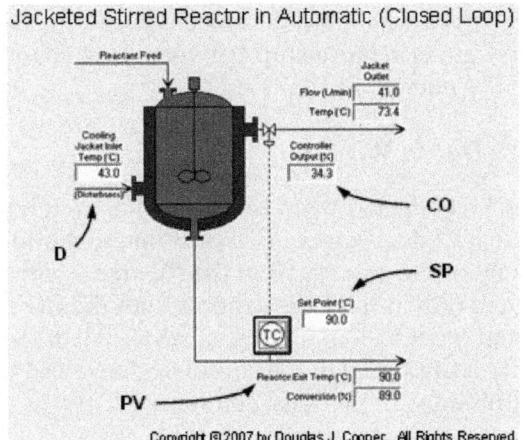

As labeled in the graphic, the important variables for this case study include:

CO = signal to valve that adjusts cooling jacket liquid flow rate (controller output, %)

PV = reactor exit stream temperature (measured process variable, °C)

SP = desired reactor exit stream temperature (set point, °C)

D = temperature of cooling liquid entering the jacket (major disturbance, °C)

Controller Design and Tuning Recipe

As with any control project, we follow our controller design and tuning recipe:

1. Establish the design level of operation (the normal or expected values for set point and major disturbances)
2. Bump the process and collect controller output (CO) to process variable (PV) dynamic process data around this design level
3. Approximate the process data behavior with a first order plus dead time (FOPDT) dynamic model
4. Use the model parameters from step 3 in rules and correlations to complete the controller design and tuning.

Step 1: Design Level of Operation (DLO)

Our DLO are presented in this chapter and are summarized:

Design PV and SP = 90 °C with approval for brief dynamic testing of ±2 °C

Design D = 43 °C with spikes up to 50 °C

Step 2: Collect Data at the DLO

The point of bumping a process is to generate and collect dynamic process data. To be of value, this data should be collected near our DLO and must clearly reveal the cause and effect relationship between how changes in the CO signal force a response in the measured PV.

Data Must Be Wire Out to Wire In

The data must be collected from the controller's viewpoint, since the controller will be making all decisions once in automatic mode. The controller only knows about the state of the process from the PV signal arriving on the "wire in" from the sensor. It can only impact the process with the CO signal it sends on the "wire out" to the final control element (*e.g.*, a valve). All devices, mechanisms and instruments that affect the signal in the complete "wire-out to wire-in" CO to PV loop must be accounted for in the recorded data.

Process Should Be Steady

To further isolate the pure cause and effect relationship, we should wait until the CO, PV and major disturbances have settled out and are reasonably constant before bumping the process. This provides us confidence that observed PV responses during the dynamic test are a direct result of the CO bumps. In the perfect world, this steady state will be at (or at least near) our DLO.

Center Data Around the DLO

Like most processes with streams comprised of gases, liquids, powders, slurries and melts, the jacketed stirred reactor displays a nonlinear or changing behavior as operating level changes. In fact, it is this nonlinear character that leads us to specify a design level of operation in the first place. To the extent possible in our manufacturing environment, we should collect our process data centered around the DLO. A model fit of such data (step 3) will then average out the nonlinear effects and provide a controller equally balanced to address process movement both up and down.

The PV Response Should Dominate the Noise

The CO bump must be far enough and fast enough to force a clear "cause and effect" response in the PV. And this PV response must clearly dominate the measurement noise.

Step 3: Fit a FOPDT Dynamic Model to Process Data Using Software

Two popular open loop (manual mode) methods for generating dynamic process response (bump test) data around the DLO include the step test and the doublet test. The primary disturbance (D) of interest in this study is cooling jacket inlet temperature. Not shown in the plots below is that D is steady at its design value of 43 °C during the bump tests.

Practitioner's Note: While the process graphic above shows the jacketed stirred reactor in automatic mode (closed loop), the step and doublet tests presented below are performed when the controller is in manual mode (open loop).

Step Test: Step tests have value because we can analyze the plot data by hand to compute the first order plus dead time (FOPDT) model parameters Kp, Tp and θp. It is important that we have proper values for these model parameters because they are used in the rules and correlations of step 4 from the recipe to complete the controller design and tuning (examples here and here).

One disadvantage of a step test is that it is conducted in manual mode. As shown in the plot below, the PV is away from the DLO for an extended period of time. This can create profitability concerns from off-spec production and perhaps even safety concerns if constraints become violated.

To collect process data in manual mode that will "average out" the nonlinear effects around our design level of operation, we start the test at steady state with the PV on one side of the DLO. Then, as shown in the plot below, we step the CO so that the measured PV moves across to settle on the other side of the DLO.

We acknowledge that it may be unrealistic to attempt such a precise step test in some production environments. But we should understand that the motivation is to obtain an approximating FOPDT model of the dynamic response behavior that averages out the changing nature of the process as it moves across the expected range of operation.

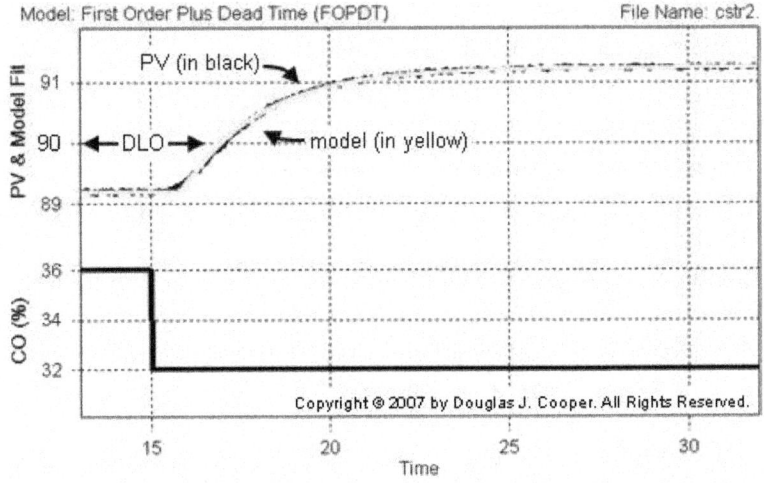

FOPDT Model Params: Kp = −0.51   Tp = 2.2   θp = 0.78

A FOPDT model fit of the CO to PV data using commercial software is shown as the yellow trace in the plot above. It clearly tracks the measured PV data quite closely. The computed model parameters are listed below the plot.

Such results can be obtained in a few mouse clicks, and the visual confirmation that "model equals data" gives us confidence that we indeed have a meaningful description of the dynamic process behavior. This, in turn, gives us confidence that the subsequent controller design will provide the performance we desire.

Doublet Test: A doublet test, again performed below in manual mode, is two CO pulses made in rapid succession and in opposite direction. The second pulse is implemented as soon as the process has shown a clear response to the first pulse that dominates the noise in the PV. It is not necessary to wait for the process to respond to steady state for either pulse.

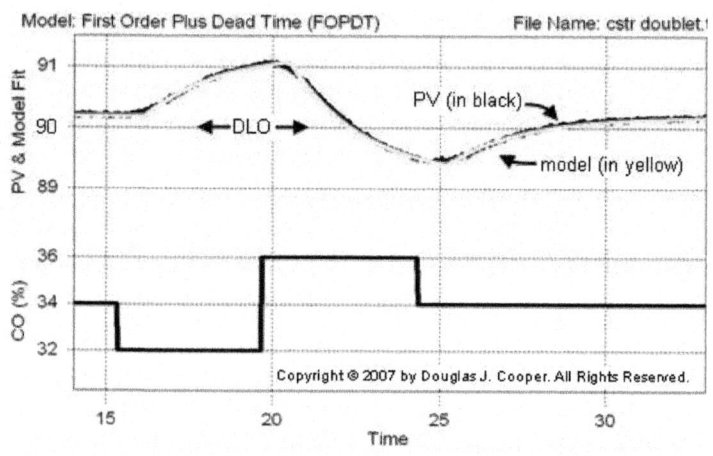

FOPDT Model Params: $K_p = -0.51$ $T_p = 2.2$ $\Theta_p = 0.84$

A FOPDT model fit of the CO to PV data using commercial software is shown as the yellow trace in the plot above. It also tracks the measured PV data quite closely.

A doublet test offers attractive benefits, including that:

it starts from and quickly returns to the DLO,

it produces data above and below the DLO to "average out" the nonlinear effects,

the PV stays close to the DLO, minimizing off-spec production and safety concerns.

For these reasons, a doublet is preferred by many practitioners for open loop testing.

The FOPDT Model

By approximating the dynamic behavior of the jacketed reactor process with a first order plus dead time (FOPDT) model, we quantify those essential features that are fundamental to control.

Aside: the FOPDT dynamic model is a linear, ordinary differential equation describing how PV(t) responds over time to changes in CO(t):

$$Tp\frac{dPV(t)}{dt} + PV(t) = Kp \cdot CO(t - \theta p)$$

Where in this case study: t [=] min, PV(t) [=] °C, CO(t – θp) [=] %

The FOPDT model describes how the PV will respond to a change in CO with the:

> Process gain, Kp, that describes the direction and how far the PV will travel,
>
> Time constant, Tp, that states how fast the PV moves after it begins its response,
>
> Dead time, θp, that is the delay that occurs from when CO is changed until when the PV begins its response.

Based on both the step and doublet test model fits shown above, we conclude that, when operating near the DLO, the jacketed reactor process dynamics are described:

- Kp = – 0.5 °C/ %
- Tp = 2.2 min
- θp = 0.8 min

We will use these parameter values in our subsequent control studies of the jacketed stirred reactor.

EXPLORING THE FOPDT MODEL WITH A PARAMETER SENSITIVITY STUDY

Quantifying Dynamic Process Behavior

Step 3 of the PID controller design and tuning recipe is to approximate process bump test data by fitting it with a first order plus dead time (FOPDT) dynamic model.

Data from a proper bump test is rich in dynamic information that is characteristic of our controller output (CO) to measured process variable (PV) relationship. A FOPDT model is a convenient way to quantify (assign numerical values to) key aspects of this CO to PV relationship for use in controller design and tuning.

In particular, when the CO changes:

> Process gain, Kp, describes the direction and how far the PV moves
>
> Time constant, Tp, describes how fast the PV responds
>
> Dead time, θp, describes how much delay occurs before the PV first begins to move

In the investigation below, we isolate and study each FOPDT parameter individually to establish its contribution to the model response curve. By the end of the study, we (hopefully) will have established the power and utility of this model in describing a broad range of common process behaviors.

Aside: the FOPDT dynamic model is a linear, ordinary differential equation describing how PV(t) responds over time to changes in CO(t):

$$Tp\frac{dPV(t)}{dt} + PV(t) = Kp \cdot CO(t - \theta p)$$

Where in this case study: t [=] min, PV(t) [=] °C, CO(t − θp) [=] %

The Base Case Model

The identical bump test data that was used in the jacketed stirred reactor study presented here. The CO step and PV response data are shown as black curves. The response of the computed FOPDT model is shown as a yellow curve.

The parameter values used to compute all of the FOPDT model responses are listed at the bottom of each plot. Throughout the study, the model parameter units are:

Kp [=] °C/%, Tp [=] min, θp [=] min

The plot below shows a FOPDT model fit as computed by commercial software. We call this fit "good" or "descriptive" because we can visually see that the computed yellow model closely tracks the measured PV data.

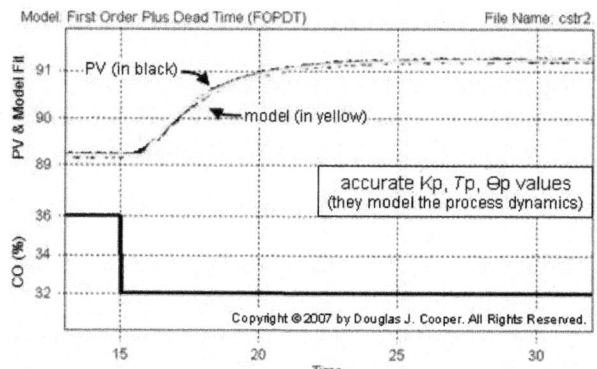

FOPDT Model Params: Kp = −0.51   Tp = 2.2   θp = 0.78

If no significant disturbances occurred during the bump test to corrupt the response, then we can have confidence in our data set. And if the FOPDT model reasonably describes the data, as it does above, then a controller designed and tuned using the parameters from the model will perform as we expect. This, in fact, is the very reason we fit the model in the first place.

Kp is the "How Far" Variable

Process gain, Kp, describes the direction and how far the PV moves in response to a change in CO.

The direction is established with the sign of Kp. We recognize that in this case, Kp must be negative because when the CO moves in one direction, the PV moves in the other. We keep the negative sign throughout this study and focus on how the absolute size of Kp impacts model response behavior.

Below we compute and display the FOPDT model response using the base case time constant, *Tp*, and dead time, θp. The yellow response curve below is different because the process gain used in the model has been increased in absolute size from the base case value of Kp = – 0.51 up to Kp = – 0.61.

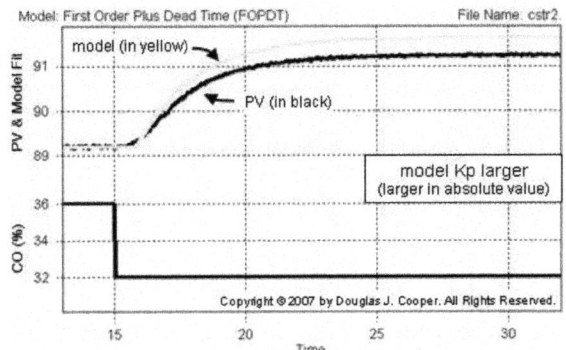

FOPDT Model Params: Kp = -0.61   Tp = 2.2   Θp = 0.78

With a larger model Kp, the "how far" response of the yellow curve clearly increases. Though perhaps not easy to see, the "how fast" and "how much delay" aspects of the response have not been impacted by the change in Kp.

Below we compute the yellow FOPDT model response using the base case *Tp* and θp, only this time, the process gain used in the model has been decreased from the base case Kp = -0.51 down to Kp = – 0.41 (decreased in absolute size).

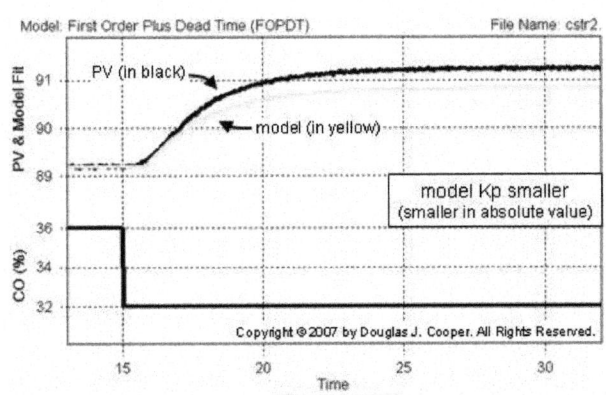

FOPDT Model Params: Kp = -0.41   Tp = 2.2   Θp = 0.78

With a smaller Kp, we see that the computed model curve undershoots the PV data. We thus establish that process gain, Kp, dictates how far the PV travels in response to changes in CO.

Tp is the "How Fast" Variable

Process time constant, Tp, describes how fast the PV moves in response to changes in CO. The plot at the top of this chapter establishes that when $Tp = 2.2$, the speed of response of the model matches that displayed by the PV data.

Below we compute and display the FOPDT model response curve using the base case gain, Kp, and dead time, θp. The curve is different because the time constant used to compute the yellow model response has been increased from the base case $Tp = 2.2$ up to $Tp = 4.0$.

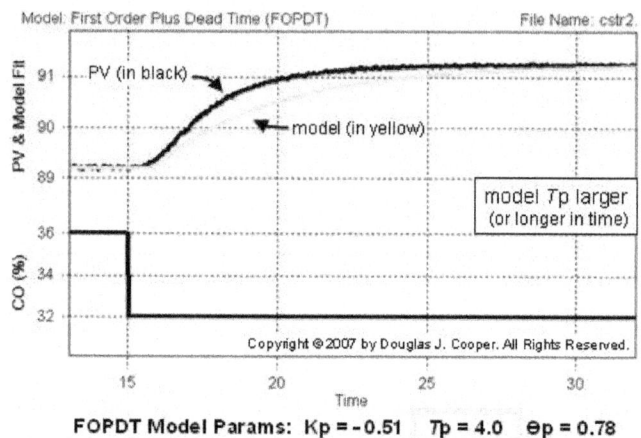

FOPDT Model Params: Kp = -0.51 Tp = 4.0 θp = 0.78

Recall that for step test data like that above, Tp is the time that passes from when the PV shows its first clear movement until when it reaches 63% of the total DPV change that is going to occur.

Since Tp marks the passage of time, a larger time constant describes a process that will take longer to complete its response. Put another way, a process with a larger Tp moves slower in response to changes in CO.

A very important observation in the above plot is that the yellow model curve, even though it moves more slowly than in the base case, ultimately reaches and levels out at the black PV data line. This is because the model response was generated using the base case Kp = -0.51.

If we accept that the size of Kp alone dictates "how far" a response will travel, then this result is consistent with our understanding.

Below we show the yellow FOPDT model response using the base case Kp and θp. The response curve is different because the model computation uses a process time constant that has been decreased from the base case $Tp = 2.2$ down to $Tp = 1.0$.

A smaller Tp means a process will take a shorter amount of time to complete its response. Thus, a process with a smaller time constant is one that moves faster in response to CO changes.

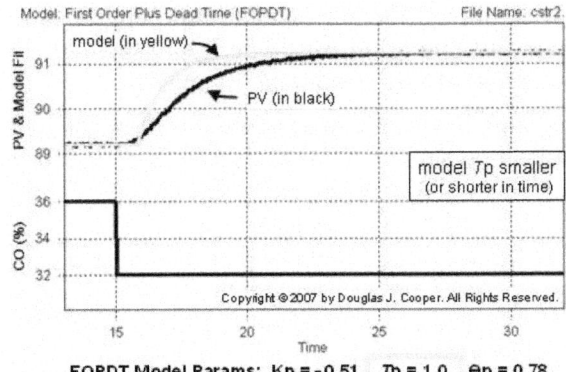

FOPDT Model Params: Kp = -0.51   Tp = 1.0   Өp = 0.78

Again note that the yellow model response curve levels out at the black PV data line because the base case value of Kp = – 0.51 was used in computing "how far" the model response will travel.

Өp is the "With How Much Delay" Variable

If we examine all of the plots above we can observe that the "how much delay" behavior of the yellow model response curves are all the same, regardless of changes in Kp or Tp.

That is, the time delay that passes from when the CO step is made until when the model shows its first clear response to that action is the same regardless of the process gain and/or time constant values used.

This is because it is Өp alone that dictates the "how much delay" model response behavior.

We compute and display the FOPDT model response curve using the base case process gain, Kp, and time constant, Tp. The yellow response is different in the plot below because the model dead time has been increased from the base case value of Өp = 0.78 up to Өp = 2.0.

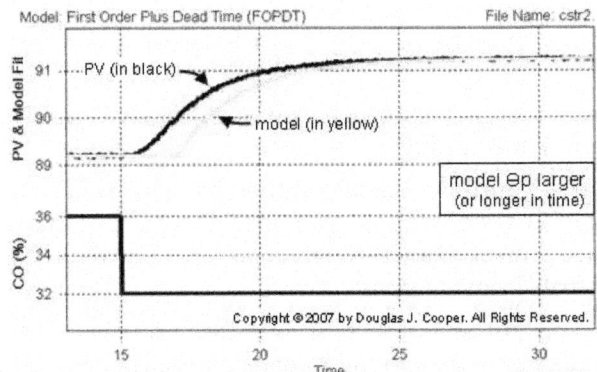

FOPDT Model Params: Kp = -0.51   Tp = 2.2   Өp = 2.0

As the plot above shows, increasing dead time simply shifts the yellow model response curve without changing its shape in any fashion. A larger dead time means a longer delay before the response first begins.

The "how far" and "how fast" shape remains identical in the model response curve above because it is Kp and Tp that dictate these behaviors.

Below we show the yellow FOPDT model response using the base case Kp and Tp. The response curve is different because the model computation uses a dead time value that has been decreased from the base case $\theta p = 0.78$ down to $\theta p = 0$.

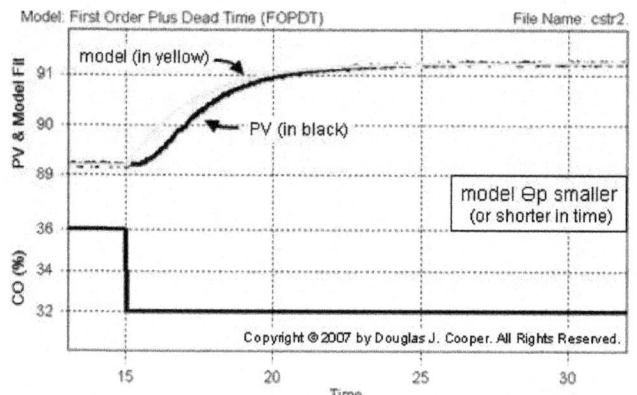

FOPDT Model Params: Kp = -0.51   Tp = 2.2   Θp = 0.0

If we specify no delay (*i.e.*, $\theta p = 0$), then the computed yellow model response curve starts to respond the instant that the CO step is made.

Conclusions

The parameter of the FOPDT model describes a unique characteristic of dynamic process behavior:

- Process gain, Kp, describes the direction and how far the PV moves.
- Time constant, Tp, describes how fast the PV responds.
- Dead time, θp, describes how much delay occurs before the PV first begins to move.

Knowing how a PV will behave in response to a change in CO is fundamental to controller design and tuning.

For example, a controller must be tuned to make small corrective actions if every CO change causes a large PV response (if Kp is large). And if Kp is small, then the controller must be tuned to make large CO corrective actions whenever the PV starts to drift from set point.

If the time constant of a process is long (if a process responds slowly), then the controller must not be issuing new corrective actions in rapid fire. Rather, the controller must permit previous CO actions to show their impact before computing further actions.

Chapter 3

PID Controller Design and Tuning Process

DESIGN AND TUNING RECIPE MUST CONSIDER
NONLINEAR PROCESS BEHAVIOR

Processes with streams comprised of gases, liquids, powders, slurries and melts tend to exhibit variations in behavior as operating level changes. This, in fact, is the very nature of a nonlinear process. For this reason, our recipe for controller design and tuning begins by specifying our design level of operation.

Controller Design and Tuning Recipe:

1. Establish the design level of operation (DLO), which is the normal or expected values for set point and major disturbances.
2. Bump the process and collect controller output (CO) to process variable (PV) dynamic process data around this design level.
3. Approximate the process data behavior with a first order plus dead time (FOPDT) dynamic model.
4. Use the model parameters from step 3 in rules and correlations to complete the controller design and tuning.

Nonlinear Behavior of the Gravity Drained Tanks

The dynamic behavior of the gravity drained tanks process is reasonably intuitive. Increase or decrease the inlet flow rate into the upper tank and the liquid level in the lower tank rises or falls in response.

One challenge this process presents is that its dynamic behavior is nonlinear. That is, the process gain, K_p; time constant, T_p; and/or dead time, θ_p; changes as operating level changes. This is evident in the open loop response plot below.

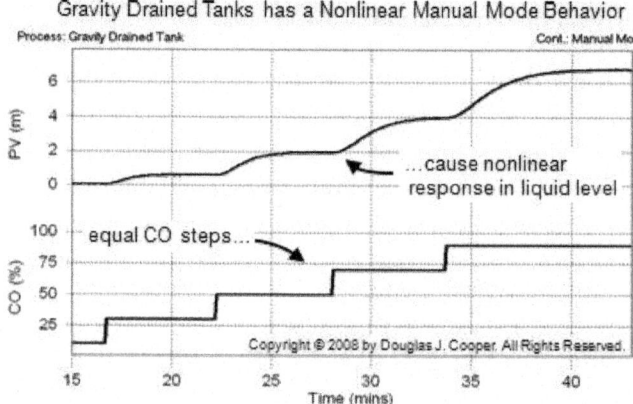

The CO is stepped in equal increments, yet the response behavior of the PV changes as the level in the tank rises. The consequence of nonlinear behavior is that a controller designed to give desirable performance at one operating level may not give desirable performance at another level.

Nonlinear Behavior of the Heat Exchanger

Nonlinear process behavior has important implications for controller design and tuning. Consider, for example, our heat exchanger process under PI control.

When tuned for a moderate response as shown in the first set point step from 140 °C to 155 °C in the plot below, the process variable (PV) responds in a manner consistent with our design goals. That is, the PV moves to the new set point (SP) reasonably quickly but does not overshoot the set point.

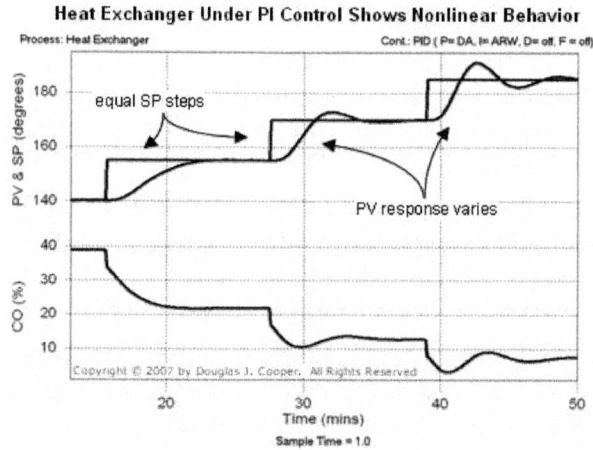

The consequence of a nonlinear process character is apparent as the set point steps continue to higher temperatures. In the third set point step from 170 °C to 185 °C, the same controller that had given a desired moderate performance now produces a PV response with a clear overshoot and some oscillation.

Such a change in performance with operating level may be tolerable in some applications and unacceptable in others.

Nonlinear behavior should not catch us by surprise. It is something we can know about our process in advance. And this is why we should choose a design level of operation as a first step in our controller design and tuning procedure.

Step 1: Establish the Design Level of Operation (DLO)

Because, as shown in the examples above, processes have process gain, K_p; time constant, T_p; and/or dead time, θ_p values that change as operating level changes, and these FOPDT model parameter values are used to complete the controller design and tuning procedure, it is important that dynamic process test data be collected at a pre-determined level of operation.

Defining this design level of operation (DLO) includes specifying where we expect the set point (SP) and measured process variable (PV) to be during normal operation, and the range of values the SP and PV might typically assume. This way we know where to explore the dynamic process behavior during controller design and tuning.

The DLO also considers our major disturbances (D). We should know the normal or typical values for our major disturbances. And we should be reasonably confident that thedisturbances are quiet so we may proceed with a bump test to generate and record dynamic process data.

Step 2. Collect Dynamic Process Data Around the DLO

The next step in our recipe is to collect dynamic process data as near as practical to our design level of operation. We do this with a bump test, where we step or pulse the CO and collect data as the PV responds.

It is important to wait until the CO, PV and D have settled out and are as near to constant values as is possible for our particular operation before we start a bump test. The point of bumping a process is to learn about the cause and effect relationship between the CO and PV.

With the process at steady state, we are starting with a clean slate. As the PV responds to the CO bumps, the dynamic cause and effect behavior is isolated and evident in the data. On a practical note, be sure the data capture routine is enabled before the initial bump is implemented so all relevant data is collected.

Two popular open loop (manual mode) methods are the step test and the doublet test.

For either method, the CO must be moved far enough and fast enough to force a response in the PV that dominates the measurement noise.

Also, our bump should move the PV both above and below the DLO during testing. With data from each side of the DLO, the model (step 3) will be able to average out the nonlinear effects.

Step Test

To collect data that will "average out" to our design level of operation, we start the test at steady state with the PV on one side of (either above or below) the DLO. Then, as shown in the plot below, we step the CO so that the measured PV moves across to settle on the other side of the DLO.

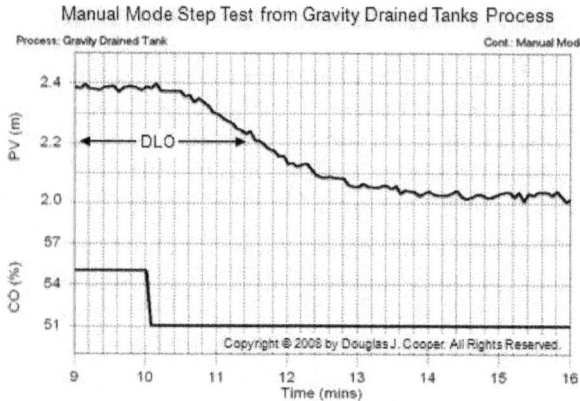

We can either start high and step the CO down (as shown above), or start low and step the CO up. Both methods produce dynamic data of equal value for our design and tuning recipe.

Doublet Test

A doublet test, as shown below, is two CO pulses performed in rapid succession and in opposite direction. The second pulse is implemented as soon as the process has shown a clear response to the first pulse that dominates the noise in the PV. It is not necessary to wait for the process to respond to steady state for either pulse.

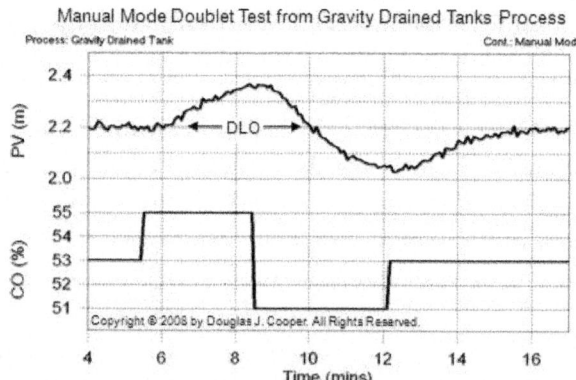

The doublet test offers attractive benefits, including that it starts from and quickly returns to the DLO, it produces data both above and below the design level to "average out" the nonlinear effects, and the PV always stays close to the

DLO, thus minimizing off-spec production. Such data does require commercial software for model fitting, however.

Step 3: Fit a FOPDT dynamic model to Process Data

In fitting a first order plus dead time (FOPDT) model, we approximate those essential features of the dynamic process behavior that are fundamental to control. We need not understand differential equations to appreciate the articles on on this site, but for completeness, the first order plus dead time (FOPDT) dynamic model has the form:

$$Tp\frac{dPV(t)}{dt} + PV(t) = Kp \cdot CO(t - \theta p)$$

Where:

$PV(t)$ = measured process variable as a function of time

$CO(t - \theta p)$ = controller output signal as a function of time and shifted by θp

θp = process dead time

t = time

When the FOPDT dynamic model is fit to process data, the results describe how PV will respond to a change in CO via the model parameters. In particular:

Process gain, Kp, describes the direction and how far PV will travel,

Time constant, Tp, states how fast PV moves after it begins its response,

Dead time, θp, is the delay from when CO changes until when PV begins to respond.

An example study that compares dynamic process data from the heat exchanger with a FOPDT model prediction can be found here. Comparisons between data and model for the gravity drained tanks can be found here and here.

Step 4: Use the Model Parameters to Complete the Design and Tuning

In step 4, the three FOPDT model parameters are used in correlations to compute controller tuning values. For example, the chart below lists internal model control (IMC) tuning correlations for the PI controller and dependent ideal PID controller, anddependent ideal PID with CO filter forms:

	Controller Gain Kc	Reset Time Ti	Deriv Time Td	Filter Const α
PI	$\dfrac{1}{Kp}\dfrac{Tp}{(\theta p + Tc)}$	Tp		
Ideal PID	$\dfrac{1}{Kp}\left(\dfrac{Tp + 0.5\theta p}{Tc + 0.5\theta p}\right)$	$Tp + 0.5\theta p$	$\dfrac{Tp\theta p}{2Tp + \theta p}$	
PID w/ CO Filter	$\dfrac{1}{Kp}\left(\dfrac{Tp + 0.5\theta p}{Tc + \theta p}\right)$	$Tp + 0.5\theta p$	$\dfrac{Tp\theta p}{2Tp + \theta p}$	$\dfrac{Tc(Tp + 0.5\theta p)}{Tp(Tc + \theta p)}$

The closed loop time constant, Tc, in the IMC correlations is used to specify the desired speed or quickness of our controller in responding to a set point change or rejecting a disturbance. The closed loop time constant is computed:

aggressive performance: Tc is the larger of $0.1 \cdot Tp$ or $0.8 \, \theta p$

moderate performance: Tc is the larger of $1 \cdot Tp$ or $8 \, \theta p$

conservative performance: Tc is the larger of $10 \cdot Tp$ or $80 \, \theta p$

Use the Recipe – It is Best Practice

The FOPDT dynamic model of step 3 also provides us the information we need to decide other controller design issues, including:

Controller Action

Before implementing our controller, we must input the proper direction our controller should move to correct for growing errors. Some vendors use the term "reverse acting" and "direct acting." Others use terms like "up-up" and "up-down" (as CO goes up, then PV goes up or down). This specification is determined solely by the sign of the process gain, Kp.

Loop Sample Time, T

Process time constant, Tp, is the clock of a process. The size of Tp indicates the maximum desirable loop sample time. Best practice is to set loop sample time, T, at 10 times per time constant or faster ($T \leq 0.1Tp$). Faster may provide modestly improved performance. Slower than five times per time constant leads to significantly degraded performance.

Dead Time Problems

As dead time grows greater than the process time constant (when $\theta p > Tp$), controller performance can benefit from a model based dead time compensator such as the Smith predictor.

Model Based Control

If we choose to employ a Smith predictor, or perhaps a dynamic feed forward element, a multivariable decoupler, or any other model based controller, we need a dynamic model of the process to enter into the control computer. The FOPDT model from step 3 of the recipe is usually appropriate for this task.

A CONTROLLER'S "PROCESS" GOES FROM WIRE OUT TO WIRE IN

A controller seeks to maintain the measured process variable (PV) at set point (SP) in spite of unplanned and unmeasured disturbances. Since $e(t) = SP - PV$, this is equivalent to saying that a controller seeks to maintain controller error, e(t), equal to zero.

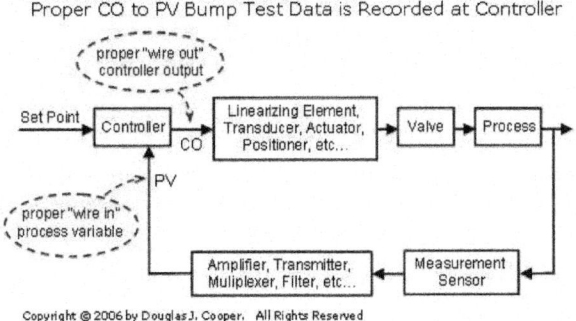

A controller repeats a measurement-computation-action procedure at every loop sample time, T. Starting at the far right of the control loop block diagram above:

A sensor measures a temperature, pressure, concentration or other property of interest from our process.

The sensor signal is transmitted to the controller. The pathway from sensor to controller might include: a transducer, an amplifier, a scaling element, quantization, a signal filter, a multiplexer, and other operations that can add delay and change the size, sign, and/or units of the measurement.

After all electronic and digital operations, the result terminates at our controller as the **"wire in"** measured process variable (PV) signal.

This "wire in" process variable is subtracted from set point in the controller to compute error, e(t) = SP − PV, which is then used in an algorithm (examples here and here) to compute a controller output (CO) signal.

The computed CO signal is transmitted on the **"wire out"** from the controller on a path to the final control element (FCE).

Similar to the measurement path, the signal from the controller to FCE might include filtering, scaling, linearization, amplification, multiplexing, transducing and other operations that can add delay and change the size, sign, and/ or units of our original CO signal.

After any electronic and digital operations, the signal reaches the valve, pump, compressor or other FCE, causing a change in the manipulated variable (a liquid or gas stream flow rate, for example).

The change in the manipulated variable causes a change in our temperature, pressure, concentration or other process property of interest, all with the goal of making e(t) = 0.

Design Based on CO to PV Dynamics

The steps of the controller design and tuning recipe include: bumping the CO signal to generate CO to PV dynamic process data, approximating this test data with a first order plus dead time (FOPDT) model, and then using the model parameters in rules and correlations to complete the controller design and tuning.

The recipe provides a proven basis for controller design and tuning that avoids wasteful and expensive trial-and-error experiments. But for success, controller design and tuning must be based on process data as the controller sees it.

The controller only knows about the state of the process from the PV signal arriving on the "wire in" *after* all operations in the signal path from the sensor. It can only impact the state of the process with the CO signal it sends on the "wire out" *before* any such operations are made in the path to the final control element.

Complete the Circuit

Sometimes we find ourselves unable to proceed with an orderly controller design and tuning. Perhaps our controller interface does not make it convenient to directly record process data. Maybe we find a vendor's documentation to be so poorly written as to be all but worthless. There are a host of complications that can hinder progress.

Being resourceful, we may be tempted to move the project forward by using portable instrumentation. It seems reasonable to collect, say, temperature in a vessel during a bump test by inserting a spare thermocouple into the liquid. Or maybe we feel we can be more precise by standing right at the valve and using a portable signal generator to bump the process rather than doing so from a remote control panel.

An approach cuts out or short circuits the complete control loop pathway. External or portable instrumentation will not be recording the actual CO or PV as the controller sees it, and the data will not be appropriate for controller design or tuning.

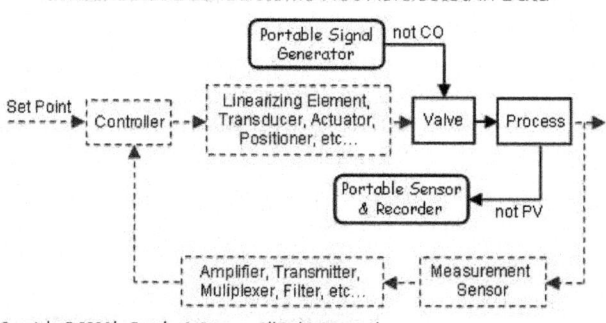

Influence of Dashed Items Not Refelected in Data

Every Item Counts

But please recognize that it can be problematic to leave out even a single step in the complete signal pathway.

A simple scaling element that multiplies the signal by a constant value, for example, may seem reasonably unimportant to the overall loop dynamics. But this alone can change the size and even the sign of Kp, thus having dramatic impact on best tuning and final controller performance.

From a controller's view, the complete loop goes from "wire out" to "wire in" as shown below.

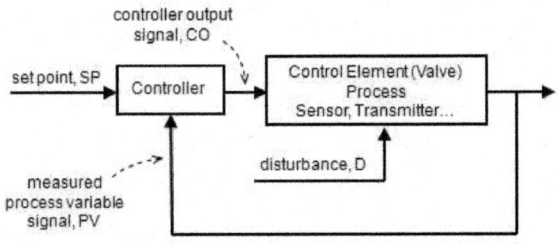

Controller's View: Wire Out (CO) to Wire In (PV) Loop

Every item in the loop counts. Always use the complete CO to PV data for process control analysis, design and tuning.

Pay Attention to Units

The signals can appear in a control loop in electronic units (*e.g.*, volts, mA), in engineering units (*e.g.* °C, Lb/hr), as percent of scale (*e.g.*, 0% to 100%), or as discrete or digital counts (*e.g.* 0 to 4095 counts).

It is critical that we remain aware of the units of a signal when working with a particular instrument or device. All values entered and computations performed must be consistent with the form of the data at that point in the loop.

THE NORMAL OR STANDARD PID ALGORITHM

The question arises quite often, "What is the normal or standard form of the PID (proportional-integral-derivative) algorithm?"

The answer is both simple and complex. Before we explore the answer, consider the screen displays shown below:

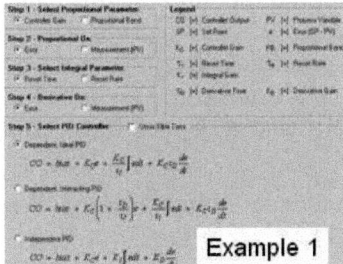

Example 1

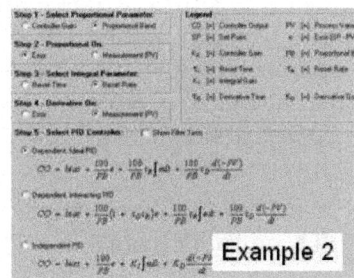

Example 2

As shown in the screen displays:

- There are three popular PID algorithm forms.
- Each of the three algorithms has tuning parameters and algorithm variables that can be cast in different ways.

So your vendor might be using one of dozens of possible algorithm forms. And if you add afilter term to your controller, the number of possibilities increases substantially.

The Simple Answer

Any of the algorithms can deliver the same performance as any of the others. There is no control benefit from choosing one form over another. They are all standard or normal in that sense.

If you are considering a purchase, select the vendor that serves your needs the best and don't dwell on the specifics of the algorithm. Some things to consider include:

- compatibility with existing controllers and associated hardware and software
- cost
- ease of installation and maintenance
- reliability
- your operating environment (is it clean? cool? dry?)

A More Complete Answer

Most of the different controller algorithm forms can be found in one vendor's product or another. Some vendors even use different forms within their own product lines.

And while the various forms are equally capable, each must be tuned (values for the adjustable parameters must be specified) using tuning correlations specifically designed for that particular control algorithm.

Commercial software makes it straightforward to get desired performance from any of them. But it is essential that you know your vendor and controller model number to ensure a correct match between controller form and computed tuning values. The alternative to an orderly design methodology is a "guess and test" approach. While used by some practitioners, such trial and error tuning squanders valuable production time, consumes more feedstock and utilities than is necessary, generates additional waste and off-spec product, and can even present safety concerns.

In most articles on Controlguru.com, we use some variation of the dependent, ideal PID controller form:

$$CO = CO_{bias} + Kc \cdot e(t) + \frac{Kc}{Ti} \int e(t)dt + Kc \cdot Td \frac{de(t)}{dt}$$

Where:

CO = controller output signal
CO_{bias} = controller bias
e(t) = current controller error, defined as SP – PV
SP = set point
PV = measured process variable
Kc = controller gain, a tuning parameter
Ti = reset time, a tuning parameter
Td = derivative time, a tuning parameter

To reinforce that the controllers all are equally capable, we occasionally use variations of the dependent, interacting form:

$$CO = CO_{bias} + Kc\left(1 + \frac{Td}{Ti}\right)e(t) + \frac{Kc}{Ti}\int e(t)dt + Kc \cdot Td\frac{de(t)}{dt}$$

or variations of the independent PID form:

$$CO = CO_{bias} + Kc \cdot e(t) + Ki\int e(t)dt + Kd\frac{de(t)}{dt}$$

Final Thoughts

Some of the subtle differences in algorithm form that we can exploit to improve control performance.

For example, derivative on error behaves different from derivative on measured PV. This is true for all of the algorithms. Derivative on error can "kick" after set point steps and this is rarely considered desirable behavior. Thus, derivative on PV is recommended for industrial applications.

And if you are considering programming the controller yourself, it is not the algorithm form that is the challenge. The big hurdle is properly accounting for the anti-reset windup and jacketing logic to allow bumpless transition between operating modes.

THE P-ONLY CONTROL ALGORITHM

General Control Loop Block Diagram

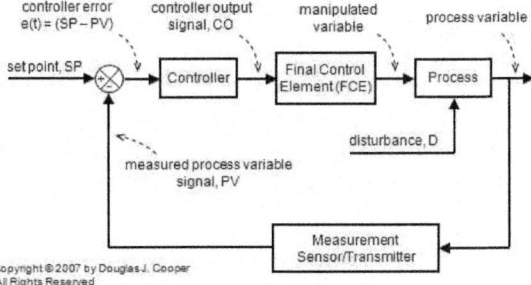

The simplest algorithm in the PID family is a proportional or P-Only controller. Like all automatic controllers, it repeats a measurement-computation-action procedure at every loop sample time, T, following the logic flow shown in the block diagram below:

Starting at the far right of the control loop block diagram above:

- A sensor measures and transmits the current value of the process variable, PV, back to the controller (the 'controller wire in')

- Controller error at current time t is computed as set point minus measured process variable, or e(t) = SP – PV

- The controller uses this e(t) in a control algorithm to compute a new controller output signal, CO

- The CO signal is sent to the final control element (*e.g.* valve, pump, heater, fan) causing it to change (the 'controller wire out')

- The change in the final control element (FCE) causes a change in a manipulated variable

- The change in the manipulated variable (*e.g.* flow rate of liquid or gas) causes a change in the PV

The goal of the controller is to make e(t) = 0 in spite of unplanned and unmeasured disturbances. Since e(t) = SP – PV, this is the same as saying a controller seeks to make PV = SP.

The P-Only Algorithm

The P-Only controller computes a CO action every loop sample time T as:

$$CO = CO_{bias} + Kc \cdot e(t)$$

Where:

CO_{bias} = controller bias or null value

Kc = controller gain, *a tuning parameter*

e(t) = controller error = SP – PV

SP = set point

PV = measured process variable

Design Level of Operation

Real processes display a nonlinear behavior, which means their apparent process gain, time constant and/or dead time changes as operating level changes and as major disturbances change. Since controller design and tuning is based on these Kp, Tp and θp values, controllers should be designed and tuned for a pre-defined level of operation.

When designing a cruise control system for a car, for example, would it make sense for us to perform bump tests to generate dynamic data when the car is traveling twice the normal speed limit while going down hill on a windy day? Of course not.

Bump test data should be collected as close as practical to the design PV when the disturbances are quiet and near their typical values. Thus, the design level of operation for a cruise control system is when the car is traveling at highway speed on flat ground on a calm day.

Definition: the design level of operation (DLO) is where we expect the SP and PV will be during normal operation while the important disturbances are quiet and at their expected or typical values.

Understanding Controller Bias

Let's suppose the P-Only control algorithm shown above is used for cruise control in an automobile and CO is the throttle signal adjusting the flow of fuel to the engine.

Let's also suppose that the speed SP is 70 and the measured PV is also 70 (units can be mph or kph depending on where you live in the world). Since PV = SP, then e(t) = 0 and the algorithm reduces to:

$$CO = CO_{bias} + Kc \cdot (0) = CO_{bias}$$

If CO_{bias} is zero, then when set point equals measurement, the above equation says that the throttle signal, CO, is also zero. This makes no sense. Clearly if the car is traveling 70 kph, then some baseline flow of fuel is going to the engine.

This baseline value of the CO is called the bias or null value. In this example, CO_{bias} is the flow of fuel that, in manual mode, causes the car to travel the design speed of 70 kph when on flat ground on a calm day.

Definition: CO_{bias} is the value of the CO that, in manual mode, causes the PV to steady at the DLO while the major disturbances are quiet and at their normal or expected values.

A P-Only controller bias (sometimes called null value) is assigned a value as part of the controller design and remains fixed once the controller is put in automatic.

Controller Gain, Kc

The P-Only controller has the advantage of having only one adjustable or tuning parameter, Kc, that defines how active or aggressive the CO will move in response to changes in controller error, e(t).

For a given value of e(t) in the P-Only algorithm above, if Kc is small, then the amount added to CO_{bias} is small and the controller response will be slow or sluggish. If Kc is large, then the amount added to CO_{bias} is large and the controller response will be fast or aggressive.

Thus, Kc can be adjusted or tuned for each process to make the controller more or less active in its actions when measurement does not equal set point.

P-Only Controller Design

All controllers from the family of PID algorithms (P-Only, PI, PID) should be designed and tuned using our proven recipe:

1. Establish the design level of operation (the normal or expected values for set point and major disturbances).

2. Bump the process and collect controller output (CO) to process variable (PV) dynamic process data around this design level.

3. Approximate the process data behavior with a first order plus dead time (FOPDT) dynamic model.

4. Use the model parameters from step 3 in rules and correlations to complete the controller design and tuning.

The Internal Model Control (IMC) tuning correlations that work so well for PI andPIDcontrollers cannot be derived for the simple P-Only controller form. The next best choice is to use the widely-published integral of time-weighted absolute error (ITAE) tuning correlation:

Moderate P-Only:

$$Kc = \frac{0.2}{Kp}\left(\frac{Tp}{\theta p}\right)^{1.22}$$

This correlation is useful in that it reliably yields a moderate Kc value. In fact, some practitioners find that the ITAE Kc value provides a response performance so predictably modest that they automatically start with an aggressive P-Only tuning, defined here as two and a half times the ITAE value:

Aggressive P-Only: Kc = 2.5 (Moderate Kc)

Reverse Acting, Direct Acting and Control Action

Time constant, Tp, and dead time, θp, cannot affect the sign of Kc because they mark the passage of time and must always be positive. The above tuning correlation thus implies that Kc must always have the same sign as the process gain, Kp.

When CO increases on a process that has a positive Kp, the PV will increase in response. The process is direct acting. Given this CO to PV relationship, when in automatic mode (closed loop), if the PV starts drifting too high above set point, the controller must decrease CO to correct the error.

This "opposite to the problem" reaction is called *negative feedback* and forms the basis of stable control.

A process with a positive Kp is direct acting. With negative feedback, the controller must be reverse acting for stable control. Conversely, when Kp is negative (a reverse acting process), the controller must be direct acting for stable control.

Since Kp and Kc always have the same sign for a particular process and stable control requires negative feedback, then:

- direct acting process (Kp and Kc positive) → use a reverse acting controller
- reverse acting process (Kp and Kc negative) → use a direct acting controller

In most commercial controllers, a positive value of the Kc is always entered. The sign (or action) of the controller is then assigned by specifying that the controller is either reverse or direct acting to indicate a positive or negative Kc respectively.

If the wrong control action is entered, the controller will quickly drive the final control element (*e.g.*, valve, pump, compressor) to full on/open or full off/closed and remain there until the proper control action entry is made.

Proportional Band

Some manufacturers use different forms for the same tuning parameter. The popular alternative to Kc found in the marketplace is proportional band, PB.

In many industry applications, both the CO and PV are expressed in units of percent. Given that a controller output signal ranges from a minimum (CO_{min}) to maximum (CO_{max}) value, then:

$$PB = (CO_{max} - CO_{min})/Kc$$

When CO and PV have units of percent and both range from 0% to 100%, the much published conversion between controller gain and proportional band results:

$$PB = 100/Kc$$

Many case studies on this site assign engineering units to the measured PV because plant software has made the task of unit conversions straightforward. If this is true in your plant, take care when using these conversion formula.

Implementation Issues

Implementation of a P-Only controller is reasonably straightforward, but this simple algorithm exhibits a phenomenon called "offset." In most industrial applications, offset is considered an unacceptable weakness. We explore P-Only control, offset and other issues for the heat exchanger and the gravity drained tanks processes.

P-ONLY CONTROL OF THE HEAT EXCHANGER SHOWS OFFSET

The general proportional only (P-Only) algorithm structure and considered important design and tuning issues associated with implementation.

Here we investigate the capabilities of the P-Only controller on our heat exchanger process and highlight some key features and weaknesses of this simple algorithm.

The heat exchanger process used in this study is shown below.

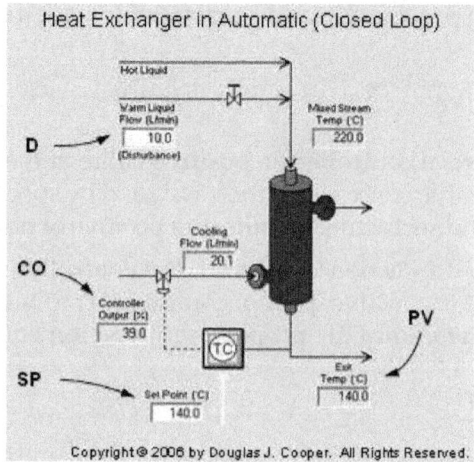

As with all controller implementations, best practice is to follow the four-step design and tuning recipe as we proceed with the study:

Step 1: Design Level of Operation (DLO)

Real processes display a nonlinear behavior. That is, their process gain (Kp), time constant (Tp) and/or dead time (θp) changes as operating level changes and as major disturbances change. Since the rules and correlations we use are based on these Kp, Tp and θp values, controllers should be designed and tuned for a specific level of operation.

The first step in the controller design recipe is to specify our design level of operation (DLO). This includes stating where we expect the set point, SP, and measured process variable, PV, to be during normal operation. Hopefully, these will be the same values as this is the point of a controller.

We also should have some sense of the range of values the SP and PV might assume so we can explore the nature of the process dynamics across that range.

For the heat exchanger, we specify that the SP and PV will normally be at 138 °C, and during production, they may range from 138 to 140 °C. Thus, we can state:

Design PV and SP = 138 °C with range of 138 to 140 °C

We also should know normal or typical values for our major disturbances and be reasonably confident that they are quiet so we may proceed with a bump test. As shown in the graphic above, the heat exchanger process has only one major disturbance variable (D), a side stream labeled Warm Liquid Flow. We specify that the expected or design value for this stream is:

Expected warm liquid flow disturbance, D = 10 L/min

We assume that D remains quiet and at this normal design value throughout the study.

Step 2: Collect Data at the DLO

The next step in the design recipe is to collect dynamic process data as near as practical to our design level of operation. We have previously collected and documentedheat exchanger step test data that matches our design conditions.

Step 3: Fit an FOPDT Model to the Design Data

Here we document a first order plus dead time (FOPDT) model approximation of the heat exchanger step test data from step 2:

- Process gain (how far), K_p = – 0.53 °C/%
- Time constant (how fast), T_p = 1.3 min
- Dead time (how much delay), θ_p = 0.8 min

Step 4: Use the Parameters to Complete the Design

The P-Only controller computes a controller output (CO) action every loop sample time T as:

$$CO = CO_{bias} + K_c \cdot e(t)$$

Where:

CO_{bias} = controller bias or null value

K_c = controller gain, a tuning parameter

$e(t)$ = controller error defined as SP – PV

- *Computing controller error, e(t):* Set point (SP) is something we enter into the controller. The PV measurement comes from our sensor (our wire in). With SP and PV values known, controller error can be computed at every loop sample time T as: $e(t)$ = SP – PV.
- *Determining Bias Value:* CO_{bias} is the value of the CO that, in manual mode, causes the PV to steady at the DLO while the major disturbances are quiet and at their normal or expected values.

The plot below shows that CO_{bias} can be located with an ordered search. That is, we move CO up and down while in manual mode until the PV settles at the design value of 138 °C while the major disturbances are quiet and at their normal or expected values.

Such a manipulation of our process may be impractical or impossible in production situations. The plot is useful, however, because it helps us visualize how the baseline (bias) value of the CO is linked to the design PV.

When we explore PI control of the heat exchanger, commercial controllers use a bumpless transfer method to automatically provide a value for CO_{bias}.

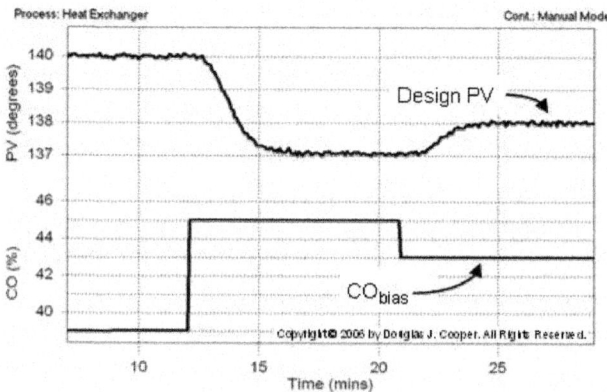

The plot above shows that when CO is held constant at 43% with the disturbances at their normal values, the PV steadies at the design value of 138 °C. Thus:

$$CO_{bias} = 43\%$$

- *Computing Controller Gain:* For the simple P-Only controller, we compute Kc with the integral of time-weighted absolute error (ITAE) tuning correlation: Moderate P-Only:

$$Kc = \frac{0.2}{Kp}\left(\frac{Tp}{\theta p}\right)^{1.22}$$

This correlation is useful in that it reliably yields a moderate Kc value.

Aside: Dead time, θp, is in the denominator in the correlation, so it cannot equal zero. Otherwise, Kc will approach infinity, a fairly useless result. Consider that all controllers measure, act, then wait until next sample time before measuring again. This "measure, act, wait" procedure has a delay (or dead time) of one sample time, T, built naturally into its structure. Thus, by definition, the minimum dead time (θp, min) in a control loop is the loop sample time, T. Dead time can certainly be larger than sample time and it usually is, but it cannot be smaller.

Whether by software or graphical analysis, if we compute a θp that is less than T, we must set $\theta p = T$ everywhere in our tuning correlations. More information about the importance of sample time to controller design and tuning can be found in this other chapter.

→ Best Practice Rule:

when using the FOPDT model for controller tuning, θp,min = T

Using our FOPDT model values from step 3, we compute:

$$Kc = \frac{0.2}{-0.53}\left(\frac{1.3}{0.8}\right)^{1.22} = -0.7\%/°C$$

And our moderate P-Only controller becomes:

$$CO = 43\% - 0.7\ e(t)$$

Implement and Test

To explore how controller gain, Kc, impacts P-Only controller behavior, we test the controller with this ITAE controller gain value. Since the Kc value tends to be moderate, we also study more active or aggressive controller behavior when we double Kc and then double it again:

$$2Kc = -1.4\ \%/°C$$

$$4Kc = -2.8\ \%/°C$$

The performance of the P-Only controller in tracking set point changes is pictured below for the ITAE Kc and its multiples. Note that the warm liquid disturbance flow, though not shown, remains constant at 10 L/min throughout the study.

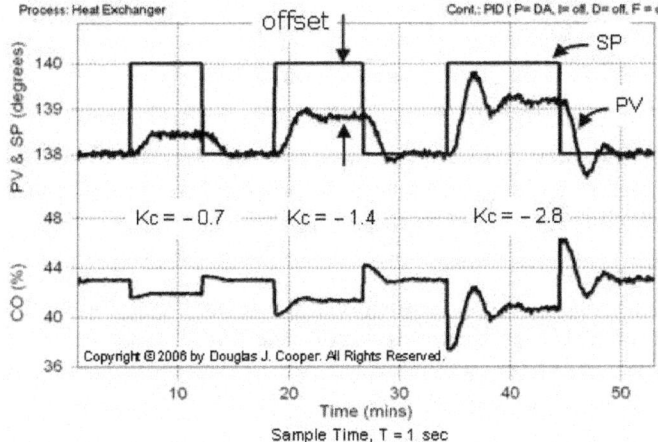

Whenever the set point is at the design level of 138 °C, then PV equals SP.

Each of the three times the SP is stepped away from the DLO, however, the PV settles out at a value short of the set point. The simple P-Only controller is not able to eliminate this "offset," or sustained error between the PV and SP. We talk more about offset below.

Kc and Controller Activity

The plot above shows the performance of the P-Only controller using three different values of Kc.

One point of this study is to highlight that as Kc increases, the activity of the controller output, CO, increases. The CO trace at the bottom of the plot shows this increasingly active behavior, seen as more dramatic moves in response to the same set point step, as Kc increases across the plot.

Thus, we establish that controller gain, Kc, is responsible for the general, and especially the initial, activity in a controller response. This "response activity related to Kc" behavior carries over to PI and PID controllers.

We also see that as Kc increases across the plot, the offset (difference between SP and final PV) decreases but the oscillatory nature of the response increases.

Offset – The Big Disadvantage of P-Only Control

The biggest advantage of P-Only control is that there is only one tuning parameter to adjust, so it is relatively easy to achieve a "best" final tuning. The disadvantage is that this simple control algorithm permits offset.

Offset occurs in most processes under P-Only control when the set point and/or disturbances are at any value other than that used to determine CO_{bias}.

To understand why offset occurs, let's work our way through the P-Only equation:

$CO = 43\% - 0.7 \cdot e(t)$

and recognize that:

when PV equals SP, then error is zero: $e(t) = 0$

if $e(t)$ is zero, then CO equals the CO_{bias} value of 43%

if CO is steady at 43%, then the PV settles to 138 °C. We know this is true because that's how CO_{bias} was determined in the first place. The first plot in this chapter shows us this.

Continuing our reasoning:

the only way CO can be different from the CO_{bias} value of 43% is if something is added or subtracted from the 43%

the only way we have something to add or subtract from the 43% is if the error $e(t)$ is not zero

if $e(t)$ is not zero, then PV cannot equal SP, and we have offset.

Possible Applications?

If P-Only controllers permit offset, do they have any place in the process world? Actually, yes.

One example is a surge or swing tank designed to smooth flows between two units. It does not matter what specific liquid level the tank maintains. The level can be at 63% or 36% and we are happy. Just as long as the tank never empties completely or fills so much that it overflows.

A P-Only controller can serve this function. Put the set point at a level of 50% and let the offset happen. We can have the controller implemented quickly and keep it tuned with little effort.

P-ONLY DISTURBANCE REJECTION OF THE GRAVITY DRAINED TANKS

We looked at the structure of the P-Only algorithm and we considered some design issues associated with implementation. We also studied the set point tracking (or servo) performance of this simple controller for the heat exchangerprocess.

Here we investigate the capabilities of the P-Only controller for liquid level control of thegravity drained tanks process. Our objective in this study is disturbance rejection (or regulatory control) performance.

Gravity Drained Tanks Process

A graphic of the gravity drained tanks process is shown below:

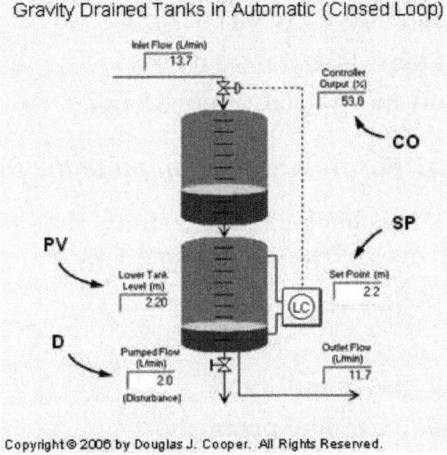

Gravity Drained Tanks in Automatic (Closed Loop)

The measured process variable (PV) is liquid level in the lower tank. The controller output (CO) adjusts the flow into the upper tank to maintain the PV at set point (SP).

The disturbance (D) is a pumped flow out of the lower tank. It's draw rate is adjusted by a different process and is thus beyond our control. Because it runs through a pump, D is not affected by liquid level, though the pumped flow rate drops to zero if the tank empties.

We begin by summarizing the previously discussed results of steps 1 through 3 of ourdesign and tuning recipe as we proceed with our P-Only control investigation:

Step 1: Determine the Design Level of Operation (DLO)

Our primary objective is to reject disturbances as we control liquid level in the lower tank. Our design level of operation (DLO) for this study is:

- design PV and SP = 2.2 m with range of 2.0 to 2.4 m
- design D = 2 L/min with occasional spikes up to 5 L/min

Step 2: Collect Process Data around the DLO

When CO, PV and D are steady near the design level of operation, we bump the CO far enough and fast enough to force a clear dynamic response in the PV that dominates the signal and process noise.

Step 3: Fit a FOPDT Model to the Dynamic Process Data

The third step of the recipe is to describe the overall dynamic behavior of the process with an approximating first order plus dead time (FOPDT) dynamic model. We define the model parameters and present details of the model fit of step test data here. A model fit of doublet test data using commercial software confirms these values:

- process gain (how far), K_p = 0.09 m/%
- time constant (how fast), T_p = 1.4 min
- dead time (how much delay), θ_p = 0.5 min

Step 4: Use the FOPDT Parameters to Complete the Design

Following the heat exchanger P-Only study, the P-Only control algorithm computes a CO action every loop sample time T as:

$CO = CO_{bias} + K_c \cdot e(t)$

Where:

CO_{bias} = controller bias or null value

K_c = controller gain, a tuning parameter

$e(t)$ = controller error, defined as SP – PV

- *Sample Time, T:* Best practice is to set the loop sample time, T, at one-tenth the time constant or faster (*i.e.*, $T \le 0.1 T_p$). Faster sampling may provide modestly improved performance. Slower sampling can lead to significantly degraded performance.

    In this study, $T \le (0.1)(1.4$ min$)$, so T should be 8 seconds or less. We meet this specification with the common vendor sample time option:

    - sample time, T = 1 sec

- *Control Action (Direct/Reverse):* The gravity drained tanks has a positive K_p. That is, when CO increases, PV increases in response. When in automatic mode (closed loop), if the PV is too high, the controller must decrease the CO to correct the error (read more here). Since the controller must move in the direction opposite of the problem, we specify:

    - controller is reverse acting

- *Dead Time Issues:* If dead time is greater than the process time constant ($\theta_p > T_p$), control becomes increasingly problematic and a Smith predictor can offer benefit. For this process, the dead time is smaller than the time constant, so:

- dead time is small and thus not a concern

- *Computing Controller Error, e(t):* Set point, SP, is manually entered into a controller. The measured PV comes from the sensor (our wire in). Since SP and PV are known values, then at every loop sample time, T, controller error can be directly computed as:

  - error, $e(t) = SP - PV$

- *Determining Bias Value,* $CO_{bias:}$ CO_{bias} is the value of CO that, in manual mode, causes the PV to remain steady at the DLO when the major disturbances are quiet and at their normal or expected values. Our doublet plots establish that when CO is at 53%, the PV is steady at the design value of 2.2 m, thus:

  - controller bias, $CO_{bias} = 53\%$

- *Controller Gain,* Kc: For the simple P-Only controller form, we use the integral of time-weighted absolute error (ITAE) tuning correlation:

  Moderate P-Only:

$$Kc = \frac{0.2}{Kp}\left(\frac{Tp}{\theta p}\right)^{1.22}$$

Aside: Regardless of the values computed in the FOPDT fit, best practice is to set θp no smaller than sample time, T (or $\theta p \geq T$) in the control rules and correlations. In this gravity drained tanks study, our FOPDT fit produced a θp much larger than T, so the "dead time greater than sample time" rule is met.

Using our FOPDT model values from step 3, we compute:

$$Kc = \frac{0.2}{0.09}\left(\frac{1.4}{0.5}\right)^{1.22} = 8\ \%/m$$

And our moderate P-Only controller becomes:

P-Only controller: $CO = 53\% + 8 \cdot e(t)$

Implement and Test

To explore how controller gain impacts P-Only performance, we test the controller with the above Kc = 8 %/m. Since the correlation tends to produce moderate performance values, we also explore increasingly aggressive or active P-Only tuning by doubling Kc (2Kc = 16 %/m) and then doubling it again (4Kc = 32 %/m).

The ability of the P-Only controller to reject step changes in the pumped flow disturbance, D, is pictured below (click for a large view) for the ITAE value of Kc and its multiples. Note that the set point remains constant at 2.2 m throughout the study.

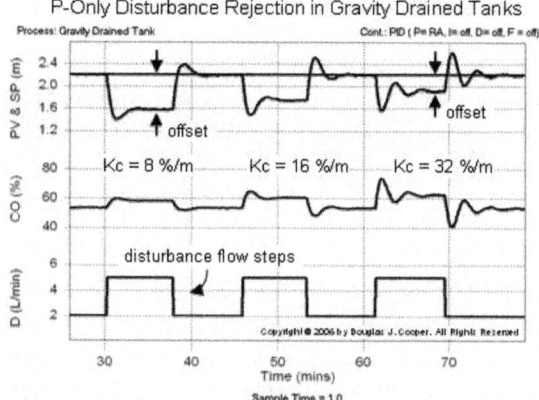

Whenever the pumped flow disturbance, D, is at the design level of 2 L/min (*e.g.*, when time is less than 30 min) then PV equals SP.

The three times that D is stepped away from the DLO, however, the PV shifts away from the set point. The simple P-Only controller is not able to eliminate this "offset," or sustained error between the PV and SP. This behavior reinforces that both set point and disturbances contribute to defining the design level of operation for a process.

The figure shows that as Kc increases across the plot:

- the activity of the controller output, CO, increases,
- the offset (difference between SP and final PV) decreases, and
- the oscillatory nature of the response increases.

Offset, or the sustained error between SP and PV when the process moves away from the DLO, is a big disadvantage of P-Only control. Yet there are appropriate uses for this simple controller.

While not our design objective, presented below is the set point tracking ability of the controller when the disturbance flow is held constant:

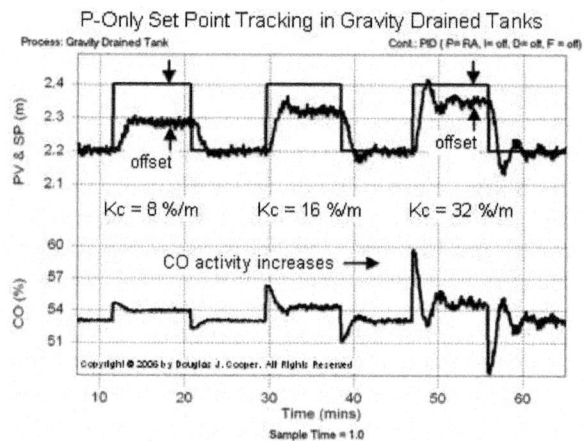

As Kc increases across the plot, the same performance observations made above apply here: the activity of CO increases, the offset decreases, and the oscillatory nature of the response increases.

Aside: it may appear that the random noise in the PV measurement signal is different in the two plots above, but it is indeed the same. Note that the span of the PV axis in the two plots differs by a factor of four. The narrow span of the set point tracking plot greatly magnifies the signal traces, making the noise more visible.

Proportional Band

Different manufacturers use different forms for the same tuning parameter. The popular alternative to controller gain found in the marketplace is proportional band, PB.

If the CO and PV have units of percent and both can range from 0 to 100%, then the conversion between controller gain and proportional band is:

$$PB = 100/Kc$$

Thus, as Kc increases, PB decreases. This reverse thinking can challenge our intuition when switching among manufacturers.

Many examples on this site assign engineering units to the measured PV because plant software has made the task of unit conversions straightforward. If this is true in your plant, take care when using this formula.

Integral Action

Integral action has the benefit of eliminating offset but presents greater design challenges.

CONTROLLER GAIN IS DIMENSIONLESS IN COMMERCIAL SYSTEMS

In modern plants, process variable (PV) measurement signals are typically scaled to engineering units before they are displayed on the control room HMI computer screen or archived for storage by a process data historian. This is done for good reasons.

When operations staff walk through the plant, the assorted field gauges display the local measurements in engineering units to show that a vessel is operating, for example, at a pressure of 25 psig (1.7 barg) and a temperature of 140 °C (284 °F).

It makes sense, then, that the computer screens in the control room display the set point (SP) and PV values in these same familiar engineering units because:

- It helps the operations staff translate their knowledge and intuition from their field experience over to the abstract world of crowded HMI computer displays.
- Familiar units will facilitate the instinctive reactions and rapid decision making that prevents an unusual occurrence from escalating into a crisis situation.

- The process was originally designed in engineering units, so this is how the plant documentation will list the operating specifications.

Controlguru.com Articles Compute Kc With Units

Like a control room display, the Controlguru.com e-book presents PV values in engineering units. In most articles, these PVs are used directly in tuning correlations to compute controller gains, Kc. As a result, the Kc values also carry engineering units.

The benefit of this approach is that controller gain maintains the intuitive familiarity that engineering units provide. The difficulty is that commercial controllers are normally configured to use a dimensionless Kc (or dimensionless proportional band, PB).

To address this issue, we explore below how to convert a Kc with engineering units into the standard dimensionless (%/%) form.

The conversion formula presented at the end of this chapter is reasonably straightforward to use. But it is derived from several subtle concepts that might benefit from explanation. Thus, we begin with a background discussion on units and scaling, and work our way toward our Kc conversion formula goal.

From Analog Sensor to Digital Signal

There are many ways to measure a process variable and move the signal into the digital world for use in a computer based control system. Below is a simplified sketch of one approach.

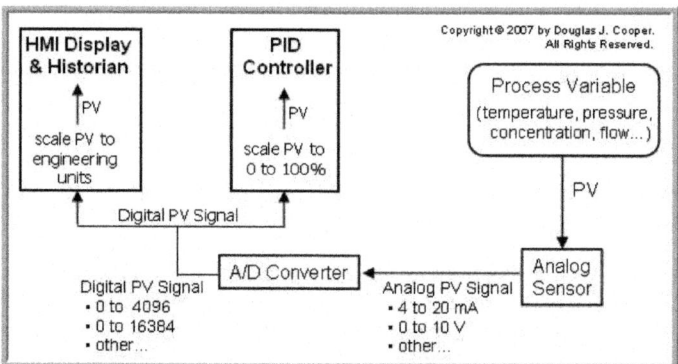

Other operations in the pathway from sensor to control system not shown in the simplified sketch might include a transducer, an amplifier, a transmitter, a scaling element, a linearizing element, a signal filter, a multiplexer, and more.

The central issue for this discussion is that the PV signal arrives at the computers and controllers in a raw digital form. The continuous analog PV measurement has beenquantized (broken into) a range of discrete increments or digital integer "counts" by an A/D (analog to digital) converter.

More counts dividing the span of a measurement signal increases the resolution of the measurement when expressed as a digital value. The ranges offered by most vendors result from the computer binary 2^n form where n is the number of bits of resolution used by the A/D converter.

Example: a 12 bit A/D converter digitizes an analog signal into $2^{12} = 4096$ discrete increments normally expressed to range from 0 to 4095 counts.

A 13 bit A/D converter digitizes an analog signal into $2^{13} = 8192$ discrete increments normally expressed to range from 0 to 8191 counts.

A 14 bit A/D converter digitizes an analog signal into $2^{14} = 16384$ discrete increments normally expressed to range from 0 to 16383 counts.

Example: if a 4 to 20 mA (milliamp) analog signal range is digitized by a 12 bit A/D converter into 0 to 4095 counts, then the resolution is:

$$(20 - 4 \text{ mA}) \div 4095 \text{ counts} = 0.00391 \text{ mA/count}$$

A signal of 7 mA from an analog range of 4 to 20 mA changes to digital counts from the 12 bit A/D converter as:

$$(7 - 4 \text{ mA}) \div 0.00391 \text{ mA/count} = 767 \text{ counts}$$

A signal of 1250 counts from a 12 bit A/D converter corresponds to an input signal of 8.89 mA from an analog range of 4 to 20 mA as:

$$4 \text{ mA} + (1250 \text{ counts}) \cdot (0.00391 \text{ mA/count}) = 8.89 \text{ mA}$$

Scaling the Digital PV Signal to Engineering Units for Display

During the configuration phase of a control project, the minimum and maximum (or zero and span) of the PV measurement must be entered. These values are used to scale the digital PV signal to engineering units for display and storage.

Example: if a temperature range of 100 °C to 500 °C is digitized into 0 to 8191 counts by a 13 bit A/D converter, the signal is scaled for display and storage by setting the minimum digital value of 0 counts = 100 °C, and maximum digital value of 8191 counts = 500 °C

Each digital count from the 13 bit A/D converter gives a resolution of:

$$(500 - 100 \text{ °C}) \div 8191 \text{ counts} = 0.0488 \text{ °C/count}$$

A signal of 175 °C from an analog range of 100 °C to 500 °C changes to digital counts from the 13 bit A/D converter as:

$$(175 - 100 \text{ °C}) \div 0.0488 \text{ °C/count} = 1537 \text{ counts}$$

A signal of 1250 counts from the 13 bit A/D converter corresponds to an input signal of 161 °C from an analog range of 100 °C to 500 °C as:

$$100 \text{ °C} + (1250 \text{ counts}) \cdot (0.0488 \text{ °C/count}) = 161 \text{ °C}$$

The intuition and field knowledge of the operations staff is maintained by using engineering units in control room displays and when storing data to a historian.

For this same reason, modern control software uses engineering units when passing variables between the function blocks used for calculations and decision-making. Calculation and decision functions are easier to understand, document and debug when the logic is written using floating point values in common engineering units.

Scaling the Digital PV Signal for Use by the PID Controller

Most commercial PID controllers use a controller gain, Kc (or proportional band, PB) that is expressed as a standard dimensionless %/%.

Note: Controller gain in commercial controllers is often said to be unitless or dimensionless, but Kc actually has units of (% of CO signal)/(% of PV signal). In a precise mathematical world, these units do not cancel, though there is little harm in speaking as though they do.

Prior to executing the PID controller calculation, the PV signal must be scaled to a standard 0% to 100% to match the "dimensionless" Kc. This happens every loop sample time, T, regardless of whether we are measuring temperature, pressure, flow, or any other process variable.

To perform this scaling, the minimum and maximum PV values in engineering units corresponding to the 0% to 100% standard PV range must be entered during setup and loop configuration.

Example: if a temperature range of 100 °C to 500 °C is digitized into 0 to 8191 counts by a 13 bit A/D converter, the signal is scaled for the PID control calculation by setting the minimum digital value of 0 counts = 0%, and the maximum digital value of 8191 counts = 100%.

Each digital count from the 13 bit A/D converter gives a resolution of:

$$(100 - 0\%) \div 8191 \text{ counts} = 0.0122\%/\text{count}$$

A signal of 1537 counts (175 °C) from a 13 bit A/D converter would translate to a signal of 18.75% as:

$$0\% + (1537) \cdot (0.0122\%/\text{value}) = 18.75\%$$

A signal of 1250 counts (161 °C) from a 13 bit A/D converter would translate to a signal of 15.25% as:

$$0\% + (1250) \cdot (0.0122\%/\text{value}) = 15.25\%$$

Control Output is 0% to 100%

The controller output (CO) from commercial controllers normally default to a 0% to 100% digital signal as well. Digital to analog (D/A) converters begin the transition of moving the digital CO values into the appropriate electrical current and voltage required by the valve, pump or other final control element (FCE) in the loop.

Note: While CO commonly defaults to a 0% to 100% signal, this may not be appropriate when implementing the outer primary controller in a cascade. The outer primary CO1 becomes the set point of the inner secondary controller, and signal scaling must match. For example, if SP2 is in engineering units, the CO1 signal must be scaled accordingly.

Care Required When Using Engineering Units For Controller Tuning

It is quite common to analyze and design controllers using data retrieved from our process historian or captured from our computer display. Just as with the articles in this e-book, this means the computed Kc values will likely be scaled in engineering units.

The sketch below highlights that scaling from engineering units to a standard 0% to 100% range used in commercial controllers requires careful attention.

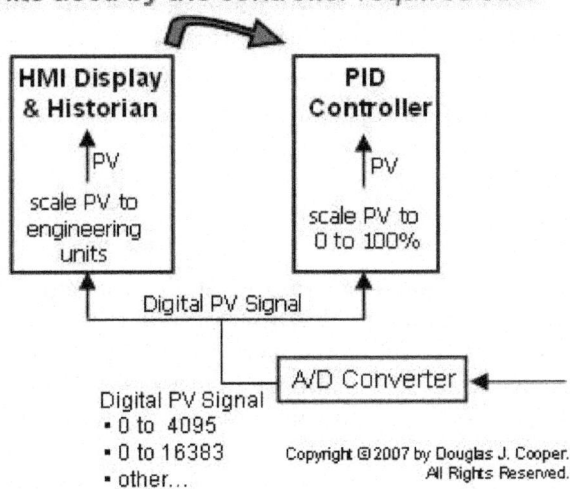

The conversion of PV in engineering units to a standard 0% to 100% range requires knowledge of the maximum and minimum PV values in engineering units. These are the same values that are entered into our PID controller software function block during setup and loop configuration. The general conversion formula is:

where:

PVmax = maximum PV value in engineering units

PVmin = minimum PV value in engineering units

PV = current PV value in engineering units

Example: a temperature signal ranges from 100 °C to 500 °C and we seek to scale it to a range of 0% to 100% for use in a PID controller. We set:

PVmin = 100 °C and PVmax = 500 °C

A temperature of 175 °C converts to a standard 0% to 100% range as:

$$[(175 - 100 °C) \div (500 - 100 °C)] \cdot (100 - 0\%) = 18.75\%$$

A temperature of 161 °C converts to a standard 0% to 100% range as:

$$[(161 - 100 °C) \div (500 - 100 °C)] \cdot (100 - 0\%) = 15.25\%$$

Applying Conversion to Controller Gain, Kc

The basis for the formula used to convert Kc from engineering units into dimensionless (%/%):

Example: the moderate Kc value in our P-Only control of the heat exchangerstudy is Kc = – 0.7 %/ °C. For this process,

PVmax = 250 °C and PVmin = 0 °C

$$Kc = (- 0.7 \%/ °C) \cdot [(250 - 0 °C) \div (100 - 0\%)]$$
$$= - 1.75 \%/\%$$

Example: the moderate value for Kc in our P-Only control of the gravity drained tanksstudy is Kc = 8 %/ °C For this process,

PVmax = 10 m and PVmin = 0 m

$$Kc = (8 \%/ m) \cdot [(10 - 0 m) \div (100 - 0\%)]$$
$$= 0.8 \%/\%$$

Final Thoughts

Textbooks are full of rule-of-thumb guidelines for estimating initial Kc values for a controller depending on whether, for example, it is a flow loop, a temperature loop or a liquid level loop. While we have great reservations with such a "guess and test" approach to tuning, it is important to recognize that such rules are based on a Kc that is expressed in a dimensionless (%/%) form.

Chapter 4

INDUSTRY AUTOMATIC CONTROL FUNDAMENTALS

FUNCTION OF AUTOMATIC CONTROL

The basic idea of a feedback control loop is most easily understood by imagining what an operator would have to do if automatic control did not exist.

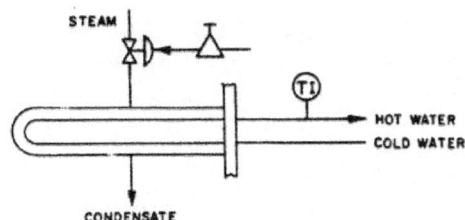

Fig. : Heat Exchanger.

A common application of automatic control found in many industrial plants, a heat exchanger which uses steam to heat cold water. In manual operation, the amount of steam entering the heat exchanger depends on the air pressure to the valve which is set on the manual regulator. To control the temperature manually, the operator would watch the indicated temperature, and by comparing it with the desired temperature, be would open or close the valve to admit more or less steam.

When the temperature had reached the desired value, the operator would simply hold that output to the valve to keep the temperature constant. Under automatic control, the temperature controller performs the same function. The measurement signal to the controller from the temperature transmitter is continuously compared to the set point signal entered into the controller. Based on a comparison of the signals, the automatic controller can tell whether the measurement signal is above or below the setpoint and move the valve accordingly until the measurement (temperature) comes to its final value.

The Feedback Loop

This simple feedback control loop serves to illustrate the four major elements of any feedback control loop.

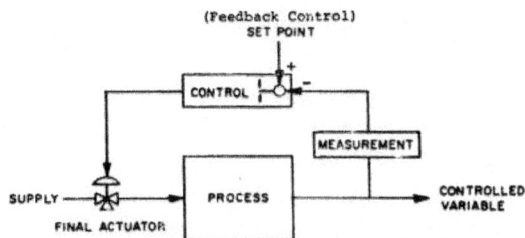

Fig. : Automatic Control Loop.

The Measurement

Measurement must be made to indicate the current value of the variable controlled by the loop. Common measurements used in industry include flow rate, pressure, level, temperature, analytical measurements such as pH, ORP and conductivity and many others particular to specific industries.

The Final Actuator

For every process there must be some final actuator, which regulates the supply of energy or material to the process and changes the measurement signal. Most often this is some kind of valve, but it might also be a belt or motor speed, louver position, *etc.*

The Process

The kinds of processes found in industrial plants are as varied as the materials they produce. They range from the simple and commonplace, such as loops to control flow rate, to the large and complex such as distillation columns in the petro-chemical industry.

The Automatic Controller

The last element of the loop is the automatic controller. Its job is to control the measurement. To "control" means to keep the measurement within acceptable limits. The mechanisms inside the automatic controller will not be considered. Therefore, the principles to be discussed can be applied equally well to both pneumatic and electronic controllers and to the controllers from any manufacturer. All automatic controllers use the same general responses, although the internal mechanisms and the definitions given for these responses may be slightly different from one manufacturer to another.

One basic concept is that for the automatic feedback control to exist, the automatic control loop must be closed. This means that information must be continuously passed around the loop. The controller must be able to move the valve, the valve must be able to affect the measurement, and the measurement signal must be reported to the controller. If this path is broken at any point, the loop is said to be open. As soon as the loop is opened, as for example, when the automatic controller is placed on manual, the automatic unit in the controller is no longer able to move the valve. Thus signals from the controller in response to changing-measurement conditions do not affect the valve and automatic control does not exist.

Controlling the Process

In performing the control function, the automatic controller uses the difference between the set point and measurement signals to develop the output signal to the valve. The accuracy and responsiveness of these signals is a basic limitation on the ability of the controller to correctly control the measurement. If the transmitter does not send an accurate signal, or if there is a lag in the measurement signal, the ability of the controller to manipulate the process will be degraded.

At the same time, the controller must receive an accurate set point signal. In controllers using pneumatic or electronic set point signals generated within the controller, miscalibration of the set point transmitter will necessarily result in the automatic control unit in the controller bringing the measurement to the wrong value. The ability of the controller to accurately position the valve is yet another limitation. If there is friction in the valve, the controller may not be able to move the valve to a specific stem position to produce a specific flow and this will show up as a difference between measurement and set point.

Repeated attempts to exactly position the valve may lead to hunting in the valve and in the measurement. or, if the controller is only able to move the valve very slowly, the ability of the controller to control the process will he degraded. One way to improve the response of control valves is to use a valve positioner, which acts as a feedback controller to position the valve at the exact position corresponding the controller output signal. Positioners, however, should be avoided in favor of volume boosters on fast responding loops such as flow arid liquid pressure.

To control the process, the change in output from the controller must be in such a direction as to oppose any change in the measurement value.

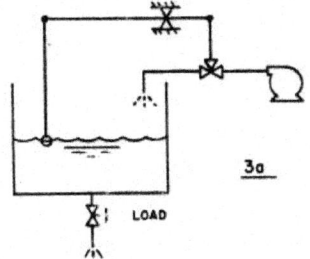

Fig. : Proportional Control Action.

A direct connected valve to control level in a tank at midscale. As the level in the tank rises, the float acts to reduce the flow rate coming in thus, the higher the liquid level the more the flow will be shut off. In the same way, as the level falls, the float will open up the valve to add more liquid to the tank. The response of this system is shown graphically.

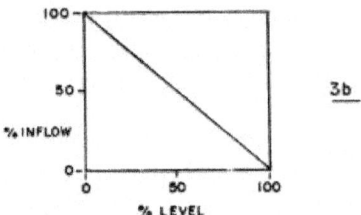

As the level goes from 0% to 100%, the valve goes from fully open to fully closed. The function of an automatic controller is to produce this kind of opposing response over varying ranges; in addition, other responses are available to more efficiently control the process.

Selecting Controller Action

Depending on the action of the valve, increases in measurement may require either increasing or decreasing outputs for control. All controllers can be switched between direct and reverse action.

· Direct action means that, when the controller sees an increasing signal from the transmitter, its output will increase.

· Reverse action means increasing measurement signals cause the controller output to decrease.

To determine which of these responses is correct, an analysis must be made of the loop. The first step is to determine the action of the valve.

For safety reasons the valve must shut if there is a failure in the plant air supply. Therefore, this valve must be air open, or fail close. Second, consider the effect of a change in measurement. For increasing temperature the steam flow to the heat exchanger should be reduced, therefore, the valve must close. To close this valve, the signal from the automatic controller to the valve must decrease.

Therefore, this controller requires reverse, or increase/decrease, action. If direct action is selected increasing signals from the transmitter will result in a larger steam. flow, causing the temperature to increase further. The result would be a run-away temperature. The same thing will happen on any decrease in temperature, causing a falling temperature. Incorrect selection of the action of the controller always results in an unstable control loop as soon as the controller is put into automatic.

Assuming that the proper action is selected on the controller, how does the controller know when the proper output has been reached. For example, to keep the level constant, a controller must manipulate the flow in to equal the flow out, any difference will cause the level to change. In other words, the flow in, or supply must balance the flow out, or demand. The controller performs its job by maintaining this balance in steady state, and acting to restore this balance between supply and demand whenever it is upset.

Upsets

Any one of three events could occur which would require a different flow to maintain the level in the tank. First, if the position of the output hand valve were opened slightly, then more flow would leave the tank, causing the level to fall. This is a change in demand, and to restore balance, the inlet flow valve has to be opened to supply a greater flow rate. A second type of unbalance condition is a change in the set point. Maintaining any other level besides midscale in the tank would cause a different flow out; this change in demand would require a different input valve position. The third kind of upset is a change in the supply. If the pressure output of the pump were to increase, even though the inlet valve remained in the same position, the increased pressure would cause a greater flow, which would at first cause the level to begin to rise. Sensing the increased measurement, the level controller would have to close the valve on the inlet to hold the level at a constant value. In the same way, any controller applied to the heat exchanger must balance the supply of heat added by the steam with the heat taken away by the water. The temperature can only remain constant if the flow of heat in equals the flow of heat out.

Process Characteristics and Controllability

The automatic controller uses changes in the position of the final actuator to control the measurement signal, moving the actuator to oppose any change it sees in the measurement signal. The controllability of any process is a function of how well the measurement signal responds to these changes in the controller output; for good control the measurement should begin to respond quickly, but then not change too rapidly. Because of the tremendous number of applications of automatic control, characterizing a process by what it does, or by industry, is an almost hopeless task. However, all processes can be described by the relationship between their inputs and outputs. The temperature response of the heat exchanger when the control valve is opened by manually increasing the controller output signal.

At first, there is no immediate response at the temperature indication, then the temperature begins to change; it rises steeply at first, and approaches a final, constant level. The process can be characterized by the two elements of its response. The first element is the dead time, or the time before the measurement begins respond, in this example, the dead time arises because the heat in the steam must be conducted to the water before it can affect the temperature, and then to the transmitter before the change can be seen. Dead time is a function of the physical dimensions of a process and such things as belt speeds and mixing rates. Second, the capacity of a process is the material or energy which has to enter or leave the process to change the measurements. It is, for example, the gallons necessary to change level, the BTU's necessary to change temperature, or the standard cubic feet of gas necessary to change pressure.

The measure of a capacity is its response to a step input. Specifically, the size of a capacity is measured by its time constant, which is defined as the time necessary to complete 63% of its total response. The time constant is a function of the size of the process and the rate of material or energy transfer. For this example, the larger the tank, and the smaller the flow rate of the steam, the longer the time constant. These numbers can be as short as a few seconds, or as long as several hours. Combined with dead time, they define how long it takes the measurement signal to respond to changes in the valve position. A process will begin to respond quickly, but then not change too rapidly, if its dead time is small and its capacity is large. In short, the larger the time constant of capacity compared to the dead time, the better the controllability of the process.

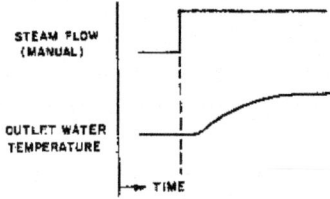

Fig. : Process Response Time.

CONTROLLER RESPONSES

The first and most basic characteristic of the controller response has been shown to be either direct or reverse action. Once this distinction has been made, several types of responses are used to control a process. These are:

1. On/off control, or two position control

2. Proportional control

3. Integral (reset)

4. Derivative

On/off control.

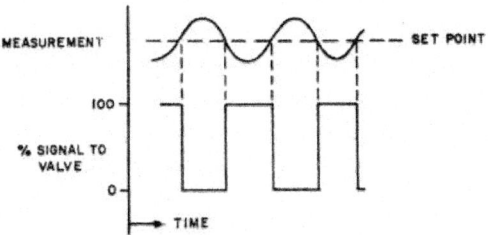

Fig. : On-off Action.

for a reverse acting controller and an air-to-close valve. An on/off controller only has two outputs, either full maximum or full minimum. For this system it has been determined that when the measurement falls below the set point, the valve must be closed to cause it to increase. Thus, whenever the signal to the automatic controller is below the set point, the controller output will be 100%. As the measurement crosses the set point the controller output goes to 0%. This eventually causes the measurement to decrease and as the measurement crosses the set point again, the output goes to maximum. This cycle will continue indefinitely, because the controller cannot balance the supply against the load.

This continuous oscillation may, or may not, be acceptable depending upon the amplitude and length of the cycle. Rapid cycling causes frequent upsets to the plant supply system and excessive valve wear. The time of each cycle depends on the dead time in the process because the dead time determines how long it takes for the measurement signal to reverse its direction once it crosses the set point and the output of the controller changes. The amplitude of the signal depends on how rapidly the measurement signal changes during each cycle. On large capacity process, such as temperature vats, the large capacity causes a very long time constant. Therefore, the measurement can change only very slowly.

The result is that the cycle occurs within a very narrow band around the set point, and this control may be quite acceptable, if the cycle is not too rapid. By far the most common type of control used in industry is on/off control. However if the process measurement is more responsive to changes in the supply, the amplitude and frequency of the cycle begins to increase. At some point, this cycle will become unacceptable and some form of proportional control must be applied.

In order to study the remaining three modes of automatic control open loop responses will be used. Open loop means that only the response of the controller will be considered.

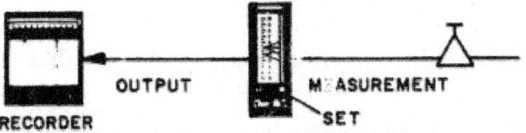

Fig. : Open Loop Controller Response.

An automatic controller with an artificial signal from a manual regulator introduced as the measurement. The set point is introduced normally and the output is recorded. With this arrangement, the specific controller responses to any desired change in measurement can be observed.

An automatic controller with an artificial signal from a manual regulator introduced as the measurement. The set point is introduced normally and the output is recorded. With this arrangement, the specific controller responses to any desired change in measurement can be observed.

Proportional Action - Proportional Controllers

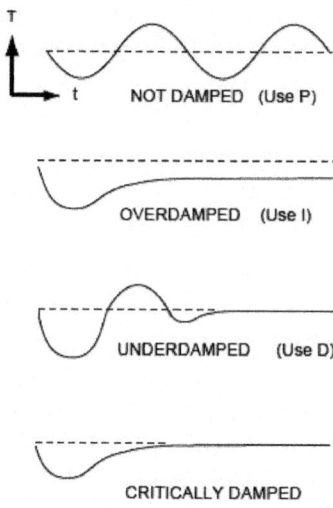

Fig. : The effects of adding P, I, and D actions to a controller.

Home heating systems, air conditioners, and refrigerators ordinarily have their temperature regulating done by "on-off" (sometimes called by engineers "bangbang") switches of some kind. For example, if a temperature is too low, and a heater is therefore turned on, there is a time delay until the heat begins to spread throughout the system and it reaches the temperature sensor ("thermostat"). When this sensor finally does turn off the heat source, it will be too late to prevent some excessive heat from continuing to spread through the system, so the temperature will overswing and temporarily become too high. This is shown in the top diagram of the figure ("not damped").

It is an oscillation, like electronic oscillations in LC circuits where the inductor causes overswing of voltage in the capacitor, even after the transistor has turned off the input. In general, it is partly caused by a time lag (or "phase difference") between the power source and the sensor.

There are two convenient ways to decrease the overswing effect in heating systems. The one used in most home systems is called an "anticipator." A very small heater is placed right next to the sensor, so when the heat is turned on, the sensor responds sooner than it would otherwise, thus decreasing the time lag. It has also been used in laboratory and factory equipment. While this makes the problem less bad, it does not completely eliminate overswing.

A better way to attack the problem, used in most modern engineering temperature controllers, is called "proportional control." In this system, the amount of heat (or cooling effect in a refrigerator) is decreased as the temperature gets closer and closer to the desired value. In other words, the power applied is proportional to the "error signal" that the sensor is indicating. This proportionality is the "P" in modern systems which are referred to as "PID controllers."

The Effect of P

The proportionality (which usually requires fairly complex electronics) has the overall effect of "damping" the controller system, and this decreases the overswing. Sometimes the knob on a controller that adds this function is labeled "damping," but most often it is called "proportional band (PB)." If the "band" is made narrower, there is a steeper gradient of power increase as the temperature goes down; if the band is set by the operator to be wider, there is a more gradual application of power (but over a wider temperature range).

The Effect of I

The damping can be used too much, and in fact it is difficult to avoid this and still stop the oscillations. When it is "overdamped", it will "settle" at the wrong temperature, somewhat "offset" from the true desired value. The cure for this problem is to use "integration." This makes use of a simple computer, usually analog instead of digital, which adds up (integrates) the error signals (desired temperature minus the sensor temperature), repeated at various times, and it slowly changes the amount of heat output to cut this integrated error down to zero.

Sometimes it is expressed in units of time (usually from 30 seconds to 2 minutes) over which the integration is carried out before it is then repeated. The integration knob on a PID controller sometimes is labeled "reset," because it changes the "setting" temperature to an artificially modified value, in order to slowly drift the temperature to what is really desired. It should be noted that this knob only has an effect on errors that exist for a long time, not on short-term "upsets."

The Effect of D

Sometimes the overall effect of a controller, in spite of the various correction factors, still overswings when a short blast of cold air occasionally occurs, *etc.*

To minimize this, a "derivative" computer can be used. This is sensitive to the slope of the measured temperature versus time, and if it is fast, it allows more corrective effect to be applied. The knob is sometimes labeled "D" and sometimes "rate." It does not affect long-term "offset" errors.

PID controllers are used for much more than temperature settings. More and more automation machinery is controlled by such equipment, which prevents the motion of robot arms from oscillating, and minimizes cumulative error, *etc.*

Some factory workers fail to understand the principles outlined above and simply "twiddle the knobs" almost randomly, hoping to get good control by luck.

NOTE: Usually the best sequence of setting these controllers is to adjust P first, then I, and then D, taking considerable time to let things reach a constant value before making the next change.

Proportional response is the basis for the three mode controller. If the other two, integral (reset) and derivative, are present, they are added to the proportional response. "Proportional" means the that the present change in the output of the controller is some multiple of the percent change in the measurement

This multiple is called the "gain" of the controller. For some controllers, proportional action is adjusted by such a "gain" adjustment, while for others a "proportional band" adjustment is used. Both have the same purposes and effect.

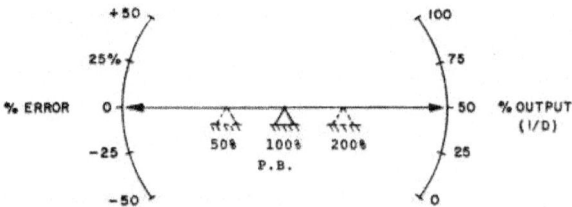

Fig. : Proportional Action.

The response of a proportional controller by an input/output pointer pivoting on one oil these positions. With the pivot in the center between the input and the output graph, 100% change in measurement is required to obtain 100% change in output, or full valve travel.A controller adjusted to respond in this way is said to have a 100% proportional band. When the pivot is moved to the right hand position, the measurement input would have to change by 200% in order to obtain full output change from 0% to 100%, this is called a 200% proportional band.

Finally, if the pivot were in the left hand position and if the measurement moved only over 50% of the scale the output would change over 100% of the scale. This is called a 50% proportional band. Thus, the smaller the proportional band, the smaller amount the measurement must change to cause full valve travel. Or, in other words the smaller the proportional band, the greater the output change for the same size measurement change.

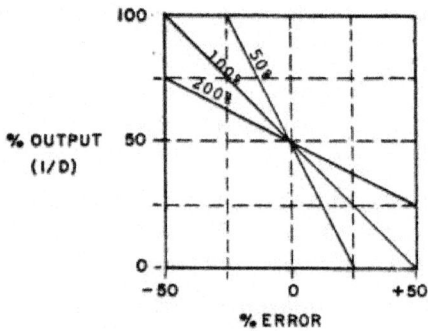

Fig. : Proportional Action.

This graph shows how the controller output will respond as a measurement deviates from set point. Each line on the graph represents a particular adjustment of the proportional band. Two basic properties of proportional control can be observed from this graph:

1. For every value of proportional band whenever the measurement equals the set point, the output is 50%.

2. Each value of the proportional band defines a unique relationship between measurement and output. For every measurement value there is a specific output value. For example; using the 100% proportional band line, whenever the measurement is 25% above the set point, the output from the controller must be 25%. The output from the controller can be 25% only if the measurement is 25% above setpoint. In the same way, whenever the output from the controller is 25%, the measurement will be 25% above set point. In other words there is one specific output value for every measurement value.

For any process control loop only one value of the proportional band is the best. As the proportional band is reduced, the controller response to any change in measurement becomes greater and greater. At some point depending upon the characteristic of each particular process, the response in the controller will be large enough to drive the measurement back in the opposite direction so far as to cause constant cycling the measurement.

This proportional band value, known as the ultimate proportional band, is a limit on the adjustment of the controller in that loop. On the other hand, if too wide a proportional band is used, the controller response to any change in measurement is too small and the measurement is not controlled as tightly as possible. The determination of the proper proportional band for any application is part of the tuning procedure for that loop. Proper adjustment of the proportional band can be observed by the response of the measurement to an upset.

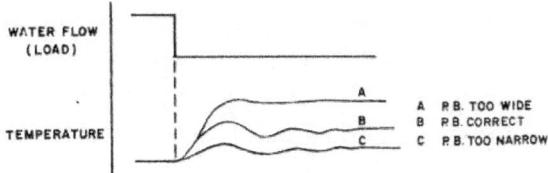

Fig. : Closed Loop Response.

Figure shows several examples of varying proportional band for the heat exchanger.

Ideally, the proper proportional band will produce quarter amplitude damping in which each half cycle is 1/2 the amplitude of the previous half cycle. The proportional band which will cause one quarter wave damping will be smaller, thereby yielding tighter control over the measured variable, as the dead time in the process decreases and the capacity increases.

One consequence of the application of proportional control to the basic control loop is offset. Offset means that the controller will maintain the measurement at a value different from the set point. This is most easily seen by referring. Note that if the load valve is opened, flow will increase through the valve and the level will begin to fall. In order to maintain the level, the supply valve would have to open. But note that because of the proportional action of the linkage the increased open position can only be achieved at a lowered level. In other words, in order to restore the balance between the flow in and the flow out, the level must stabilize at a value below the set point.

This difference, which will be maintained by the control loop, is called offset, and is characteristic of the application of proportional-only control feedback loops. The acceptability of proportional-only control depends on whether or not this offset can be tolerated. Since the error necessary to produce any output decreases with proportional band, the narrower the proportional band, the less the offset will be. For large capacity, small dead time applications accepting a very narrow proportional band, proportional-only control will probably be satisfactory since the measurement will remain within a small percentage band around the set point.

If it is essential that there be no steady state difference between measurement and set point under all load conditions, an additional function must be added to the controller.

INTEGRAL ACTION (RESET)

This function is called Integral Action or reset. The open loop response of the reset mode is shown in

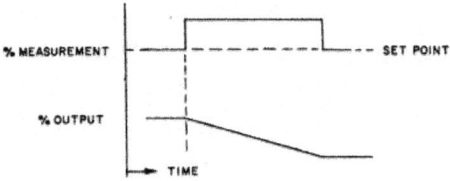
Fig. : Integral Action.

Figure which indicates a step change in the artificial measurement away from the set point at some instant in time.

So long as the measurement was at the set point, there is no change in the output due to the reset mode in the controller. However, when any error exists between measurement and set point, the reset will cause the output to begin to change and to continue to change so long as the error exists. This function, then, causes the output to change until the proper output necessary to hold the measurement at the set point at various loads is achieved. This response is added to the proportional response of the controller.

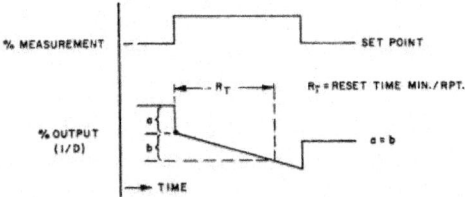
Fig. : Proportional Plus Integral Action.

The step change in the measurement first causes a proportional response, and then a reset response which is added to the proportional. The more reset action there is in the controller, the more quickly the output changes due to the reset response. The reset adjustment determines how rapidly the output changes as a function of time. Among the various brands of controllers, the amount of reset action is measured in one of two ways; either in minutes per repeat, or the number of repeats per minute.

For those controllers measuring reset in minutes per repeat, the reset time is the amount of time necessary for the reset mode to repeat the open loop response caused by proportional mode for a step change in error. Thus, for these controllers, the smaller the reset number, the greater the action of the reset mode. On those controllers which measure reset action in repeats per minute, the adjustment in indicates how many repeats of the proportional action are generated by the reset mode in one minute. Thus, for these controllers the higher the reset number the greater the reset action. The proper amount of reset action depends upon how fast the measurement can respond to the additional valve travel it causes.

The controller must not drive the valve faster than the dead time in the process, allowing the measurement to respond, or else the valve will go to its limits before the measurement can be brought back to the set point. The valve will then remain in its extreme position until the measurement crosses the set point whereupon the controller will drive the valve to its opposite extreme where will stay until the measurement crosses the set point in the opposite direction.

The result will be a reset cycle in which the valve travels from one extreme to another as the measurement oscillates around the set point. When reset is applied in controllers on batch processes where the measurement is away from the set point for long periods between batches, the reset may drive the output to its maximum resulting in "reset wind?up". When the next batch is started, the output will not come off its maximum until the measurement crosses the setpoint, causing large overshoots. This problem can be prevented by including a "batch switch" in the controller.

DERIVATIVE ACTION

The third response found on controllers is the derivative mode. Whereas proportional response responds to the size of the error and reset responds to the size and time duration of the error, the derivative mode responds to how quickly the error is changing.

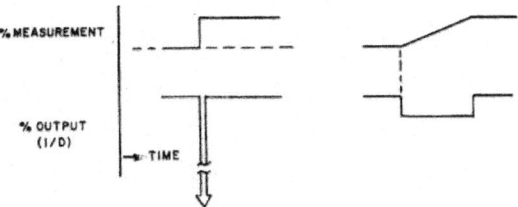

Fig. : Derivative Action.

The first is a response to a stop change of the measurement away from the set point. For a step, the measurement is changing infinitely fast, and the derivative mode in the controller causes a very large change in the output, or spike, which dies immediately because the measurement has stopped changing after the step. The second response shows the response of the derivative mode to a measurement which is changing at a constant rate. The derivative output is proportional to the rate of change of this error. The greater the rate of change, the greater the output due to derivative. The derivative holds this output so long as the measurement is changing. As soon as the measurement stops changing, whether or not it is at the set point, above or below it, the response due to derivative will cease. Among all brands of controllers, derivative response is commonly measured in minutes.

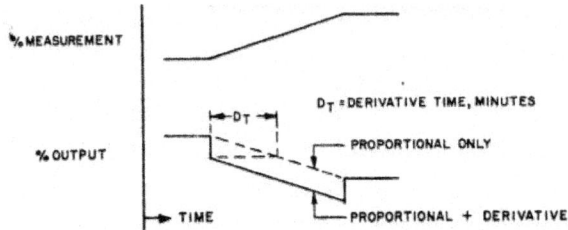

Fig. : Proportional Plus Derivative Action.

The derivative time in minutes is the time that the open loop proportional plus derivative response is ahead of the response due to proportional alone. Thus, the greater the derivative number the greater the derivative response. Changes in the error are the result of changes in either the set point, or the measurement, or both. To avoid a large output spike caused by step changes in the set point, most modern controllers apply derivative action only to changes in the measurement.

Derivative action in controllers helps to control processes with especially large time constants and significant dead time; derivative is unnecessary on those processes which respond fairly quickly to valve motion, and cannot be used at all on process with noise in the measurement signal, such as flow, since the derivative in the controller will respond to the very rapid changes in measurement which it sees in the noise. This will cause large and rapid variations in the controller output, which will keep the valve constantly moving up and down, wearing the valve and causing the measurement to cycle.

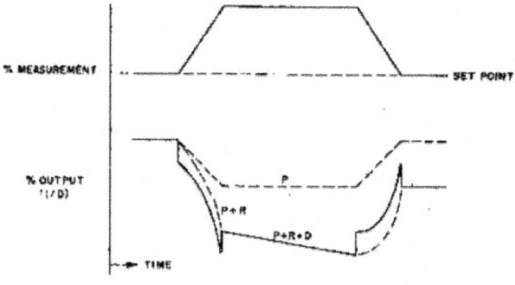

Fig. : F + T + D Action.

The combined proportional, reset, and derivative response to a simulated heat exchanger temperature measurement which deviates from the set point due to a load change. When the measurement begins to deviate from the set point, the first response from the controller is a derivative response proportional to the rate of change of measurement which opposes the movement of the measurement away from the set point. This derivative response is combined with the proportional response, and-in addition, as the reset in the controller sees the error increase, it drives the valve farther still.

This action continues until the measurement stops changing, when derivative response ceases. Since there is still an error, the measurement continues to change due to reset, until the measurement begins to move back towards the set point. As soon as the measurement begins to move back toward the set point, there is a derivative response proportional to the rate of change in the measurement opposing the return of the measurement toward the set point. The reset response continues because there is still error, although its contribution decreases with the error.

Also, the output due to proportional is changing. Thus, the measurement comes back towards the set point. As soon as the measurement reaches the set point and-stops changing, derivative response again ceases and the proportional output is back to 50%. With the measurement back at the set point, there is no longer any changing response due to reset. However, the output is at a new value. This new value is the result of the reset action during the time that the measurement was away from the set point, and compensates for the load change which caused the original upset.

INTRODUCTION TO PROGRAMMABLE LOGIC CONTROLLERS

The development of Programmable Logic Controllers (PLCs) was driven primarily by the requirements of automobile manufacturers who constantly changed their production line control systems to accommodate their new car models. In the past, this required extensive rewiring of banks of relays - a very expensive procedure.

In the 1970s, with the emergence of solid-state electronic logic devices, several auto companies challenged control manufacturers to develop a means of changing control logic without the need to totally rewire the system. The Programmable Logic Controller (PLC) evolved from this requirement. (PLC™ is a registered trademark of the Allen-Bradley Co. but is now widely used as a generic term for programmable controllers.) A number of companies responded with various versions of this type of control.

The PLCs are designed to be relatively "user-friendly" so that electricians can easily make the transition from all-relay control to electronic systems. They give users the capability of displaying and trouble-shooting ladder logic on a cathode ray tube (CRT) that showed the logic in real time. The logic can be "rewired" (programmed) on the CRT screen, and tested, without the need to assemble and rewire banks of relays.

The existing push-buttons, limit switches, and other command components continue to be used, and become input devices to the PLC. In like manner, the contactors, auxiliary relays, solenoids, indicating lamps, *etc.*, become output devices controlled by the PLC.

The ladder logic is contained as software (memory) in the PLC, replacing the inter-wiring previously required between the banks of relays. If one understands the interface between the hardware and the software, the transition to PLCs is relatively easy to accomplish. This approach to control allows "laymen" to use the control without necessarily being expert computer programmers.

The following introduction to PLCs should be considered generic in content. While each PLC manufacturer may have unique addressing systems, or varying instruction sets, you will find that the similarities will outnumber the differences. A typical program appears on the CRT as a ladder diagram, with contacts, coils, and circuit branching, very similar to that which appears on an equivalent schematic for relay logic.

This page is a help to try to understand the transition from relays to PLC control, rather than trying to teach the details of designing and programming a specific brand of equipment.

Manufacturers have numerous programming schools, from basic to advanced training, and you should consider attending them if you plan to become a proficient programmer.

PLC Hardware

Programmable controllers have a modular construction. They require a power supply, control processor unit (CPU), input/output rack (I/O), and assorted input and output modules. Systems range in size from a compact "shoe-box" design with limited memory and I/O points, to systems that can handle thousands of I/O, and multiple, inter-connected CPUs. A separate programming device is required, which is usually an industrial computer terminal, a personal computer, or a dedicated programmer.

Industrial controllers market :

- **Allen Bradley** Programmable Logic Controllers
- **Siemens** Programmable Logic Controllers

- Mitsubishi Programmable Logic Controllers

- **Omron** Programmable Logic Controllers

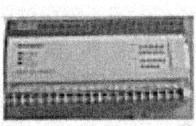 

- **GE** Programmable Logic Controllers
- **Klockner Moeller** Programmable Logic Controllers.

Power Supply

The internal logic and communication circuitry usually operates on 5 and 15 volt DC power. The power supply provides filtering and isolation of the low voltage power from the AC power line. Power supply assemblies may be separate modules, or in some cases, plug-in modules in the I/O racks. Separate control transformers are often used to isolate inputs and CPU from output devices. The purpose is to isolate this sensitive circuitry from transient disturbances produced by any highly inductive output devices.

CPU

This unit contains the "brains" of the PLC. It is often referred to as a microprocessor or sequencer. The basic instruction set is a high level program, installed in Read Only Memory (ROM). The programmed logic is usually stored in Electrically Erasable Permanent Read Only Memory (EEPROM). The CPU will save everything in memory, even after a power loss. Since it is "electrically erasable', the logic can be edited or changed as the need arises. The programming device is connected to the CPU whenever the operator needs to monitor, trouble-shoot, edit, or program the system, but is not required during the normal running operations.

I/O rack

This assembly contains slots to receive various input and output modules. The rack can be local, combined with the CPU and power supply, or remote. Each rack is given a unique address so that the CPU can recognize it. Within each rack, the slots have unique addresses. Power and communication cables are required for remote installations. The replaceable I/O modules plug into a back-plane that communicates directly with the CPU or through the cable assembly. Field wiring terminates on "swing arms" that plug into the face of the I/O modules. This allows a quick change of I/O modules without disconnecting the field wiring.

Every module terminal also has a unique address.

I/O modules are available in many different configurations, and voltages, (AC and DC). Special modules are available to read analog signals and produce analog outputs, provide communication capabilities, interface with motion control systems, *etc.* The input modules provide isolation from the "real world" control voltages, and give the CPU a continuous indication of the on/off status of each input termination. Inputs sense the presence of voltages at their terminals, and therefore usually have very low current requirements.

Output modules receive commands from the CPU and switch isolated power on and off at the output terminals. Output modules must be capable of switching currents required by the load connected to each terminal, so more attention must be given to current capacity of output modules and their power supply.

Programming Devices

Every brand of PLC has its own programming hardware. Sometimes it is a small hand-held device that resembles an oversized calculator with a liquid crystal display (LCD).

Computer-based programmers typically use a special communication board, installed in an industrial terminal or personal computer, with the appropriate software program installed.

Computer-based programming allows "off-line" programming, where the programmer develops his logic, stores it on a disk, and then "down-loads" the program to the CPU at his convenience. In fact, it allows more than one programmer to develop different modules of the program.

Programming can be done directly to the CPU if desired. When connected to the CPU the programmer can test the system, and watch the logic operate as each element is intensified in sequence on the CRT when the system is running. Since a PLC can operate without having the programming device attached, one device can be used to service many separate PLC systems. The programmer can edit or change the logic "on-line" in many cases.

Trouble shooting is greatly simplified, once you understand the addressing system. Every I/O point has a corresponding address in the CPU memory.

The Control Loop

An automatic control has the task of bringing the output signal x of a process-controlled-system to its predetermined value as well as to keep this value despite the influence from disturbances z.

In a digital automatic control, the regulating variable x is periodically collected and compared with the reference variable input w.

The difference is an error signal $xd = w-x$, which is then processed in the controller to a correcting variable y. This correcting variable acts as a feedback in the process-controlled-system.

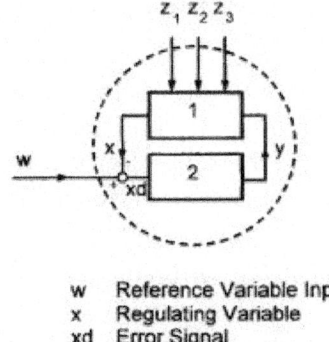

w   Reference Variable Input
x   Regulating Variable
xd   Error Signal
y   Correcting Variable
z   Disturbance

1   Process-controlled-system
2   Control System

Fig. : Control loop and signal flow diagram.

Sensors and Measuring Transducers

The regulating variable can be any physical quantity. Generally in process technology such variables are pressure, temperature, medium level and flow, amongst others.

Some sensors can be directly connected to the controller such as resistance- and temperature-sensors. In other instances, such as with measuring transducers whose output is an electric quantity, the connection must be done between the sensor and controller. Generally the industrial controllers are designed for measuring transducers with standard signal outputs **(0/4 - 20 mA)**.

• Industrial transducers.

Actuators and Actuating Drives

In most heat and process technology applications, the correcting variable **y** acts via a valve, a flap or via any other mechanic adjusting mechanism. There are three drive types to operate such actuators:

Industrial controllers market :

• **Allen Bradley** Programmable Logic Controllers

• **Siemens** Programmable Logic Controllers

• **Mitsubishi** Programmable Logic Controllers

• **Omron** Programmable Logic Controllers

• **GE** Programmable Logic Controllers

• **Klockner Moeller** Programmable Logic Controllers

• **Newport** controllers

- Process controllers for the industry.

- **Yokogawa** Products

- **Electric actuating drive** consisting of an electromotor and set of gears. This drive works as an integral element and is actuated via a three-step controller. Additionally, there are electric actuating drives with an integrated (series connected) position controller. These work as a proportional element and are actuated via continuous controllers.

- **Pneumatic actuating drive** with compressed air and with an electropneumatic position controller or with an electropneumatic transducer. These work as a proportional element and are actuated via continuous controllers.

- **Hydraulic actuating drive** with an electrically operated oil pump and electrohydraulic position controller. These also work as a proportional element and are actuated via continuous controllers.

Electric actuating drives with alternating or three phase current motors are robust require low maintenance and are economical.

Pneumatic actuating drives are faster than electric ones and are explosion-proof. However they are not really suitable for a big actuating power.

Hydraulic actuating drives are fast and also suitable for a big actuating power. They are, however, more expensive than pneumatic or electric actuating drives.

With these three types of actuating drives, automatic controls are constantly carried out. Relays, contactors or thyristors; are used as actuators in discontinuous temperature control loops for electrical heating and/or cooling.

The Controller

In the front-end circuit the regulating variable **x** is compared with the reference variable input **w** and the error signal **xd** is determined. The error signal is then converted with or without a time response to an output signal. The output signal of the amplifier can immediately represent the correcting variable **y** if, for example, actuators working as proportional elements or actuating drives are controlled by the output signal.

In the electric actuating drives the correcting variable **y** occurs only after the actuating drive operates. The necessary adjusting increments are obtained as a pulse interval modulated signal from the controller output.

Depending on the design of the circuit, the controller works as a **Proportional - (P)**, **Proportional-Differential (PD)**, Proportional-Integral- (PI) or **Proportional-Integral-Differential (PID)** controller.

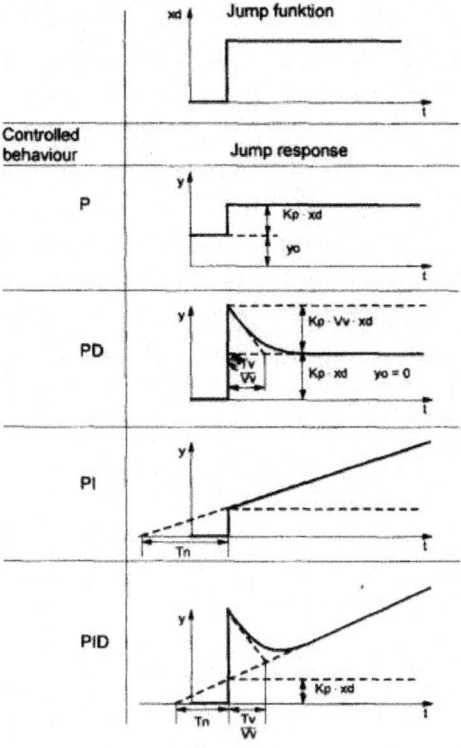

Fig. : Jump responses with different time responses.

When a jump function is sent to the controller inlet, the respective jump responses are developed.

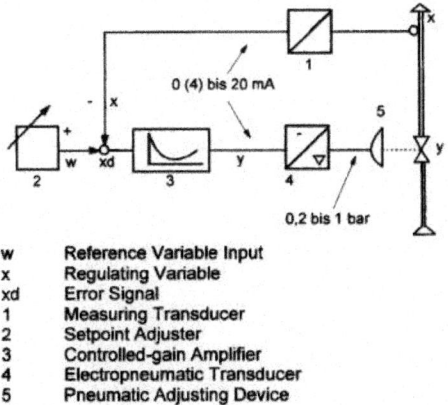

w	Reference Variable Input
x	Regulating Variable
xd	Error Signal
1	Measuring Transducer
2	Setpoint Adjuster
3	Controlled-gain Amplifier
4	Electropneumatic Transducer
5	Pneumatic Adjusting Device

Fig. : Function diagram – Continuous controller.

Characteristic quantities of a P- and PD-controller are the proportional gain **Kp** and the operating point **yo**. The operating point is defined as the *output signal, which has a zero error signal.*

Unlike the P- or PD-controller, in the PI-controller a permanent error signal independent from the operating point, adjustment of the reference variable input and change of the disturbance is avoided by means of the integral part of the controller. The parameter of the integral part is the *integral action time* **Tn**.

With a PID-controller it is possible to achieve an improvement of the dynamic quality due to its additional **D-part**. The **D-part** is determined by the rate amplification **Vv** and the rate time **Tv**.

The controller output signals must be matched on the actuating drives. Two controller types have been set for the most important actuating drives.

A **continuous controller** is normally used in pneumatic and hydraulic actuating drives and a three-step controller in electric actuating drives.

The continuous controller is mainly used in plants with pneumatic actuating drives.

The controller output signal **0(4) to 20 mA** constantly works on the adjusting device via an electropneurnatic transducer.

Unlike the continuous controller, **discontinuous controllers** don't have a constant output signal. Instead the correcting variable can only, be on or off, *i.e.* for example voltage on/off.

However, it is also possible to regulate a process with a discontinuous controller. Discontinuous controllers change the operating ratio instead of the value of the output signal.

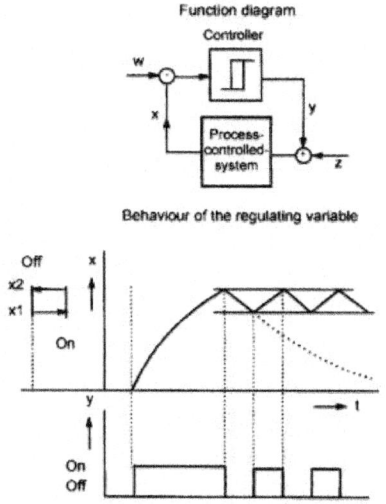

Fig. : Two-step controller without feedback.

Discontinuously switching two-step controllers are employed in the actuation from relays, contactors or thyristors installed in electrical heating or cooling systems.

The two-step controller switches when the regulating variable is outside the area between x1 and x2. A steady-state vibration results, whose frequency depends on the dead time of the process system and on the switching hysteresis of the controller.

Since in most cases the obtained controlled results don't meet the requirements, the sampling frequency is increased which reduces the amplitude of the controlled oscillation. By these means, it is often possible to obtain with a two-step controller a controlled result of a P- or PI-controller.

In process systems with small dead times, the sampling frequency can be very high. This results in a high contact load of the relay in the controller output. If the sampling frequency is lowered due to an increase of the switching difference, the control accuracy lowers again.

If a portion of the output signal is returned to the input and combined with the deviation, the characteristics of the two-step controller are fundamentally changed. The deviation is significantly reduced and a time response is obtained - like with a continuous controller.

A PID-two-step controller can be developed if for example an impulse-interval-transformer with an adjustable period is placed after the controlled-gain amplifier.

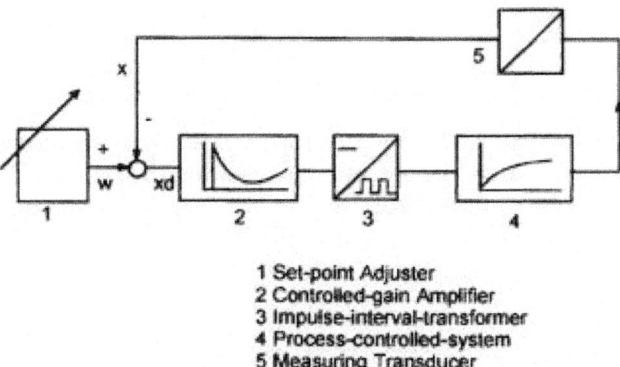

1 Set-point Adjuster
2 Controlled-gain Amplifier
3 Impulse-interval-transformer
4 Process-controlled-system
5 Measuring Transducer

Fig. : PID-two-step controller.

In principle, three-stop controllers consist of two interconnected two-step controllers. With these controllers the cooling / heating effect is achieved when the setpoint is respectively exceeded / fallen short of.

In these applications, two-step controllers are used whose correcting variable is split into two parts. Additionally, two outputs are assigned. Between the two parts there is an adjustable dead zone. In each part the pulse-width repetition rate runs through 0 to 100%. Such controllers are designated as three-step controllers.

Actuating drives also have three switching modes: Open, close and stop. In cases where an electrical motor is employed as a drive for the right- and left-rotation, the actuating drive is controlled by a three-step controller.

Such actuating drives need a certain period of time until the desired damper position is reached. In case the controller does not give any further signals, the actuating drives remain in the reached position. These controllers, are designated **three-step controllers**.

The three-step controller switches the electromotor of the actuating drive to right rotation, stop position and left-rotation with relays or static switches. Additionally, the controller can influence the adjusting speed of the adjusting device due to the different pulse-width repetition rates.

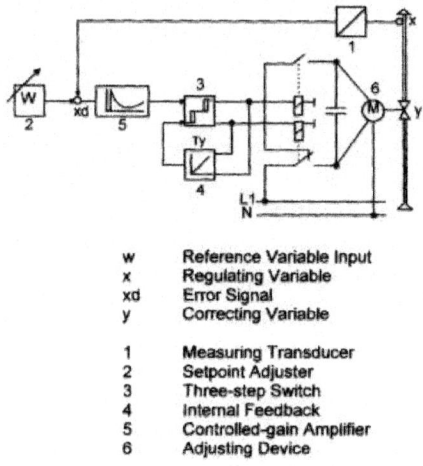

w	Reference Variable Input
x	Regulating Variable
xd	Error Signal
y	Correcting Variable
1	Measuring Transducer
2	Setpoint Adjuster
3	Three-step Switch
4	Internal Feedback
5	Controlled-gain Amplifier
6	Adjusting Device

Fig. : Tree-step controller Function diagram.

The step response created by these impulses in the actuator is similar to a step response of a continuous Pl-controller. Therefore, the parameters **Kp** and **Tn** are also used in the description of the step response from three-step controllers. One uses the expression quasicontinuous control.

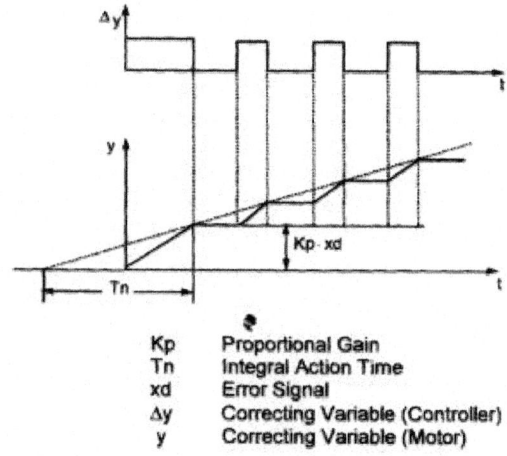

Kp	Proportional Gain
Tn	Integral Action Time
xd	Error Signal
Δy	Correcting Variable (Controller)
y	Correcting Variable (Motor)

Fig. : Step response and characteristic quantities.

The operation margin is adjustable to achieve, for example, a suppression of the disturbing signal and, thus, a stabilizing effect.

The feedback of the correcting variable y can occur in two forms in three-step controllers (TSC): as an output signal of the position encoder (connected to the motor shaft) with external feedback of the correcting variable (TSC Extern) or via an internal copy of the correcting variable (TSC Intern).

The integral component of the adjusting device is simulated with an integrator with adjustable floating time (parameter Ty). The integrator replaces the position-feedback, To prevent the internal integrator and the PID-output from operating in the saturation zone over the time, both variables are decreased by the same value if necessary. To avoid an integral saturation, the slew rate of the I-part is limited by the series connected controlled-gain amplifier.

The position controller has an adjustable minimum interpulse length **Ton** and a minimum interpulse period **Toff**. The interpulse length **Ton** gives as a result an operating threshold Ae as follows:

Switch on: Aee $=$ $2\dfrac{Te}{Ty} \cdot 100\%$

Switch off: Aea $=$ $\dfrac{Te}{Ty} \cdot 100\%$

Hysteresis: Aee – Aea $=$ $\dfrac{Te}{Ty} \cdot 100\%$

Interval: Aa $=$ $\dfrac{Ta}{Ty} \cdot 100\%$

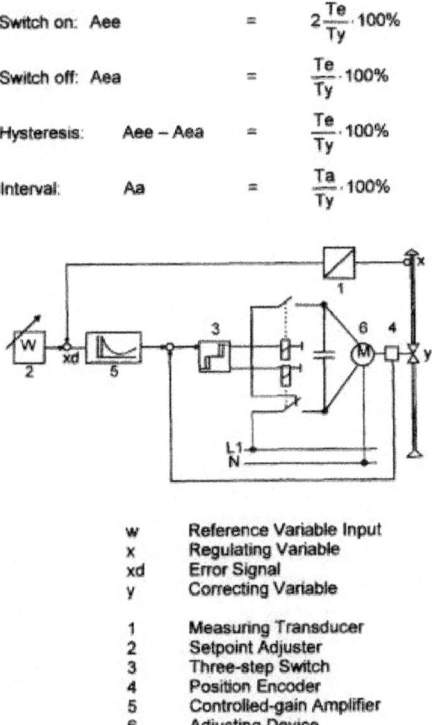

w	Reference Variable Input
x	Regulating Variable
xd	Error Signal
y	Correcting Variable

1	Measuring Transducer
2	Setpoint Adjuster
3	Three-step Switch
4	Position Encoder
5	Controlled-gain Amplifier
6	Adjusting Device

Fig. : Function diagram of a TSC extern.

In the **TSC Extern**" mode the continuous controller output signal is compared with the correcting variable y of the adjusting device. The output signal divergence is fed to a three-step controller (with a PD- feedback structure) which controls the right-/ left-rotation of the actuator.

This way, it is possible to achieve a correcting value limitation with the parameters **ya** and **ye** as well as an absolute-value setting for a safety correcting value **ys**.

The parameters **Toff** and Ton are also instrumental in the output structure to adjust the minimum interpulse period and length. Additionally, these parameters together with **Ty** help in the optimization of the position control loop.

$$\text{Switch on: } Aee = 6\frac{Te}{Ty} \cdot 100\%$$

$$\text{Switch off: } Aea = 3\frac{Te}{Ty} \cdot 100\%$$

$$\text{Hysteresis: } Aee - Aea = 3\frac{Te}{Ty} \cdot 100\%$$

$$\text{Interval: } Aa = \frac{Ta}{Ty} \cdot 100\%$$

Chapter 5

INTEGRAL ACTION AND PI CONTROL

Like the P-Only controller, the Proportional-Integral (PI) algorithm computes and transmits a controller output (CO) signal every sample time, T, to the final control element (*e.g.*, valve, variable speed pump). The computed CO from the PI algorithm is influenced by the controller tuning parameters and the controller error, e(t).

PI controllers have two tuning parameters to adjust. While this makes them more challenging to tune than a P-Only controller, they are not as complex as the three parameter PID controller.

Integral action enables PI controllers to eliminate offset, a major weakness of a P-only controller. Thus, PI controllers provide a balance of complexity and capability that makes them by far the most widely used algorithm in process control applications.

The PI Algorithm

While different vendors cast what is essentially the same algorithm in different forms, here we explore what is variously described as the dependent, ideal, continuous, position form:

$$CO = CO_{bias} + Kc \cdot e(t) + \frac{Kc}{Ti} \int e(t)dt$$

Where:

CO = controller output signal (the wire out)

CO_{bias} = controller bias or null value; set by bumpless transfer as explained below

e(t) = current controller error, defined as SP – PV

SP = set point

PV = measured process variable (the wire in)

Kc = controller gain, a tuning parameter

Ti = reset time, a tuning parameter

The first two terms to the right of the equal sign are identical to the P-Only controller.

The integral mode of the controller is the last term of the equation. Its function is to integrate or continually sum the controller error, e(t), over time.

Some things we should know about the reset time tuning parameter, Ti:

- It provides a separate weight to the integral term so the influence of integral action can be independently adjusted.
- It is in the denominator so smaller values provide a larger weight to (*i.e.* increase the influence of) the integral term.
- It has units of time so it is always positive.

Function of the Proportional Term

As with the P-Only controller, the proportional term of the PI controller, Kc·e(t), adds or subtracts from CO_{bias} based on the size of controller error e(t) at each time t.

As e(t) grows or shrinks, the amount added to CO_{bias} grows or shrinks immediately and proportionately. The past history and current trajectory of the controller error have no influence on the proportional term computation.

The idea for a set point response. The error used in the proportional calculation is shown on the plot:

- At time t = 25 min, e(25) = 60–56 = 4
- At time t = 40 min, e(40) = 60–62 = –2

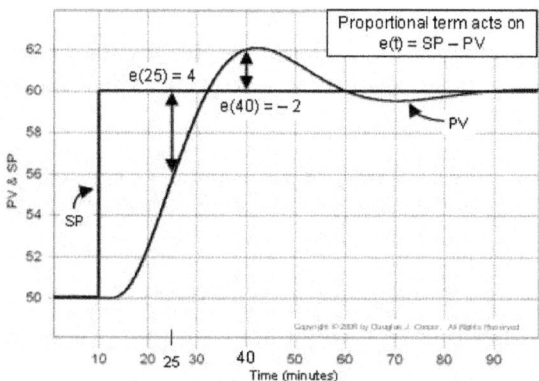

Recalling that controller error e(t) = SP – PV, rather than viewing PV and SP as separate traces as we do above, we can compute and plot e(t) at each point in time t.

Below is the identical data to that above only it is recast as a plot of e(t) itself. Notice that in the plot above, PV = SP = 50 for the first 10 min, while in the error plot below, e(t) = 0 for the same time period.

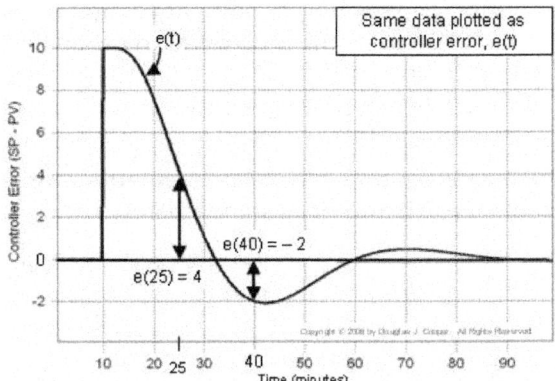

This plot is useful as it helps us visualize how controller error continually changes size and sign as time passes.

Function of the Integral Term

While the proportional term considers the current size of e(t) only at the time of the controller calculation, the integral term considers the history of the error, or how long and how far the measured process variable has been from the set point over time.

Integration is a continual summing. Integration of error over time means that we sum up the complete controller error history up to the present time, starting from when the controller was first switched to automatic.

Controller error is e(t) = SP − PV. In the plot below, the integral sum of error is computed as the shaded areas between the SP and PV traces.

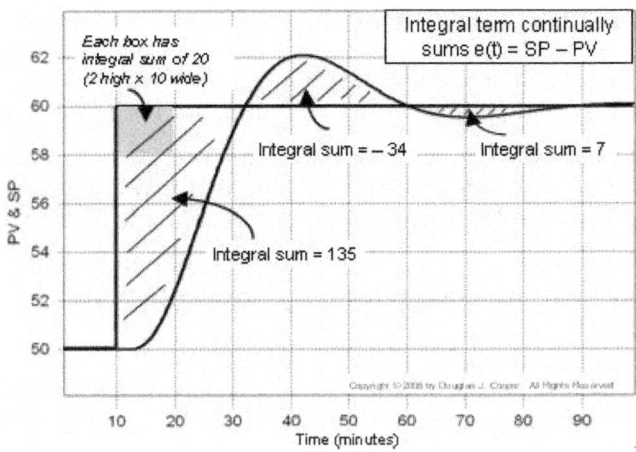

Each box in the plot has an integral sum of 20 (2 high by 10 wide). If we count the number of boxes contained in the shaded areas, we can compute the integral sum of error.

So when the PV first crosses the set point at around t = 32, the integral sum has grown to about 135. We write the integral term of the PI controller as:

$$\frac{Kc}{Ti}\int_0^{32} e(t)dt = \frac{Kc}{Ti}(135)$$

Since it is controller error that drives the calculation, we get a direct view the situation from a controller error plot as shown below:

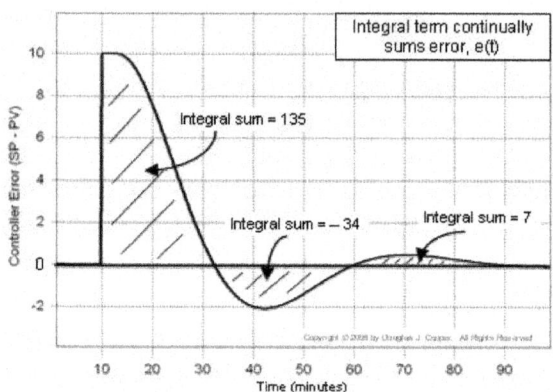

Note that the integral of each shaded portion has the same sign as the error. Since the integral sum starts accumulating when the controller is first put in automatic, the total integral sum grows as long as e(t) is positive and shrinks when it is negative.

At time t = 60 min on the plots, the integral sum is 135 – 34 = 101. The response is largely settled out at t = 90 min, and the integral sum is then 135 – 34 + 7 = 108.

Integral Action Eliminates Offset

The previous sentence makes a subtle yet very important observation. The response is largely complete at time t = 90 min, yet the integral sum of all error is not zero.

In this example, the integral sum has a final or residual value of 108. It is this residual value that enables integral action of the PI controller to eliminate offset.

The most processes under P-only control experience offset during normal operation. Offset is a sustained value for controller error (*i.e.*, PV does not equal SP at steady state).

We recognize from the P-Only controller:

$$CO = CO_{bias} + Kc.e(e(t)$$

that CO will always equal CO_{bias} unless we add or subtract something from it.

The only way we have something to add or subtract from CO_{bias} in the P-Only equation above is if e(t) is not zero. It e(t) is not steady at zero, then PV does not equal SP and we have offset.

However, with the PI controller:

$$CO = CO_{bias} + Kc \cdot e(t) + \frac{Kc}{Ti} \int e(t)dt$$

we now know that the integral sum of error can have a final or residual value after a response is complete. This is important because it means that e(t) can be zero, yet we can still have something to add or subtract from CO_{bias} to form the final controller output, CO.

So as long as there is any error (as long as e(t) is not zero), the integral term will grow or shrink in size to impact CO. The changes in CO will only cease when PV equals SP (when e(t) = 0) for a sustained period of time.

At that point, the integral term can have a residual value as just discussed. This residual value from integration, when added to CO_{bias}, essentially creates a new overall bias value that corresponds to the new level of operation.

In effect, integral action continually resets the bias value to eliminate offset as operating level changes.

Challenges of PI Control

There are challenges in employing the PI algorithm:

- The two tuning parameters interact with each other and their influence must be balanced by the designer.
- The integral term tends to increase the oscillatory or rolling behavior of the process response.

Because the two tuning parameters interact with each other, it can be challenging to arrive at "best" tuning values. The value and importance of our design and tuning recipe increases as the controller becomes more complex.

Initializing the Controller for Bumpless Transfer

When we switch any controller from manual mode to automatic (from open loop to closed loop), we want the result to be uneventful. That is, we do not want the switchover to cause abrupt control actions that impact or disrupt our process

We achieve this desired outcome at switchover by initializing the controller integral sum of error to zero. Also, the set point and controller bias value are initialized by setting:

- SP equal to the current PV
- CO_{bias} equal to the current CO

With the integral sum of error set to zero, there is nothing to add or subtract from CO_{bias} that would cause a sudden change in the current controller output. With the set point equal to the measured process variable, there is no error to drive a change in our CO. And with the controller bias set to our current CO value, we are prepared by default to maintain current operation.

Thus, when we switch from manual mode to automatic, we have "bumpless transfer" with no surprises. This is a result everyone appreciates.

Reset Time Versus Reset Rate

Different vendors cast their control algorithms in slightly different forms. Some use proportional band rather than controller gain. Also, some use reset rate, Tr, instead of reset time. These are simply the inverse of each other:

$$Tr = 1 / Ti$$

No matter how the tuning parameters are expressed, the PI algorithms are all equally capable.

But it is critical to know your manufacturer before you start tuning your controller because parameter values must be matched to your particular algorithm form. Commercial software for controller design and tuning will automatically address this problem for you.

PI CONTROL OF THE HEAT EXCHANGER

We investigated P-Only control of the heat exchanger process and learned that while P-Only is an algorithm that is easy to tune and maintain, it has a severe limitation. Specifically, its simple form permits steady state error, called offset, in most processes during normal operation.

Then we moved on to integral action and PI control. We focused in that chapter on the structure of the algorithm and explored the mathematics of how the proportional and integral terms worked together to eliminate offset.

Here we test the capabilities of the PI controller on the heat exchanger process. Our focus is on design, implementation and basic performance issues. Along the way we will highlight some strengths and weaknesses of this popular algorithm.

As with all controller implementations, best practice is to follow our proven four-step design and tuning recipe as we proceed with this case study.

Step 1: Design Level of Operation (DLO)

Real processes display a nonlinear behavior. That is, their process gain, time constant and/or dead time changes as operating level changes and as major disturbances change. Since controller design and tuning is based on these process Kp, Tp and θp values, controllers should be designed and tuned for a specific level of operation.

Thus, the first step in our controller design recipe is to specify our design level of operation (DLO). This includes stating:

- Where we expect the set point, SP, and measured process variable, PV, to be during normal operation.
- The range of values the SP and PV might assume so we can explore the nature of the process dynamics across that range.

We will track along with the same design conditions used in the P-Only control study to permit a direct comparison of performance and capability. As in that study, we specify:

- Design PV and SP = 138 °C with range of 138 to 140 °C

We also should know normal or typical values for our major disturbances and be reasonably confident that they are quiet so we may proceed with a bump test. The heat exchanger process has only one major disturbance variable, and consistent with the previous study:

- Expected warm liquid flow disturbance = 10 L/min

Step 2: Collect Data at the DLO

The next step in the design recipe is to collect dynamic process data as near as practical to our design level of operation. We have previously collected and documented heat exchanger step test data that matches our design conditions.

Step 3: Fit an FOPDT Model to the Design Data

Here we document a first order plus dead time (FOPDT) model approximation of the step test data from step 2:

- Process gain (how far), Kp = –0.53 °C/%
- Time constant (how fast), Tp = 1.3 min
- Dead time (how much delay), θp = 0.8 min

Step 4: Use the Parameters to Complete the Design

One common form of the PI controller computes a controller output (CO) action every loop sample time T as:

Where:

CO = controller output signal (the wire out)

CO_{bias} = controller bias or null value; set by bumpless transfer as explained below

e(t) = current controller error, defined as SP – PV

SP = set point

PV = measured process variable (the wire in)

Kc = controller gain, a tuning parameter

Ti = reset time, a tuning parameter

Loop Sample Time, T

Best practice is to specify loop sample time, T, at 10 times per time constant or faster (T ≤ 0.1Tp). For this study, T ≤ 0.13 min = 8 sec. Faster sampling may provide modestly improved performance, while slower sampling can lead to significantly degraded performance. Most commercial controllers offer an option of T = 1.0 sec, and since this meets our design rule, we use that here.

Computing controller error, e(t)

Set point, SP, is something we enter into the controller. The PV measurement comes from our sensor (our wire in). With SP and PV known, controller error, e(t) = SP – PV, can be directly computed at every loop sample time T.

Determining Bias Value

Strictly speaking, CO_{bias} is the value of the CO that, in manual mode, causes the PV to steady at the DLO while the major disturbances are quiet and at their normal or expected values.

Bumpless Transfer

A desirable feature of the PI algorithm is that it is able to eliminate the offset that can occur under P-Only control. The integral term of the PI controller provides this capability by providing updated information that, when combined with the controller bias, keeps the process centered as conditions change.

Since integral action acts to update (or reset) our bias value over time, CO_{bias} can be initialized in a straightforward fashion to a value that produces no abrupt control actions when we switch to automatic. Most commercial controllers do this with a simple "bumpless transfer" feature. When switching to automatic, they initialize:

- SP equal to the current PV
- CO_{bias} equal to the current CO

With the set point equal to the measured process variable, there is no error to drive a change in our controller output. And with the controller bias set to our current controller output, we are prepared by default to maintain current operation.

We will use a controller that employs these bumpless transfer rules when we switch to automatic. Hence, we need not specify any value for CO_{bias} as part of our design.

Computing Controller Gain and Reset Time

Here we use the industry-proven Internal Model Control (IMC) tuning correlations. The first step in using the IMC correlations is to compute T_c, the closed loop time constant. All time constants describe the speed or quickness of a response. The closed loop time constant describes the desired speed or quickness of a controller in responding to a set point change or rejecting a disturbance.

If we want an active or quickly responding controller and can tolerate some overshoot and oscillation as the PV settles out, we want a small T_c (a short response time) and should choose *aggressive* tuning:

aggressive: T_c is the larger of $0.1 \cdot T_p$ or $0.8\, \theta_p$

Moderate tuning is for a controller that will move the PV reasonably fast while producing little to no overshoot.

moderate: T_c is the larger of $1 \cdot T_p$ or $8\, \theta_p$

If we seek a more sluggish controller that will move things in the proper direction, but quite slowly, we choose *conservative* tuning (a big or long T_c).

conservative: T_c is the larger of $10 \cdot T_p$ or $80\, \theta_p$

Once we have decided on our desired performance and computed the closed loop time constant, T_c, with the above rules, then the PI correlations for controller gain, K_c, and reset time, T_i, are:

Notice that reset time, T_i, is always set equal to the time constant of the process, regardless of desired controller activity.

Moderate Response Tuning:

For a controller that will move the PV reasonably fast while producing little to no overshoot, choose:

Moderate T_c = the larger of $1 \cdot T_p$ or $8\, \theta_p$

= larger of $1(1.3 \text{ min})$ or $8(0.8 \text{ min})$

= 6.4 min

Using this T_c and our model parameters in the tuning correlations above, we arrive at the moderate tuning values:

Aggressive Response Tuning

1. For an active or quickly responding controller where we can tolerate some overshoot and oscillation as the PV settles out, specify

    Aggressive T_c = the larger of $0.1 \cdot T_p$ or $0.8\, \theta_p$

    = larger of $0.1(1.3 \text{ min})$ or $0.8(0.8 \text{ min})$

    = 0.64 min

    and the aggressive tuning values are:

Practitioner's Note: The FOPDT model parameters used in the tuning correlations above have engineering units, so the Kc values we compute also have engineering units. In commercial control systems, controller gain (or proportional band) is normally entered as a dimensionless (%/%) value. For commercial implementations, we could:

- Scale the process data before fitting our FOPDT dynamic model so we directly compute a dimensionless Kc.

- Convert the model Kp to dimensionless %/% after fitting the model but before using the FOPDT parameters in the tuning correlations.

- Convert Kc from engineering units into dimensionless %/% after using the tuning correlations. CO is already scaled from 0 – 100% in the above example. Thus, we convert Kc from engineering units into dimensionless %/% using the formula:

For the heat exchanger, PVmax = 250 °C and PVmin = 0 °C. The dimensionless Kc values are thus computed:

- moderate Kc = (– 0.34 %/ °C)·[(250 – 0 °C) ÷ (100 – 0%)] = – 0.85 %/%
- aggressive Kc = (– 1.7%/ °C)·[(250 – 0 °C) ÷ (100 – 0%)] = – 4.2 %/%

We use Kc with engineering units in the remainder of this chapter and are careful that our PI controller is formulated to accept such values. We would be mindful if we were using a commercial control system, however, to ensure our tuning parameters are cast in the form appropriate for our equipment.

Controller Action

The process gain, Kp, is negative for the heat exchanger, indicating that when CO increases, the PV decreases in response. This behavior is characteristic of a reverse acting process. Given this CO to PV relationship, when in automatic mode (closed loop), if the PV starts drifting above set point, the controller must increase CO to correct the error. Such negative feedback is an essential component of stable controller design.

A process that is naturally reverse acting requires a controller that is direct acting to remain stable. In spite of the opposite labels (reverse acting process and direct acting controller), the details presented above show that both Kp and Kc are negative values.

In most commercial controllers, only positive Kc values can be entered. The sign (or action) of the controller is then assigned by specifying that the controller is either reverse acting or direct acting to indicate a positive or negative Kc, respectively.

If the wrong control action is entered, the controller will quickly drive the final control element (FCE) to full on/open or full off/closed and remain there until a proper control action entry is made.

Implement and Test

Below we test our two PI controllers on the heat exchanger process simulation. Shown are two set points step pairs from 138 °C up to 140 °C and back again.

The first set point steps to the left show the PI controller performance using the moderate tuning values computed above. The second set point steps to the right show the controller performance using the aggressive tuning values. Note that the warm liquid disturbance flow, though not shown, remains constant at 10 L/min throughout the study.

The asymmetrical behavior of the PV for the set point steps up compared to the steps down is due to the very nonlinear character of the heat exchanger.

If we seek tuning between moderate and aggressive performance, we would average the Kc values from the tuning rules above.

But if we believe we had collected good bump test data (we saw a clear response in the PV when we stepped the CO and the major disturbances were quiet during the test), and the FOPDT model fit appears to be visually descriptive of the data, then we have a good value for Tp and that means a good value for Ti.

If we are going to fiddle with the tuning, we can tweak Kc and we should leave the reset time alone.

Tuning Recipe Saves Time and Money

The exciting result is that we achieved our desired controller performance based on one bump test and following a controller design recipe. No trial and error was involved. Little off-spec product was produced. No time was wasted.

The method of approximating complex behavior with a FOPDT model and then following a recipe for controller design and tuning has been used successfully on a broad spectrum of processes with streams composed of gases, liquids, powders, slurries and melts. It is a reliable approach that has been proven time and again at diverse plants from a wide range of companies.

PI DISTURBANCE REJECTION OF THE GRAVITY DRAINED TANKS

When exploring the capabilities of the P-Only controller in rejecting disturbances for the gravity drained tanks process, we confirmed the observations we had made during the the P-Only set point tracking study for the heat exchanger.

In particular, the P-Only algorithm is easy to tune and maintain, but whenever the set point or a major disturbance moves the process from the design level of operation, a sustained error between the process variable (PV) and set point (SP), called offset, results.

Further, we saw in both case studies that as controller gain, Kc, increases (or as proportional band, PB, decreases):

- the activity of the controller output, CO, increases

- the oscillatory nature of the response increases
- the offset (sustained error) decreases

We explore the benefits of integral action and the capabilities of the PI controller for rejecting disturbances in the gravity drained tanks process. We have previously presented the fundamentals behind PI controland its application to set point tracking in the heat exchanger.

As with all controller implementations, best practice is to follow our proven four-stepdesign and tuning recipe. One benefit of the recipe is that steps 1-3, summarized below from our P-Only study, remain the same regardless of the control algorithm being employed. After summarizing steps 1-3, we complete the PI controller design and tuning in step 4.

Step 1: Determine the Design Level of Operation (DLO)

The control objective is to reject disturbances as we control liquid level in the lower tank. Our design level of operation (DLO):

- design PV and SP = 2.2 m with range of 2.0 to 2.4 m
- design D = 2 L/min with occasional spikes up to 5 L/min

Step 2: Collect Process Data around the DLO

When CO, PV and D are steady near the design level of operation, we bump the CO and force a clear response in the PV that dominates the noise.

Step 3: Fit a FOPDT Model to the Dynamic Process Data

We then describe the process behavior by fitting an approximating first order plus dead time (FOPDT) dynamic model to the test data from step 2. We define the model parameters and present details of the model fit of step test data here. A model fit ofdoublet test data using commercial software confirms these values:

- process gain (how far), Kp = 0.09 m/%
- time constant (how fast), Tp = 1.4 min
- dead time (how much delay), θp = 0.5 min

Step 4: Use the FOPDT Parameters to Complete the Design

We explore what is often called the dependent, ideal form of the PI control algorithm:

Where:

CO = controller output signal (the wire out)

CO_{bias} = controller bias or null value; set by bumpless transfer

e(t) = current controller error, defined as SP – PV

SP = set point

PV = measured process variable (<u>the wire in</u>)

Kc = controller gain, a tuning parameter

Ti = reset time, a tuning parameter

Aside: our observations using the dependent ideal PI algorithm directly apply to the other popular PI controller forms. For example, the integral gain, Ki, in the independent algorithm form:can be computed directly from controller gain and reset time as: Ki = Kc/Ti.

In the P-Only study, we established that for the gravity drained tanks process:

- sample time, T = 1 sec

- the controller is reverse acting

- dead time is small compared to Tp and thus not a concern in the design

Controller Gain, Kc, and Reset Time, Ti

We use our FOPDT model parameters in the industry-proven Internal Model Control (IMC) tuning correlations to compute PI tuning values.

The first step in using the IMC correlations is to compute Tc, the closed loop time constant. All time constants describe the speed or quickness of a response. Tc describes the desired speed or quickness of a controller in responding to a set point change or rejecting a disturbance.

If we want an active or quickly responding controller and can tolerate some overshoot and oscillation as the PV settles out, we want a small Tc (a short response time) and should choose aggressive tuning:

Aggressive Response: Tc is the larger of 0.1 $\cdot Tp$ or 0.8 θp

If we seek a sluggish controller that will move things in the proper direction, but quite slowly, we choose conservative tuning (a big or long Tc).

Conservative Response: Tc is the larger of 10 $\cdot Tp$ or 80 θp

Moderate tuning is for a controller that will move the PV reasonably fast while producing little to no overshoot.

Moderate Response: Tc is the larger of 1 $\cdot Tp$ or 8 θp

With Tc computed, the PI controller gain, Kc, and reset time, Ti, are computed as:

Notice that reset time, Ti, is always equal to the process time constant, Tp, regardless of desired controller activity.

Moderate Response Tuning:

1. For a controller that will move the PV reasonably fast while producing little to no overshoot, choose:

Moderate Tc = the larger of $1 \cdot Tp$ or $8 \cdot \theta p$

= larger of $1(1.4 \text{ min})$ or $8(0.5 \text{ min})$

= 4 min

Using this Tc and our model parameters in the tuning correlations above, we arrive at the moderate tuning values:

Aggressive Response Tuning:

1. For an active or quickly responding controller where we can tolerate some overshoot and oscillation as the PV settles out, specify:

   Aggressive Tc = the larger of $0.1 \cdot Tp$ or $0.8 \cdot \theta p$

   = larger of $0.1(1.4 \text{ min})$ or $0.8(0.5 \text{ min})$

   = 0.4 min

and the aggressive tuning values are:

Practitioner's Note: The FOPDT model parameters used in the tuning correlations above have engineering units, so the Kc values we compute also have engineering units. In commercial control systems, controller gain (or proportional band) is normally entered as a dimensionless (%/%) value. To address this, we could:

- Scale the process data before fitting our FOPDT dynamic model so we directly compute a dimensionless Kc.

- Convert the model Kp to dimensionless %/% after fitting the model but before using the FOPDT parameters in the tuning correlations.

- Convert Kc from engineering units into dimensionless %/% after using the tuning correlations. Since we already have Kc in engineering units, we employ the third option. CO is already scaled from 0 – 100% in the above example. Thus, we convert Kc from engineering units into dimensionless %/% using the formula:

For the gravity drained tanks, $PV_{max} = 10$ m and $PV_{min} = 0$ m. The dimensionless Kc values are thus computed:

- moderate $Kc = (3.5 \text{ %/m}) \cdot [(10 - 0 \text{ m}) \div (100 - 0\%)] = 0.35 \text{ %/%}$

- aggressive $Kc = (17 \text{ %/m}) \cdot [(10 - 0 \text{ m}) \div (100 - 0\%)] = 1.7 \text{ %/%}$

We use the Kc with engineering units in the remainder of this chapter and are careful that our PI controller is formulated to accept such values. If we were using these results in a commercial control system, we would be careful to ensure our tuning parameters are cast in the form appropriate for our equipment.

Controller Action

The process gain, Kp, is positive for the gravity drained tanks, indicating that when CO increases, the PV increases in response. This behavior is characteristic of a direct acting process. Given this CO to PV relationship, when in automatic mode (closed loop), if the PV starts drifting above set point, the controller must decrease CO to correct the error. Such negative feedback is an essential component of stable controller design.

A process that is naturally direct acting requires a controller that is reverse acting to remain stable. In spite of the opposite labels (direct acting process and reverse acting controller), the details presented above show that both Kp and Kc are positive values.

In most commercial controllers, only positive Kc values can be entered. The sign (or action) of the controller is then assigned by specifying that the controller is either reverse acting or direct acting to indicate a positive or negative Kc, respectively.

If the wrong control action is entered, the controller will quickly drive the final control element (FCE) to full on/open or full off/closed and remain there until a proper control action entry is made.

Implement and Test

The ability of the PI controller to reject changes in the pumped flow disturbance, D, is pictured below for the moderate and aggressive tuning values computed above. Note that the set point remains constant at 2.2 m throughout the study.

The aggressive controller shows a more energetic CO action, and thus, a more active PV response. However, the penalty for this increased activity is some overshoot and oscillation in the process response.

Please be aware that the terms "moderate" and "aggressive" hold no magic. If we desire a control performance between the two, we need only average the Kc values from the tuning rules above. Note, however, that these rules provide a constant reset time, Ti, regardless of our desired performance. So if we believe we have collected a good process data set, and the FOPDT model fit looks like a reasonable approximation of this data, then $Ti = Tp$ always.

While not our design objective, presented below is the set point tracking ability of the PI controller when the disturbance flow is held constant:

Again, the aggressive tuning values provide for a more active response.

Aside: it may appear that the random noise in the PV measurement signal is different in the two plots above, but it is indeed the same. Note that the span of the PV axis in each plot differs by a factor of four. The narrow span of the set point tracking plot greatly magnifies the signal traces, making the noise more visible.

Comparison With P-Only Control

The performance of a P-Only controller in addressing the same disturbance rejection and set point tracking challenge. A comparison of that study with the results presented here reveals that PI controllers:

- can eliminate the offset associated with P-Only control,
- have integral action that increases the tendency for the PV to roll (or oscillate),
- have two tuning parameters that interact, increasing the challenge to correct tuning when performance is not acceptable.

Derivative Action

The addition of the derivative term to complete the PID algorithm provides modest benefit yet significant challenges.

THE CHALLENGE OF INTERACTING TUNING PARAMETERS

Many process control practitioners tune by "intuition," fiddling their way to final tuning by a combination of experience and trial-and-error.

Some are quite good at approaching process control as art. Since they are the ones who define "best" performance based on the goals of production, the capabilities of the process, the impact on down stream units, and the desires of management, it can be difficult to challenge any claims of success.

To explore the pitfalls of a trial and error approach and reinforce that there is science to controller tuning, we consider the common dependent, ideal form of the PI controller:

$$CO = CO_{bias} + Kc \cdot e(t) + \frac{Kc}{Ti} \int e(t)dt$$

Where:

CO = controller output

e(t) = controller error = set point – process variable = SP – PV

Kc = controller gain, a tuning parameter

Ti = reset time, a tuning parameter

For this form, controller activity or aggressiveness increases as Kc increases and as Ti decreases (Ti is in the denominator, so smaller values increase the weighting on the integral action term, thus increasing controller activity).

Since Kc and Ti individually can make a controller more or less aggressive in its response, the two tuning parameters interact with each other. If current controller performance is not what we desire, it is not always clear which value to raise or lower, or by how much.

Example of Interaction Confusion

To illustrate, consider a case where we seek to balance a fairly rapid response to a set point change (a short rise time) against a small overshoot. While every process application is different, we choose to call the response plot below our desired or base case performance.

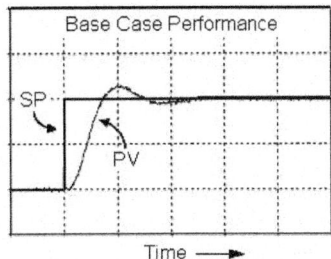

Now consider the two response plots below. These were made using the identical process and controller to that above. The only difference between the base case response above and plot A and plot B below is that different Kc and T_i tuning values were used in each one.

And now the question: what tuning adjustments are required to restore the desired base case performance above starting from each plot below? Or alternatively: how has the tuning been changed from base case performance to produce these different behaviors?

There are no tricks in this question. The "process" is a simple linear second order system with modest dead time. Controller output is not hitting any limits. The scales on the plot are identical. Everything is as it seems, except PI controller tuning is different in each case.

Study the plots for a moment before reading ahead and see if you can figure it out. Each plot has a very different answer.

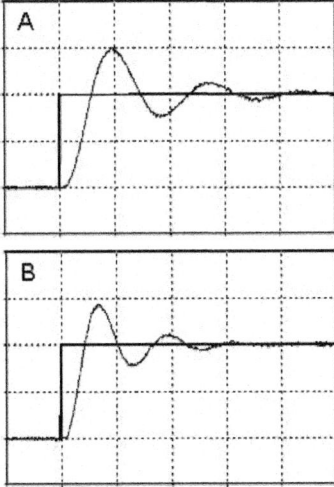

Some Hints

Before we reveal the answer, here is a hint. One plot has been made more active or aggressive in its response by doubling Kc while keeping Ti constant at the original base case value.

The other cuts Ti in half (remember, decreasing Ti makes this PI form more active) while keeping Kc at the base case value:

So we have:

• Base case = Kc and Ti

• Plot A or B = 2Kc and Ti

• Other Plot B or A = Kc and Ti/2

Still not sure? Here is a final hint: remember from our proportional action is largely responsible for the first movements in a response. We also integral action tends to increase the oscillatory or cycling behavior in the PV.

It is not easy to know the answer, even with these huge hints, and that is the point of this chapter.

The Answer

Below is a complete tuning map with the base case performance from our challenge problem in the center. The plot shows how performance changes as Kc and Ti are doubled and halved from the base case for the dependent, ideal PI controller form.

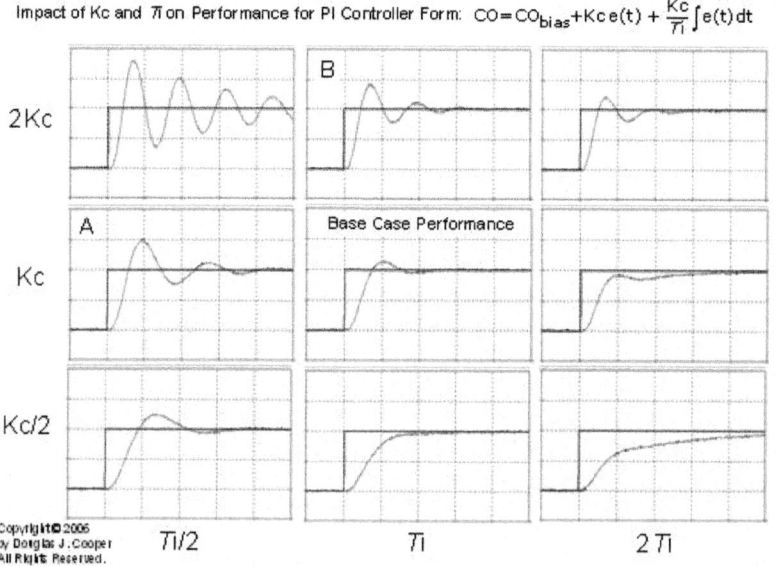

Impact of Kc and Ti on Performance for PI Controller Form: $CO = CO_{bias} + Kc\, e(t) + \dfrac{Kc}{Ti}\int e(t)\, dt$

Starting from the center and moving up on the map from the base case performance brings us to plot B. As indicated on the tuning map axis, this direction increases (doubles) controller gain, Kc, thus making the controller more active or aggressive. Moving down on the map from the base case decreases (halves) Kc, making the controller more sluggish in its response.

Moving left on the map from the base case brings us to plot A. As indicated on the tuning map axis, this direction decreases reset time (cuts it in half), again making the controller more active or aggressive. Moving right on the map from the base case increases (doubles) reset time, making the controller more sluggish in its response.

It is clear from the tuning map that the controller is more active or aggressive in its response when Kc increases and Ti decreases, and more sluggish or conservative when Kc decreases and Ti increases.

Building on this observation, it is not surprising that the upper left most plot (2Kc and Ti/2) shows the most active controller response, and the lower right most plot (Kc/2 and 2Ti) is the most conservative or sluggish response.

Back to the question. With what we now know, the answer:

• Base case = Kc and Ti

• Plot B = 2Kc and Ti

• Plot A = Kc and Ti/2

Interacting Parameters Makes Tuning Problematic

The PI controller has only two tuning parameters, yet it produces very similar looking performance plots located in different places on a tuning map.

If our instincts lead us to believe that we are at plot A when we really are at plot B, then the corrective action we make based on this instinct will compound our problem rather than solve it. This is strong evidence that trial and error is not an efficient or appropriate approach to tuning.

When we consider a PID controller with three tuning parameters, the number of similar looking plots in what would be a three dimensional tuning map increases dramatically. Trial and error tuning becomes almost futile.

We have been exploring a step by step tuning recipe approach that produces desired results without the wasted time and off-spec product that results from trial and error tuning.

If we follow this industry-proven methodology, we will improve the safety and profitability of our operation.

Interesting Observation

Before leaving this subject, we make one more very useful observation from the tuning map. This will help build our intuition and may help one day when we are out in the plant.

The right most plot in the center row (Kc, 2Ti) of the tuning map above is reproduced below.

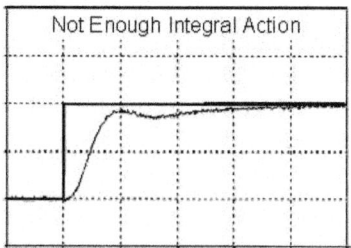

Notice how the PV shows a dip or brief oscillation on its way up to the set point? This is a classic indication that the proportional term is reasonable but the integral term is not getting enough weight in the calculation. For the PI form used in this chapter, that would mean that the reset time, Ti, is too large since it is in the denominator.

If we cover the right half of the "not enough integral action" plot, the response looks like it is going to settle out with some offset, as would be expected with a P-Only controller. When we consider the plot as a whole, we see that as enough time passes, the response completes. This is because the weak integral action finally accumulates enough weight in the calculation to move the PV up to set point.

This "oscillates on the way" pattern is a useful marker for diagnosing a lack of sufficient integral action.

PI DISTURBANCE REJECTION IN THE JACKETED STIRRED REACTOR

The control objective for the jacketed reactor is to minimize the impact on reactor operation when the temperature of the liquid entering the cooling jacket changes. As a base case study, we establish here the performance capabilities of a PI controller in achieving this objective.

The important variables for this study are labeled in the graphic:

CO = signal to valve that adjusts cooling jacket liquid flow rate (controller output, %)

PV = reactor exit stream temperature (measured process variable, °C)

SP = desired reactor exit stream temperature (set point, °C)

D = temperature of cooling liquid entering the jacket (major disturbance, °C)

We follow our industry proven recipe to design and tune our PI controller:

Step 1: Design Level of Operation (DLO)

The details of expected process operation and how this leads to our DLO are summarized:

- Design PV and SP = 90 °C with approval for brief dynamic (bump) testing of ±2 °C.

- Design D = 43 °C with occasional spikes up to 50 °C.

Step 2: Collect Process Data around the DLO

When CO, PV and D are steady near the design level of operation, we bump the process to generate CO-to-PV cause and effect response data.

Step 3: Fit a FOPDT Model to the Dynamic Process Data

We approximate the dynamic behavior of the process by fitting a first order plus dead time (FOPDT) dynamic model to the test data from step 2. The results of the modeling study are presented in detail here and are summarized:

- Process gain (direction and how far), $Kp = -0.5$ °C/%

- Time constant (how fast), $Tp = 2.2$ min

- Dead time (how much delay), $\theta p = 0.8$ min

Step 4: Use the FOPDT Parameters to Complete the Design

As in the heat exchanger PI control study, we explore what is often called the dependent, ideal form of the PI control algorithm:

Where:

CO = controller output signal (the wire out)

CO_{bias} = controller bias or null value; set by bumpless transfer

e(t) = current controller error, defined as SP – PV

SP = set point

PV = measured process variable (the wire in)

Kc = controller gain, a tuning parameter

Ti = reset time, a tuning parameter

Aside: our observations using the dependent ideal PI algorithm directly apply to the other popular PI controller forms. For example, the integral gain for the independent algorithm form, written as:can be computed as: $Ki = Kc/Ti$. The Kc is the same for both forms, though it is more commonly called the proportional gain for the independent algorithm.

1. *Sample Time,* T

2. Best practice is to set the loop sample time, T, at one-tenth the time constant or faster (*i.e.,* $T \leq 0.1Tp$). Faster sampling may provide modestly improved performance, while slower sampling can lead to significantly degraded performance.

In this study, $T \le 0.1(2.2 \text{ min})$, so T should be 13 seconds or less. We meet this with the sample time option available from most commercial vendors:

◊ sample time, $T = 1$ sec

Control Action (Direct/Reverse)

The jacketed stirred reactor process has a negative Kp. That is, when CO increases, PV decreases in response. Since a controller must provide negative feedback, if the process is reverse acting, the controller must be direct acting. That is, when in automatic mode (closed loop), if the PV is too high, the controller must increase the CO to correct the error. Since the controller must move in the same direction as the problem, we specify:

- ◊ controller is direct acting • *Dead Time Issues*
- If dead time is greater than the process time constant ($\theta p > Tp$), control becomes increasingly problematic and a Smith predictor can offer benefit. For this process, the dead time is smaller than the time constant, so:
- ◊ dead time is small and not a concern

Computing Controller Error, e(t)

- Set point, SP, is manually entered into a controller. The measured PV comes from the sensor (our wire in). Since SP and PV are known values, then at every loop sample time, T, controller error can be directly computed as:
- ◊ error, $e(t) = SP - PV$

Determining Bias Value, CObias

- CO_{bias} is the value of CO that, in manual mode, causes the PV to steady at the DLO when the major disturbances are quiet and at their normal or expected values. When integral action is enabled, commercial controllers determine the bias value with a bumpless transfer procedure.

That is, when switching to automatic, the controller initializes the SP to the current value of PV, and CO_{bias} to the current value of CO. By choosing our current operation as our design state (at least temporarily at switchover), there is no corrective action needed by the controller that will bump the process. Thus,

◊ controller bias, CO_{bias} = current CO for a bumpless transfer

- *Controller Gain, Kc, and Reset Time, Ti*

We use our FOPDT model parameters in the industry-proven Internal Model Control (IMC) tuning correlations to compute PI tuning values.

The first step in using the IMC correlations is to compute *Tc*, the closed loop time constant. *Tc* describes how active our controller should be in responding to a set point change or in rejecting a disturbance.

The performance implications of choosing Tc have been explored previously for PI control of the heat exchangerand the gravity drained tanks case studies.

In short, the closed loop time constant, Tc, is computed based on whether we seek:

- aggressive action and can tolerate some overshoot and oscillation in the PV response,
- moderate action where the PV will move reasonably fast but show little overshoot,
- conservative action where the PV will move in the proper direction, but quite slowly.

Once this decision is made, we compute Tc with these rules:

- *Aggressive Response*: Tc is the larger of $0.1 \cdot Tp$ or $0.8\ \theta p$
- *Moderate Response*: Tc is the larger of $1 \cdot Tp$ or $8\ \theta p$
- *Conservative Response*: Tc is the larger of $10 \cdot Tp$ or $80\ \theta p$

With Tc computed, the PI controller gain, Kc, and reset time, Ti, are computed as:

Notice that reset time, Ti, is always equal to the process time constant, Tp, regardless of desired controller activity.

Moderate Response Tuning:

- For a controller that will move the PV reasonably fast while producing little to no overshoot, choose:

    Moderate Tc = the larger of $1 \cdot Tp$ or $8\ \theta p$

    = larger of $1(2.2 \text{ min})$ or $8(0.8 \text{ min})$

    = 6.4 min

Using this Tc and our model parameters in the tuning correlations above, we arrive at the moderate tuning values:

Aggressive Response Tuning:

1. For an active or quickly responding controller where we can tolerate some overshoot and oscillation as the PV settles out, specify:

    Aggressive Tc = the larger of $0.1 \cdot Tp$ or $0.8\ \theta p$

    = larger of $0.1(2.2 \text{ min})$ or $0.8(0.8 \text{ min})$

    = 0.64 min

and the aggressive tuning values are:

Practitioner's Note: The FOPDT model parameters used in the tuning correlations above have engineering units, so the Kc values we compute also have engineering units. In commercial control systems, controller gain (or proportional band) is normally entered as a dimensionless (%/%) value. For commercial implementations, we could:

- Scale the process data before fitting our FOPDT dynamic model so we directly compute a dimensionless Kc.
- Convert the model Kp to dimensionless %/% after fitting the model but before using the FOPDT parameters in the tuning correlations.
- Convert Kc from engineering units into dimensionless %/% after using the tuning correlations.CO is already scaled from 0 – 100% in the above example. Thus, we convert Kc from engineering units into dimensionless %/% using the formula:

For the jacketed stirred reactor, PVmax = 250 °C and PVmin = 0 °C. The dimensionless Kc values are thus computed:

- moderate Kc = (– 0.6 %/ °C)·[(250 – 0 °C) ÷ (100 – 0%)] = – 1.5 %/%
- aggressive Kc = (– 3.1%/ °C)·[(250 – 0 °C) ÷ (100 – 0%)] = – 7.8 %/%

We use Kc with engineering units in the remainder of this chapter and are careful that our PI controller is formulated to accept such values. We would be mindful if we were using a commercial control system, however, to ensure our tuning parameters are cast in the form appropriate for our equipment.

Implement and Test

The ability of the PI controller to reject changes in the cooling jacket inlet temperature, D, is pictured below for the moderate and aggressive tuning values computed above. Note that the set point remains constant at 90 °C throughout the study.

As expected, the aggressive controller shows a more energetic CO action, and thus, a more active PV response.

While not our design objective, presented below is the set point tracking ability of the PI controller when the disturbance temperature is held constant.

The plot shows that set point tracking performance matches the descriptions used above for choosing Tc:

- Use aggressive action if we seek a fast response and can tolerate some overshoot and oscillation in the PV response.
- Use moderate action if we seek a reasonably fast response but seek little to no overshoot in the PV response.

Important => Ti Always Equals Tp

The rules provide a constant reset time, Ti, regardless of our desired performance. So if we believe we have collected a good process data set, and the FOPDT model fit looks like a reasonable approximation of this data, then we have a good estimate of the process time constant and $Ti = Tp$ regardless of desired performance.

If we are going to tweak the tuning, Kc should be the only value we adjust. For example, if we seek a performance between moderate and aggressive, we average the Kc values while Ti remains constant.

INTEGRAL (RESET) WINDUP, JACKETING LOGIC AND THE VELOCITY PI FORM

A valve cannot open more than all the way. A pump cannot go slower than stopped. Yet an improperly programmed control algorithm can issue such commands.

Herein lies the problem of integral windup (also referred to as reset windup or integral saturation). It is a problem that has been around for decades and was solved long ago. We prevent it to help those who choose to write their own control algorithm.

The PI Algorithm

To increase our comfort level with the idea that different vendors cast the same PI algorithm in different forms, we choose the independent, continuous, position PI form for this discussion:

Where:

CO = controller output signal (the wire out)

CO_{bias} = controller bias or null value

e(t) = current controller error, defined as SP − PV

SP = set point

PV = measured process variable (the wire in)

Kc = proportional gain, a tuning parameter

Ki = integral gain, a tuning parameter

Note that Kc is the same parameter in both the dependent and independent forms, though it is more typically called controller gain in the dependent form.

Every procedure and observation we have previously discussed about PI controllers applies to both forms. Both even use the same tuning correlations. To tune Ki, we compute Kc and Ti for the dependent form and then divide ($Ki = Kc/Ti$).

Integral (Reset) Windup

The integration is a continual summing. Integration of error means that we continually sum controller error up to the present time.

The integral sum starts accumulating when the controller is first put in automatic and continues to change as long as controller error exists.

If an error is large enough and/or persists long enough, it is mathematically possible for the integral term to grow very large (either positive or negative):

This large integral, when combined with the other terms in the equation, can produce a CO value that causes the final control element (FCE) to saturate. That is, the CO drives the FCE (*e.g.* valve, pump, compressor) to its physical limit of fully open/on/maximum or fully closed/off/minimum.

And if this extreme value is still not sufficient to eliminate the error, the simple mathematics of the controller algorithm, if not jacketed with protective logic, permits the integral term to continue growing.

If the integral term grows unchecked, the equation above can command the valve, pump or compressor to move to 110%, then 120% and more. Clearly, however, when an an FCE reaches its full 100% value, these last commands have no physical meaning and consequently, no impact on the process.

Control is Lost

Once we cross over to a "no physical meaning" computation, the controller has lost the ability to regulate the process.

When the computed CO exceeds the physical capabilities of the FCE because the integral term has reached a large positive or negative value, the controller is suffering from*windup*. Because windup is associated with the integral term, it is often referred to as*integral windup* or *reset windup*.

To prevent windup from occurring, modern controllers are protected by either:

- Employing extra "jacketing logic" in the software to halt integration when the CO reaches a maximum or minimum value.

- Recasting the controller into a discrete velocity form that, by its very formulation, naturally avoids windup.

Visualizing Windup

To better visualize the problem of windup and the benefit of anti-windup protection, consider the plot from our heat exchanger process.

To the left is the performance of a PI controller with no windup protection. To the right is the performance of the same controller protected by an anti-windup strategy.

For both controllers, the set point is stepped from 200 °C up to 215 °C and back again. As shown in the lower trace on the plot, the controller moves the CO to 0%, closing the valve completely, yet this is not sufficient to move the PV up to the new set point.

To the left in the plot, the impact of windup is a degraded controller performance. When the set point is stepped back to its original value of 200 °C, the windup condition causes a delay in the CO action. This in turn causes a delay in the PV response.

To the right in the plot, anti-windup protection permits the CO, and thus PV, to respond promptly to the command to return to the original SP value of 200 °C.

More Details on Windup

The plot below offers more detail. As labeled on the plot:

1. To the left for the Controller with Wind-up case, the SP is stepped up to 215 °C. The valve closes completely but is not able to move the PV all the way to the high set point value. Integration is a summing of controller error, and since error persists, the integration term grows very large.

   The sustained error permits the controller to windup (saturate). While it is not obvious from the plot, the PI algorithm is computing values for CO that ask the valve to be open –5%, –8% and more. The control algorithm is just simple math with no ability to recognize that a valve cannot be open to a negative value.

   Note that the chart shows the CO signal bottoming out at 0% while the controller algorithm is computing negative CO values. This misleading information is one reason why windup can be difficult to diagnose as the root cause of a problem from visual inspection of process data trend plots.

2. When the SP is stepped back to 200 °C, it seems as if the CO does not move at first. In reality, the control algorithm started moving the CO when the SP changed, but the values remain in the physically meaningless range of negative numbers.

   So while the valve remains fully closed at 0%, the integral sum is accumulating controller errors of opposite sign. As time passes, the integral term shrinks or "unwinds" as the running sum of errors balance out.

3. When the integral sum of errors shrinks enough, it no longer dominates the CO computation. The CO signal returns from the physically meaningless world of negative values. The valve can finally move in response.

4. To the right in the plot above, the controller is protected from windup. As a result, when the set point is stepped back to 200 °C, the CO immediately reacts with a change that is proportional to the size of the SP change. The PV moves quickly in response to the CO actions as it tracks the SP back to 200 °C.

Solution 1: Jacketing Logic on the Position Algorithm

The PI controller at the top of this chapter is called the position form because the computed CO is a specific intermediate value between full on/open/maximum and closed/off/minimum. The continuous PI algorithm is specifying the actual position (*e.g.*, 27% open, 64% of maximum) that the final control element (FCE) should assume.

1. *Simple Logic Creates Additional Problems*

2. It is not enough to have logic that simply limits or clips the CO if it reaches a maximum (COmax) or minimum (COmin) value because this does nothing to check the growth of the integral sum of errors term.

In fact, such simple logic was used in the "control with windup" plots just discussed. The CO seems stuck at 0% and we are unaware that the algorithm is actually computing negative valve positions as described in item 1 above.

- *Anti-Windup Logic Outline*
- When we switch from manual mode to automatic, we assume that we have initialized the controller using a <u>bumpless transfer</u> procedure. That is, at switchover, the integral sum of error is set to zero, the SP is set equal to the current PV, and the controller bias is set equal to the current CO (implying that COmin < CO_{bias} < COmax).Thus, there is nothing to cause CO to immediately change and "bump" our process at switchover.

One approach to creating anti-windup jacketing logic is to artificially manipulate the integral sum of error itself. With our controller properly initialized, the approach is to flip the algorithm around and *back-calculate a value for the integral sum of error* that will provide a desired controller output value (COdesired), or:

Note that COdesired can be different in different situations. For example,

- We do not want tuning parameter adjustments to cause sudden CO movements that bump our process. So if tuning values have changed, COdesired is the value of CO from the previous loop calculation cycle.
- If the PI controller computes CO values that are above COmax or below COmin, then we must be concerned about windup and COdesired is set equal to the limiting COmaxor COmin value.

The anti-windup logic followed at every loop sample time, T, is thus:

1. If tuning parameters have changed since the last loop calculation cycle, then COdesired = current CO. Back calculate the integral sum of error so CO remains unchanged from the previous sample time. This prevents sudden CO bumps due to tuning changes.
2. Update SP and PV for this loop calculation cycle.
3. compute:
4. If CO > COmax or if CO < COmin, then the anti-windup (integral desaturation) logic of step 5 is required. Otherwise, proceed to step 6.
5. If CO > COmax, then CO = COdesired = COmax. if CO < COmin, then CO = COdesired = COmin. Back calculate the integral sum of error using our selected COdesired and save it for use in the next control loop calculation cycle.
6. Implement CO

Solution 2 – Use the Velocity (Discrete) Controller Form

Rather than computing a CO signal indicating a specific position for our final control element, an alternative is to compute a signal that specifies a change, ΔCO, from current position for the FCE. As explained below, this is called the velocity or discrete controller form.

We employ the dependent algorithm for this presentation, but the derivation that follows can be applied in an analogous fashion to the independent PI form. To derive the discrete velocity form, we must first write the continuous, position form of the PI controller to include the independent variable on the controller output, showing it properly as CO(t) to reflect that it changes with time:

Please note that this controller is identical to all dependent PI forms as presented in other articles in this e-book. The only difference is we are being more mathematically precise in our expression of CO(t).

Reason for Anti-Windup Protection

Discrete velocity algorithms compute a ΔCO that signals the FCE to move a specific distance and direction from its current position. The PI controller form above, the computation does not keep track of the current FCE position, nor does it mathematically accumulate any integral sums.

In a sense, the accumulation of integration is stored in the final control element itself. If a long series of ΔCO moves are all positive, for example, the valve, pump or compressor will move toward its maximum value. And once the FCE reaches its maximum limit, any ΔCO commands to move further will have no impact because, as stated in the first sentences of this chapter, a valve cannot open more than all the way and a pump cannot go slower than stopped. It is the physical nature of the FCE itself that provides protection from over-accumulation (*i.e.*, windup).

As long as the CO never reaches COmax or COmin, the continuous position and discrete velocity forms of the PI controller provide identical performance. A properly jacketed continuous position PI controller will also provide windup protection equal to the discrete velocity form. Implicit in these statements is that sample time,T, is reasonably fast and that T and the tuning values (Kc and *Ti*) are the same when comparing implementations.

Concerns with Discrete Velocity PID

Unfortunately, the usefulness of the discrete velocity form is limited because the method suffers problems whenderivative action is included. We find that we must take the derivative of a derivative, yielding a second derivative. A second derivative applied to data that contains even modest noise can produce nonsense results.

Some vendors implement this form anyway and include a signal filter and additional logic sequences to address the problem. Thus, even with the anti-windup benefits of a discrete velocity algorithm, we find the need to jacket the algorithm with protective logic.

PID CONTROL AND DERIVATIVE ON MEASUREMENT

Like the PI controller, the Proportional-Integral-Derivative (PID) controller computes a controller output (CO) signal for the final control element every sample time T.

The PID controller is a "three mode" controller. That is, its activity and performance is based on the values chosen for three tuning parameters, one each nominally associated with the proportional, integral and derivative terms.

The PI controller is a reasonably straightforward equation with two adjustable tuning parameters. The number of different ways that commercial vendors can implement the PI form is fairly limited, and they all provide the same performance if properly tuned.

With the addition of a third adjustable tuning parameter, the number of algorithm permutations increases markedly. And there are even different forms of the PID equation itself. This creates added challenges for controller design and tuning.

Here we focus on what a derivative is, how it is computed, and what it means for control. We also explore why derivative on measurement is widely recommended for industrial practice.

We narrow our world in this chapter and focus on the dependent, ideal form of the controller. In later articles we will circle back and talk about the different algorithm forms, methods for design and tuning, algorithm limitations, and other practical issues.

The Dependent, Ideal PID Form

A popular way vendors express the dependent, ideal PID controller is:

$$CO = CO_{bias} + Kc \cdot e(t) + \frac{Kc}{Ti} \int e(t)dt + Kc \cdot Td \frac{de(t)}{dt}$$

Where:

CO = controller output signal (the wire out)

CO_{bias} = controller bias; set by bumpless transfer

e(t) = current controller error, defined as SP − PV

SP = set point

PV = measured process variable (the wire in)

Kc = controller gain, a tuning parameter

Ti = reset time, a tuning parameter

Td = derivative time, a tuning parameter

The first three terms to the right of the equal sign are identical to the PI controller we have already explored in some detail.

The derivative mode of the PID controller is an additional and separate term added to the end of the equation that considers the derivative (or rate of change) of the error as it varies over time.

The Contribution of the Derivative Term

The proportional term considers *how far* PV is from SP at any instant in time. Its contribution to the CO is based on the size of e(t) only at time t. As e(t) grows or shrinks, the influence of the proportional term grows or shrinks immediately and proportionately.

The integral term addresses *how long* and how far PV has been away from SP. The integral term is continually summing e(t). Thus, even a small error, if it persists, will have a sum total that grows over time and the influence of the integral term will similarly grow.

A derivative describes how steep a curve is. More properly, a derivative describes the slope or the rate of change of a signal trace at a particular point in time. Accordingly, the derivative term in the PID equation above considers *how fast*, or the rate at which, error is changing at the current moment.

Derivative on PV is Opposite but Equal

While the proportional and integral terms of the PID equation are driven by the controller error, e(t), the derivative computation in many commercial implementations should be based on the value of PV itself.

The derivative of e(t) is mathematically identical to the negative of the derivative of PV everywhere except when set point changes. And when set point changes, derivative on error results in an undesirable control action called *derivative kick*.

Math Note: the mathematical defense that "derivative of e(t) equals the negative derivative of PV when SP is constant" considers that, since e(t) = SP – PV, the equation below follows. That is, derivative of error equals derivative of set point minus process variable. The derivative of a constant is zero, so when SP is constant, mathematically, the derivative (or slope or rate of change) of the controller error equals the derivative (or slope or rate of change) of the measured process variable, PV, except the sign is opposite.

$$\frac{de(t)}{dt} = \frac{d(\overset{0}{SP} - PV)}{dt} = -\frac{dPV}{dt}$$

The figures below provide a visual appreciation that the derivative of e(t) is the negative of the derivative of PV.

The top plot shows the measured PV trace after a set point step. The bottom plot shows the e(t) = SP – PV trace for the same event.

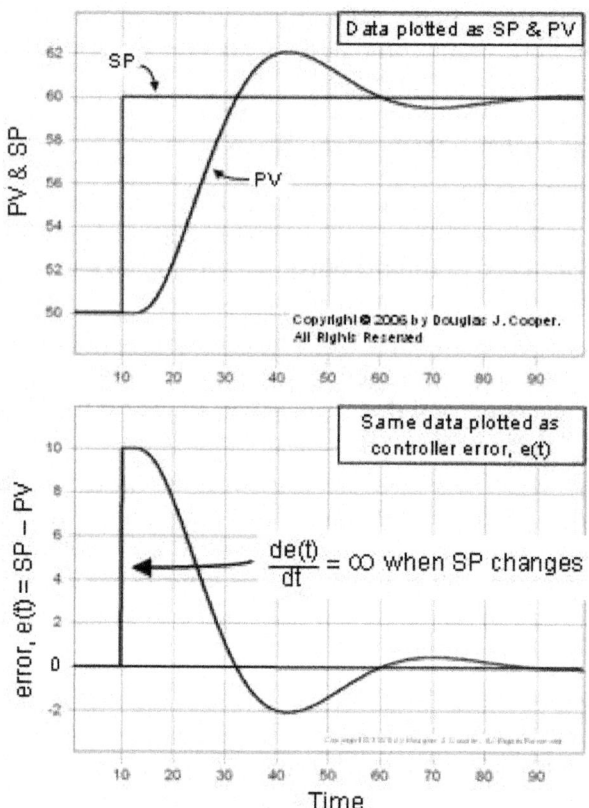

If we compare the two plots after the SP step at time t = 10, we see that the PV trace in the upper plot is an exact reflection of the e(t) trace in the lower plot. The PV trace ascends, peaks and then settles, while in a reflected pattern, the e(t) trace descends, dips and then settles.

Mathematically, this "mirror image" of trace shapes means that the derivatives (or slopes or rates of change) are the same everywhere after the SP step, except they are opposite in sign.

Derivative on PV Used in Practice

While the shape of e(t) and PV are opposite but equal everywhere after the set point step, there is an important difference at the moment the SP changes. The lower plot shows a vertical spike in e(t) at this moment. There is no corresponding spike in the PV plot.

The derivative (or slope) of a vertical spike in the theoretical world approaches infinity. In the real world it is at least a very big number. If Td is large enough to provide any meaningful weight to the derivative term, this huge derivative value will cause a large and sudden manipulation in CO. This large manipulation in CO, referred to as *derivative kick*, is almost always undesirable.

As long as loop sample time, T, is properly specified, the PV trace will follow a gradual and continuous response, avoiding the dramatic vertical spike evident in the e(t) trace.

Because derivative on e(t) is identical to derivative on PV at all times except when the SP changes, and when the set point does change, derivative on error provides information we don't want our controller to use, we substitute the "math note" equation in the yellow box above to obtain the PID with derivative on measurement controller:

$$CO = CO_{bias} + Kc \cdot e(t) + \frac{Kc}{Ti} \int e(t)dt - Kc \cdot Td \frac{dPV}{dt}$$

Derivative on PV Does Not "Kick"

Below we show the heat exchanger case study under PID control using the dependent, ideal algorithm form and moderate tuning values as computed in this chapter.

The first set point steps to the left in the plot below show loop performance when PID with derivative on error is used. The set points steps to the right present the identical scenario except that PID with derivative on measurement is used.

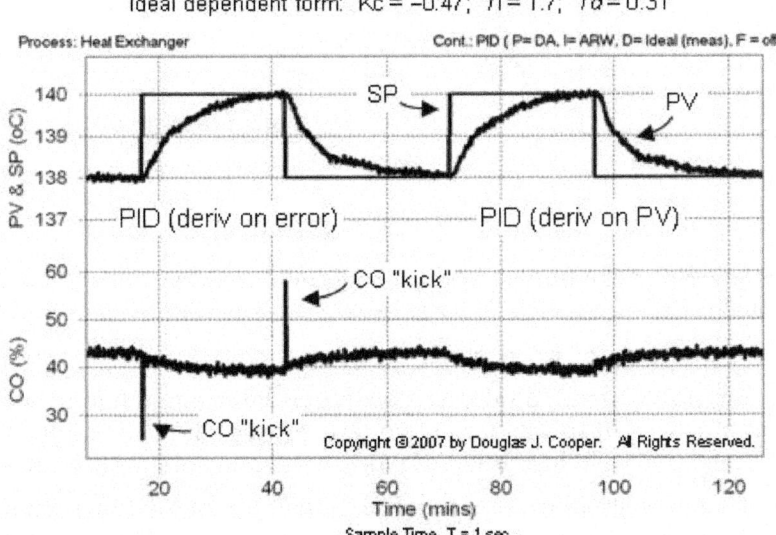

Heat Exchanger Under PID Control
Ideal dependent form: Kc = –0.47; Ti = 1.7; Td = 0.31

The "kick" that dominates the CO trace when derivative on error is used is rather dramatic and somewhat unsettling. While it exists for only a brief moment and does not impact performance in this example, we should not assume this will always be true. In any event, such action will eventually take a toll on mechanical final control elements.

We recommend that derivative on measured PV be used if our vendor provides the option (fortunately most do). The tuning values remain the same for both algorithms.

Understanding Derivative Action

A rapidly changing PV has a steep slope and this yields a large derivative. This is true regardless of whether a dynamic event has just begun or if it has been underway for some time.

In the plot below, the derivative dPV/dt describes the slope or "steepness" of PV during a process response.

Early in the response, the slope is large and positive when the PV trace is increasing rapidly. When PV is decreasing, the derivative (slope) is negative. And when the PV goes through a peak or a trough, there is a moment in time when the derivative is zero.

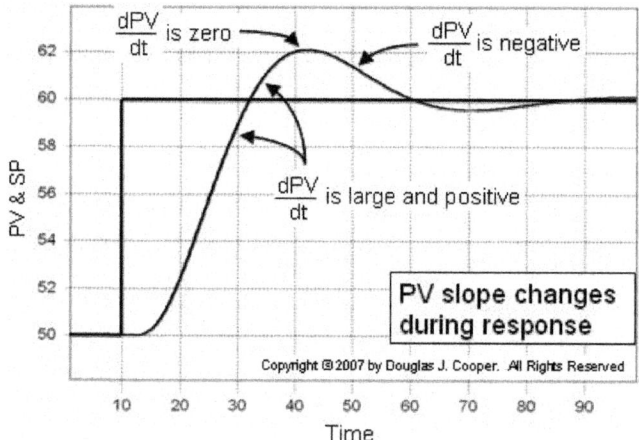

To understand the impact of this changing derivative, let's assume for discussion that:

- Controller gain, Kc, is positive.

- Derivative time, Td (always positive) is large enough to provide meaningful weight to the derivative term. After all, if Td is very small, the derivative term has little influence, regardless of the slope of the PV.

The negative sign in front of the derivative term of the PID with derivative on measurement controller (and given the above assumptions) means that the impact on CO from the derivative term will be opposite to the sign of the slope:

$$CO = CO_{bias} + Kc \cdot e(t) + \frac{Kc}{Ti} \int e(t)dt - Kc \cdot Td \frac{dPV}{dt}$$

Thus, when dPV/dt is large and positive, the derivative term has a large influence and seeks to decrease CO.

Conversely, when dPV/dt is negative, the derivative term seeks to increase CO.

It is interesting to note that the derivative term does not consider whether PV is heading toward or away from the set point (whether e(t) is positive or negative). The only consideration is whether PV is heading up or down and how quickly.

The result is that derivative action seeks to inhibit rapid movements in the PV. This could be an especially useful characteristic when seeking to dampen the oscillations in PV that integral action tends to magnify.

THE CHAOS OF COMMERCIAL PID CONTROL

The design and tuning of a three mode PID controller follows the proven recipe we have used with success for P-Only control (*e.g.*, here and here) and PI Control (*e.g.*, here, here and here). The decisions and procedures we established for steps 1-3 of the design and tuning recipe in these previous studies remain unchanged as we move on to the PID algorithm.

Step 4 of the recipe remains the same as well. But it is essential in this step that we match the rules and correlations of step 4 with the particular controller algorithm form we are using.

The challenge arises because the number of PID algorithm forms available from hardware vendors increases markedly when derivative action is included. And unfortunately, these PID algorithms are implemented in many different forms across the commercial market.

The potential for confusion by even a careful practitioner is significant. For example:

- there are three popular PID algorithm forms, and
- each of these three forms have multiple parameters that are cast in different ways.

As a result, there are literally dozens of possible PID algorithm forms. Matching each controller form with its proper design rules and correlations requires careful attention if performed without the help of software tools.

Common Algorithm Forms

Listed below are the three common PID controller forms. If offered as an option by our vendor (most do offer it), derivative on measured process variable (PV) is the recommended PID form:

- Dependent, ideal PID controller form (derivative on measurement):

$$CO = CO_{bias} + Kc \cdot e(t) + \frac{Kc}{Ti} \int e(t)dt - Kc \cdot Td \frac{dPV}{dt}$$

- Dependent, interacting form (derivative on measurement):

$$CO = CO_{bias} + Kc\left(1 + \frac{Td}{Ti}\right)e(t) + \frac{Kc}{Ti} \int e(t)dt - KcTd\frac{dPV}{dt}$$

- Independent PID form (derivative on measurement):

$$CO = CO_{bias} + Kc \cdot e(t) + Ki \int e(t)dt - Kd\frac{dPV}{dt}$$

Where for the above:

CO = controller output signal (the wire out)

CO_{bias} = controller bias; set by bumpless transfer

e(t) = current controller error, defined as SP – PV

SP = set point

PV = measured process variable (the wire in)

Kc = controller gain (also called proportional gain), a tuning parameter

Ki = integral gain, a tuning parameter

Kd = derivative gain, a tuning parameter

Ti = reset time, a tuning parameter

Td = derivative time, a tuning parameter

Tuning Parameters

Because there has been little standardization on nomenclature, the same tuning parameters can appear under different names in the commercial market. Perhaps more unfortunate, the same parameter can even have a different name within a single company's product line.

We will not attempt to list all of the different names here, though we will look at a solution. A few notes to consider:

1. The dependent forms appear most in products commonly used in the process industries, but the independent form is not uncommon.

2. The majority of DCS and PLC systems now use controller gain, Kc, for their dependent PID algorithms. There are notable exceptions, however, such as Foxboro who uses proportional band (PB = 100/Kc assuming PV and CO both range from 0 to 100%).

3. Reset time, Ti, is slightly more common for the dependent PID algorithms, though it is rarely called that in product documentation. Reset rate, defined as Tr = 1/Ti, comes in a close second. Again, the name for this parameter changes with product.

4. Most vendors use derivative time, Td, for their dependent PID algorithms, though few refer to it by that name in their product documentation.

Tune One, Tune Them All

Some good news in all this confusion is that the different forms, if tuned with the proper correlations, will perform exactly the same. No one form is better than another, it is just expressed differently.

In fact, we can show equivalence among the parameters, and thus algorithms, with these relations.

- *Proportional*

$$Kc, ideal = Kc, interact \left(1 + \frac{Td, interact}{Ti, interact}\right)$$

Kc, independent = Kc, ideal

- *Integral*

$$Ti, ideal = Ti, interact \left(1 + \frac{Td, interact}{Ti, interact}\right)$$

$$Ki, independent = \frac{Kc, ideal}{Ti, deal}$$

- *Derivative*

$$Td, ideal = \frac{Td, interact}{\left(1 + \frac{Td, interact}{Ti, interact}\right)}$$

Kd, independent = (Kc, ideal) (Td, ideal)

Though not presented here, analogous conversion relations can be developed for forms expressed using proportional band and/or reset rate.

Clarity in the Chaos

It is perhaps reasonable to hope that industrial practitioners will have an intuitive understanding of proportional, integral and derivative action. They might know the benefits each term offers and problems each presents. And experienced practitioners will know how design, tune and validate a PID implementation.

Expecting a practitioner to convert that knowledge and intuition over into the confusion of the commercial PID marketplace might not be so reasonable.

Given this, the best solution for those in the real world is to use software that lets us focus on the big picture while the software ensures that details are properly addressed.

Such productivity software should not only provide a "click and go" approach to algorithm and tuning parameter selection, but should also provide this information simply based on our choice of equipment manufacturer and product line.

For example, below is a portion of the controller manufacturer selection available in one commercial softwarepackage:

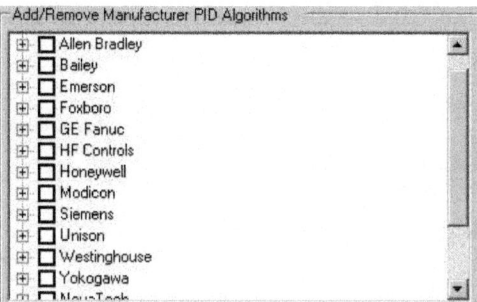

If you select Allen Bradley, Emerson, and Honeywell in the above list, the choice of PID controllers for each company is shown in the next three images:

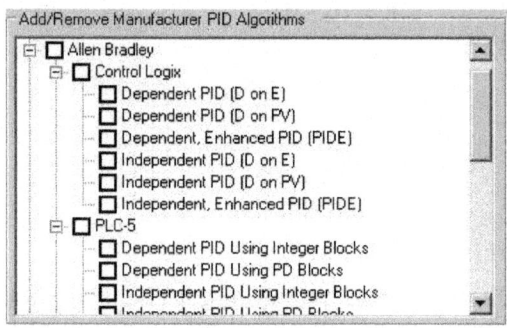

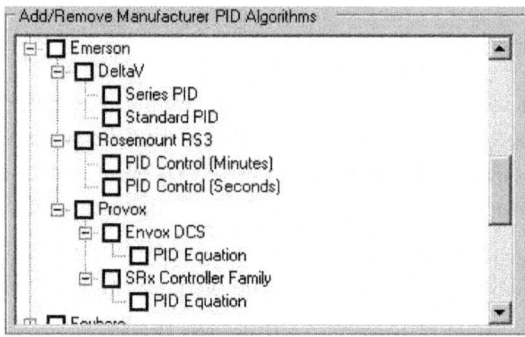

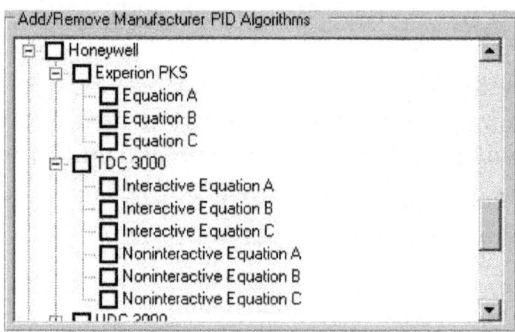

It is clear from these displays that there are different terms and many options for us to select from, all for PID control. And it may not be obvious that the different terms above refer to some version of our "basic three" PID forms.

Too much is at stake in a plant to ask a practitioner to keep track of it all. Software can get us past the details during PID controller design and tuning so we can focus on mission-critical control tasks like improving safety, performance and profitability.

Note: the Laplace domain is a subject that most control practitioners can avoid their entire careers, but it provides is a certain mathematical "elegance."Below, for example, are the three controller forms assuming derivative on error. Even without familiarity with Laplace, perhaps you will agree the three PID forms indeed look like part of the same family:

- *Dependent, Ideal (Non-interacting) Form*

$$\frac{CO(s)}{E(s)} = Kc\left(1 + \frac{1}{Ti} + Td\,s\right)$$

- *Dependent, Interacting Form*

$$\frac{CO(s)}{E(s)} = Kc\left(1 + \frac{1}{Ti\,s}\right)(Td\,s + 1)$$

- *Dependent Form*

$$\frac{CO(s)}{E(s)} = Kc + \frac{Ki}{s} + K\,d\,s$$

PID CONTROL OF THE HEAT EXCHANGER

We investigated P-Only control and then PI control of a heat exchanger. Here we explore the benefits and challenges of derivative action with a PID control study of this process. Our focus is on basic design, implementation and performance issues.

We follow the same four-step design and tuning recipe we use for all control implementations. A benefit of the recipe, beyond the fact that it is easy to use, widely applicable, and reliable in industrial applications, is that steps 1-3 of the recipe remain the same regardless of the control algorithm being employed.

Summary results of steps 1-3 from the previous heat exchanger control studies are presented below.

Step 1: Specify the Design Level of Operation (DLO)

- Design PV and SP = 138 °C with operation ranging from 138 to 140 °C
- Expected warm liquid flow disturbance = 10 L/min

Step 2: Collect Process Data around the DLO

PI control chapter referenced above for a summary.

Step 3: Fit an FOPDT Model to the Dynamic Data

The first order plus dead time (FOPDT) model approximation of the heat exchanger data from step 2 is:

- Process gain (how far), Kp = –0.53 °C/%
- Time constant (how fast), Tp = 1.3 min
- Dead time (how much delay), θp = 0.8 min

Step 4: Use the Parameters to Complete the Design

Vendors market the PID algorithm in a number of different forms, creating a confusing array of choices for the practitioner.

The preferred algorithm in industrial practice is PID with derivative on PV. The three most common of these forms each have their own tuning correlations, and they all provide identical performance as long as we take care to match each algorithm with its proper correlations during implementation.

Because the three common forms are identical in capability and performance if properly tuned, the observations and conclusions we draw from any one of these algorithms applies to the other forms.

Among the most widely used is the Dependent, Ideal (Non-interacting) form:

$$CO = CO_{bias} + Kc \cdot e(t) + \frac{Kc}{Ti} \int e(t)dt - Kc \cdot Td \frac{dPV}{dt}$$

Where:

CO = controller output signal (the wire out)

CO_{bias} = controller bias; set by bumpless transfer

e(t) = current controller error, defined as SP – PV

SP = set point

PV = measured process variable (the wire in)

Kc = controller gain, a tuning parameter

Ti = reset time, a tuning parameter

Td = derivative time, a tuning parameter

As explained in the PI control study, best practice is to set loop sample time, T, at 10 times per time constant or faster (T ≤ 0.1Tp). For this process, controller sample time, T = 1.0 sec. Also, like most commercial controllers, we employ bumpless transfer. Thus, when switching to automatic, SP is set equal to the current PV and CO_{bias} is set equal to the current CO.

Controller Gain, Reset Time & Derivative Time

We use the industry-proven Internal Model Control (IMC) tuning correlations in this study. These require specifying the closed loop time constant, Tc, that describes the desired speed or quickness of our controller in responding to a set point change or rejecting a disturbance.

Guidance for computing Tc for an aggressive, moderate or conservative controller action are listed in our PI control study and summarized as:

- *aggressive:* Tc is the larger of 0.1 ·Tp or 0.8 θp
- *moderate*: Tc is the larger of 1 ·Tp or   8 θp
- *conservative*: Tc is the larger of 10 ·Tp or 80 θp

With Tc determined, the IMC tuning correlations for the Dependent, Ideal (Non-Interacting) PID form are:

$$Kc = \frac{1}{Kp}\left(\frac{Tp + 0.5\,\theta p}{Tc + 0.5\,\theta p}\right); \; Ti = Tp + 0.5\,\theta p; \; Td = \frac{Tp\,\theta p}{2Tp + \theta p}$$

Note that, similar to the PI controller tuning correlations, only controller gain contains Tc, and thus, only Kc changes based on a desire for a more active or less active controller.

We start our study by choosing an aggressive response tuning:

Aggressive Tc = the larger of 0.1 ·Tp or 0.8 · θp

$\qquad\qquad$ = larger of 0.1 (1.3 min) or 0.8 (0.8 min)

$\qquad\qquad$ = 0.64 min

Using this Tc and our Kp, Tp and θp from Step 3 in the correlations above, we compute these aggressive PID tuning values:

Aggressive Ideal PID: Kc = –3.1 %/°C; Ti = 1.7 min; Td = 0.31 min

Controller Action

Controller gain is negative for the heat exchanger, yet most commercial controllers require that a positive value of Kc be entered. The way we indicate a negative sign is to choose the *direct acting* controller option during implementation. If the wrong control action is entered, the controller will quickly drive the final control element to full on/open or full off/closed and remain there until a proper control action entry is made.

Practitioner's Note: Controller gain, Kc, always has the same sign as the process gain, Kp. For a process with a negative Kp such as our heat exchanger, when the CO increases, the PV decreases (sometimes called up-down behavior). With the controller in automatic, if the PV is too high, the controller has to increase CO to correct for the error. The CO acts directly toward the problem, and thus, is said to be direct acting. If Kp (and thus Kc) are positive or up-up, when the PV is too high, the controller has to decrease CO to correct for the error. The CO acts in reverse of the problem and is said to be reverse acting.

Implement and Test

Below we compare the performance of two controllers side-by-side for the heat exchanger process simulation. Shown are two set point step pairs from 138 °C up to 140 °C and back again. Though not shown, the disturbance flow rate remains constant at 10 L/min throughout the study. To the left is set point tracking performance for an aggressively tuned PI controller. To the right is the set point tracking performance of our aggressively tuned PID controller.

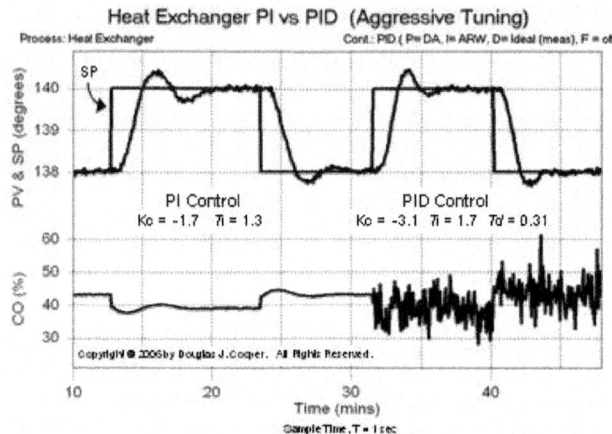

The PID controller not only displays a faster rise time (because Kc is bigger), but also a faster settling time. This is because derivative action seeks to inhibit rapid movements in the process variable. One result of this is a decrease in the rolling or oscillatory behavior in the PV trace.

Perhaps more significant, however, is the obvious difference in the CO signal trace for the two controllers. Derivative action causes the noise (random error) in the PV signal to be amplified and reflected in the control output. Such extreme control action will wear a mechanical final control element, requiring increased maintenance.

This unfortunate consequence of noise in the measured PV is a serious disadvantage with PID control.

Tune One, Tune Them All

To complete this study, we compare the Dependent, Ideal (Non-interacting) form above to the performance of the Dependent, Interacting form:

$$CO = CO_{bias} + Kc\left(1+\frac{Td}{Ti}\right)e(t) + \frac{Kc}{Ti}\int e(t)dt - KcTd\frac{dPV}{dt}$$

The IMC tuning correlations for this form are:

$$Kc = \frac{1}{Kp}\left(\frac{Tp}{Tc + 0.5\,\theta p}\right); Ti = Tp; \quad Td = 0.5\,\theta p$$

For variety, we explore moderate response tuning:

Moderate Tc = the larger of $1 \cdot Tp$ or $8 \cdot \theta p$

             = larger of 1 (1.3 min) or 8 (0.8 min)

             = 6.4 min

Using this Tc and our model parameters in the proper tuning correlations, we arrive at these moderate tuning values:

Dependent, Interacting PID: Kc = –0.36 %/°C; Ti = 1.3 min; Td = 0.40 min

Dependent, Ideal PID: Kc = –0.47 %/°C; Ti = 1.7 min; Td = 0.31 min

As shown in the plot below, we see that moderate tuning provides a reasonably fast PV response while producing no overshoot.

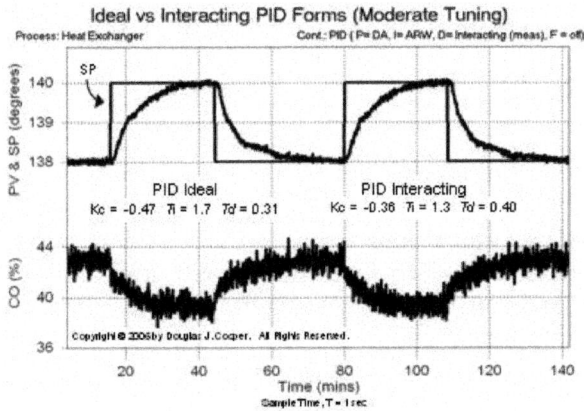

But more important, we establish that the interacting PID form and the ideal PID form provide identical performance when tuned with their own correlations.

MEASUREMENT NOISE DEGRADES DERIVATIVE ACTION

At the start of a recent Practical Process Control workshop, I asked the attendees what the "D" in PID stood for. One fellow immediately shouted from the back of the room, "Disaster?" Another piped in, "How about Danger?" When the laughter died down, another emphatically stated, "D is for Do not use." This one got a good laugh out of me. I had not heard it before and thought it was perfect. And here's why…

Benefits and Drawbacks

Derivative action has its largest influence when the measured process variable (PV) is changing rapidly (when the slope of the PV trace is steep).

The three terms of a properly tuned PID controller thus work together to provide a rapid response to error (proportional term), to eliminate offset (integral term), and to minimize oscillations in the PV (derivative term).

While this sounds great in theory, unfortunately, there are serious drawbacks to including derivative action in our controller.

A PID controller has *three* tuning parameters (three modes) that all interact and must be balanced to achieve a desired performance. It is often not at all obvious which of the three tuning parameters must be adjusted to correct behavior if performance is considered to be undesirable.

Trial and error tuning is hopeless for any but the most skilled practitioner. Fortunately, our tuning recipe provides a quick route to a safe and profitable PID performance.

A second disadvantage relates to the uncertainty in the derivative computation for processes that have noise in the PV signal.

PID Controller Form

We use the Dependent, Ideal (Non-interacting) form:

$$CO = CO_{bias} + Kc \cdot e(t) + \frac{Kc}{Ti} \int e(t)dt - Kc \cdot Td \frac{dPV}{dt}$$

Nomenclature and tuning correlations for conservative, moderate and aggressive performance.

The various PID algorithms forms provide identical performance if each algorithm is matched to its proper tuning correlations. Hence, the observations and conclusions presented below are general in nature and are not specific to a particular algorithm form.

Derivative Action Dampens Oscillations

The plot below shows the impact of derivative action on set point response performance. Because noise in the PV signal can impact performance, an idealized noise-free simulation was used to create the plot.

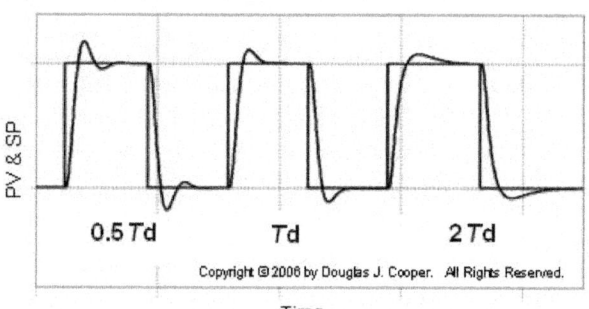

The plot shows the PV response to three pairs of set point (SP) steps. The middle response shows the base case performance of a PID controller tuned using the aggressive correlations.

For the set point steps to the right and left of the base case, the derivative time, Td, is adjusted while the controller gain, Kc, and reset time, Ti, are kept constant. This let's us isolate the impact of derivative time on performance.

It is apparent that when derivative action is cut in half to the left in the plot, the oscillating nature of the response increases. And when Td is doubled to the right, the increased derivative action inhibits rapid movement in the PV, causing the rise time and settling time to lengthen.

We saw in this tuning map from the chapter on interacting tuning parameters for PI controllers that with only two tuning parameters, performance response plots could look similar.

With the addition of Td, we now have a three dimensional tuning map with a great many similar-looking plots. If we are unhappy with our controller performance, knowing which parameter to adjust and by how much borders on the impossible. With three mode PID control, the orderly approach of our tuning recipe becomes fundamental to success.

Measurement Noise Leads to Controller Output "Chatter"

The PID control of the heat exchanger study and summarized here, the side-by-side comparison of PI vs PID control shown below illustrates one unwelcome result of adding derivative action to our controller.

The CO signal trace along the bottom of the plot clearly changes when the derivative term is added. Specifically, derivative action causes the noise (random error) in the PV signal to be amplified and reflected in the controller output.

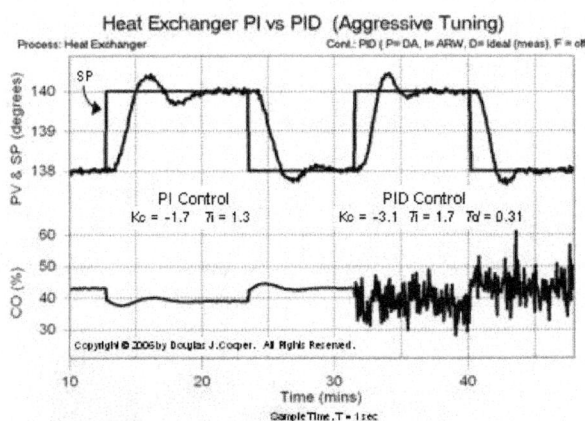

The reason for this extreme CO action or "chatter". As indicated in the plot, a noisy PV signal produces conflicting derivatives as the slope appears to dramatically alternate direction at every sample.

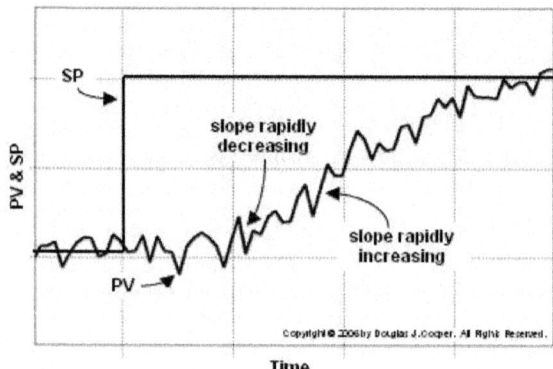

The consequence of a PV that repeatedly changes from "rapidly increasing slope" to "rapidly decreasing slope" is a derivative term that computes a series of large, alternating CO actions.

The ultimate impact of this alternating derivative computation on the total CO depends on the size of Td (the weight given to the derivative term). As Td grows larger, the "chatter" in the CO signal grows in response.

In any event, extreme control action will increase the wear on a mechanical final control element and lead to an increase in maintenance costs.

Larger Noise Means Larger Problems

The plot below shows a more subtle problem that measurement noise can cause with derivative action. In particular, this problem arises when the level of operation is near a controller output constraint (either the maximum or minimum CO).

The PID tuning values in the plot are constant throughout the experiment. As indicated on the plot, measurement noise is increased in increments across three set point tracking tests.

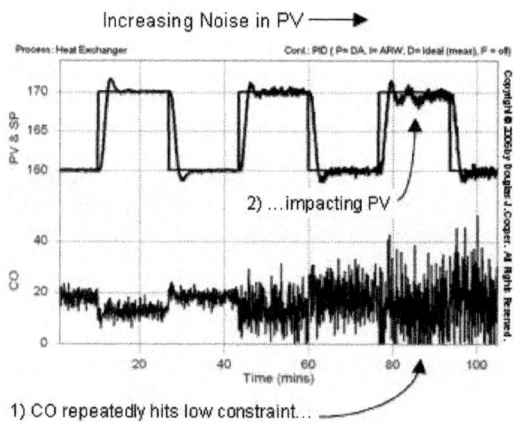

As long as measurement noise causes the derivative to alternate equally between suddenly increasing and suddenly decreasing, and the controller output can reflect this "equality in randomness" unimpeded, then controller performance is reasonably consistent in spite of increasing noise in the PV signal.

If a constraint inhibits the controller output from the "equality in randomness" symmetry, causing it to become skewed or off center, then controller performance degrades.

This is illustrated in the right most set point steps above. For this case, the controller output signal becomes so active that it repeatedly hits the minimum CO value. By constraining CO, the controller output loses its symmetry, causing the PV to wander.

What's the Solution?

One solution is to include a signal filter somewhere in the PID loop. There are several possible locations, a half dozen candidate filter algorithms, and choice of a hardware or software implementation. We explore filtering enough to see big-picture concepts and the potential benefits, though we will by no means exhaust the topic.

PID DISTURBANCE REJECTION OF THE GRAVITY DRAINED TANKS

We have explored disturbance rejection in the gravity drained tanks process using P-Only and then PI control. In the PI study, we confirmed the observations we had made in the PI control of the heat exchanger investigation.

In particular, we learned that PI controllers:

- can eliminate the offset associated with P-Only control,
- have integral action that increases the tendency for the PV to roll (or oscillate),
- have two tuning parameters that interact, making it challenging to correct tuning when performance is not acceptable.

Here we investigate the benefits and challenges of derivative action and PID control when disturbance rejectionremains our control objective.'

As with all controller implementations, we follow our four-step design and tuning recipe. A benefit of this recipe is that steps 1-3 are independent of the controller used, so our previous results from steps 1 and 2 and step 3 can be used in this PID study.

We summarize those previous results before proceeding to step 4 and the design and tuning of a PID controller.

Step 1: Determine the Design Level of Operation (DLO)

The control objective is to reject disturbances as we control liquid level in the lower tank. Our DLO for this study is:

- design PV and SP = 2.2 m with range of 2.0 to 2.4 m
- design D = 2 L/min with occasional spikes up to 5 L/min

Step 2: Collect Process Data around the DLO

When CO, PV and D are steady near the design level of operation (DLO).

Step 3: Fit a FOPDT Model to the Dynamic Process Data

We approximate the dynamic behavior of the process by fitting test data with a first order plus dead time (FOPDT) dynamic model. A fit of step test data and doublet test data yields these values:

- process gain (how far), $Kp = 0.09$ m/%
- time constant (how fast), $Tp = 1.4$ min
- dead time (how much delay), $\theta p = 0.5$ min

Step 4: Use the FOPDT Parameters to Complete the Design

The preferred PID algorithm in industrial practice employs derivative on PV, and vendors market this controller inseveral different forms. Each algorithm form has its own tuning correlations, and if we take care to match algorithm with correlation, they all provide identical capability and performance.

For tuning, we rely on the industry-proven Internal Model Control (IMC) tuning correlations. These require only one specification, the closed loop time constant (Tc), that describes the desired speed or quickness of our controller in responding to a set point (SP) change or rejecting a disturbance (D).

Our PI control study describes what to expect from an aggressive, moderate or conservative controller. Once our desired performance is chosen, the closed loop time constant is computed:

- *aggressive*: Tc is the larger of $0.1 \cdot Tp$ or $0.8 \; \theta p$
- *moderate*: Tc is the larger of $1 \cdot Tp$ or $8 \; \theta p$
- *conservative*: Tc is the larger of $10 \cdot Tp$ or $80 \; \theta p$

Because the popular PID forms perform the same if properly tuned, the observations and conclusions we draw from any one algorithms applies to the other forms.

Dependent Ideal PID

Among the most widely used algorithms is the Dependent Ideal (Non-interacting) PID form:

$$CO = CO_{bias} + Kc \cdot e(t) + \frac{Kc}{Ti} \int e(t)dt - Kc \cdot Td \frac{dPV}{dt}$$

Where:

CO = controller output signal (the wire out)

CO_{bias} = controller bias; set by bumpless transfer

e(t) = current controller error, defined as SP – PV

SP = set point

PV = measured process variable (the wire in)

Kc = controller gain, a tuning parameter

Ti = reset time, a tuning parameter

Td = derivative time, a tuning parameter

Design and Tune

In the P-Only study, we had established that for the gravity drained tanks process:

- sample time, T = 1 sec
- the controller is reverse acting
- dead time is small compared to Tp and thus not a concern in the design

After we choose a Tc based on our desired performance, the tuning correlations for the Dependent Ideal PID form are:

$$Kc = \frac{1}{Kp} \left(\frac{Tp + 0.5\,\theta p}{Tc + 0.5\,\theta p} \right); Ti = Tp + 0.5\,\theta p; \ Td = \frac{Tp\,\theta p}{2\,Tp + \theta p}$$

Similar to the PI controller tuning correlations, only controller gain contains Tc, and thus, only Kc changes based on the need for a more or less active controller.

Implement and Test

We first explore an aggressive response tuning for our ideal PID controller:

Aggressive Tc = the larger of 0.1 ·Tp or 0.8 θp

= larger of 0.1 (1.4 min) or 0.8 (0.5 min)

= 0.4 min

Using this Tc and our Kp, Tp and θp from Step 3 in the tuning correlations above, we compute these aggressive controller gain, reset time and derivative time tuning values:

Aggressive Ideal PID: Kc = 28 %/m; Ti = 1.7 min; Td = 0.21 min

The performance of this controller in rejecting changes in the pumped flow disturbance (D) for the gravity drainedtanks is shown to the right in plot. For comparison, the performance of an aggressive PI controller is shown in the plot to the left. Note that the set point (SP) remains constant at 2.2 m throughout the study.

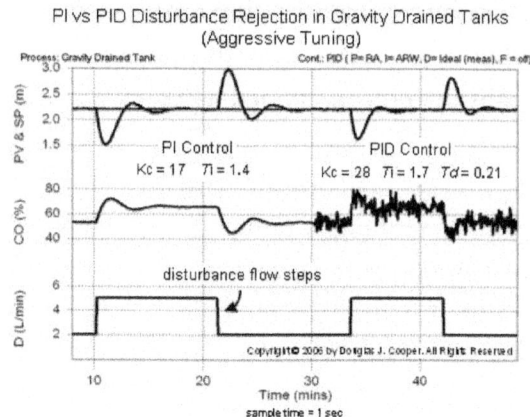

The maximum deviation of the PV from set point during the disturbance rejection event is smaller for the PID controller relative to the PI controller. The PID controller also provides a faster settling time because derivative action tends to reduce the rolling or oscillatory behavior in the PV trace.

Like the heat exchanger PID study, there is an obvious difference in the CO signal trace for the PI vs PID controllers. Derivative action causes the noise (random error) in the PV signal to be amplified and reflected in the control output (CO) signal.

Such extreme control action will cause excessive wear in a valve or other mechanical final control element, requiring increased maintenance. This consequence of noise in the measured PV can be a serious disadvantage with PID control.

Ideal vs Interacting PID

We compare the Dependent Ideal PID form above to the performance of the Dependent Interacting PID form and establish that they are identical in performance if properly tuned. The Dependent Interacting form is written:

$$CO = CO_{bias} + Kc\left(1 + \frac{Td}{Ti}\right)e(t) + \frac{Kc}{Ti}\int e(t)dt - Kc \cdot Td\frac{dPV}{dt}$$

Design and Tune

We use the same rules above to choose a Tc that reflects our desired performance. The IMC tuning correlations for the Dependent, Interacting form are then:

$$Kc = \frac{1}{Kp}\left(\frac{Tp}{Tc + 0.5\,\theta p}\right); \quad Ti = Tp; \quad Td = 0.5\,\theta p$$

As before, only controller gain contains Tc, and thus, only Kc changes based on a desire for a more or less active controller. Sample time remains for this implementation at T = 1 sec and the controller remains as reverse acting.

Implement and Test

We choose a moderate response tuning in this example:

Moderate Tc = the larger of 1 ·Tp or 8·θp

$\qquad$ = larger of 1.0 (1.4 min) or 8 (0.5 min)

$\qquad$ = 0.4 min

Using this Tc and our model parameters in the proper tuning correlations (ideal or interacting), we arrive at these moderate tuning values:

Moderate Ideal PID:$\qquad$ Kc = 4.3 %/m; $\quad$ Ti = 1.7 min; $\quad$ Td = 0.21 min

Moderate Interacting PID: Kc = 3.7 %/m; $\quad$ Ti = 1.4 min; $\quad$ Td = 0.25 min

As shown in the plot below, moderate tuning provides a reasonably fast disturbance rejection response while producing little or no oscillations as the PV settles.

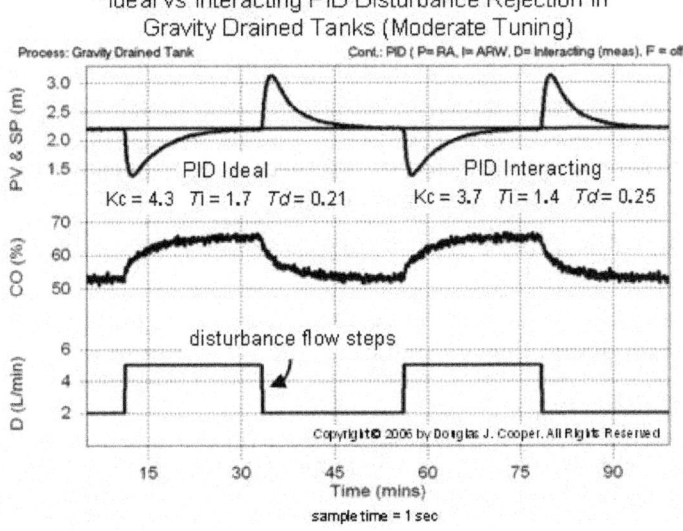

The indistinguishable behavior confirms that the two controllers indeed are identical in capability and performance if tuned with their own correlations.

Aside: Our observations using the dependent ideal and dependent interacting PID algorithms directly apply to the other popular PID controller forms.

For example, the independent PID algorithm form is written:

$$CO = CO_{bias} + Kc \cdot e(t) + Ki \int e(t)dt - Kd \frac{dPV}{dt}$$

The integral and derivative gains in the above independent form can be computed, for example, using the ideal PID correlations as:

$$Ki = Kc/Ti \text{ and } Kd = Kc \times Td.$$

Because of these mathematical identities, performance and capability observations drawn about one algorithm will apply directly to the other.

Ideal Moderate *vs* Ideal Aggressive

We compare moderate tuning side-by-side with aggressive tuning for the dependent ideal PID controller. For a different perspective, we make this comparison using a set point tracking objective.

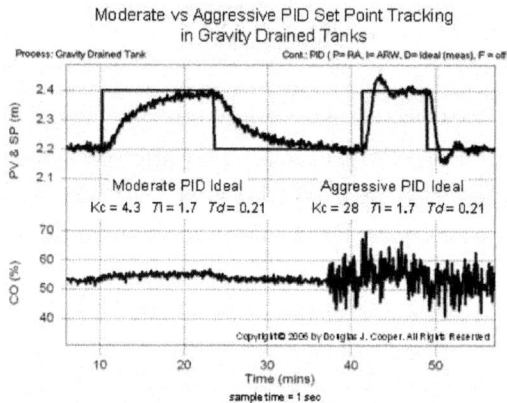

The performance of the two controllers matches the design descriptions provided here. That is, a controller tuned with:

- a moderate Tc will move the PV reasonably fast while producing little to no overshoot.

- an aggressive Tc will move the PV quickly enough to produce some overshoot and then oscillation as the PV settles out.

Need for CO Filtering

The excessive activity in the CO signal can be a problem, and a controller output (CO) signal filter is one solution. An interesting observation from the above plot is that the degree of "chatter" in the CO signal grows as controller gain, Kc, increases.

USING SIGNAL FILTERS IN OUR PID LOOP

In our study of the derivative mode of a PID controller, we explored how noise or random error in the measured process variable (PV) can degrade controller performance.

The derivative action can cause the noise in the PV measurement to be reflected and amplified in the controller output (CO) signal, producing "chatter" in the final control element (FCE). This extreme control action will increase the wear on a mechanical FCE (*e.g.*, a valve) and lead to increased maintenance needs.

Sources of Noise

Random behavior in the PV measurement arises because of signal noise and process noise.

Signal noise tends to have higher frequency relative to the characteristic dynamics of process control applications (*i.e.*, processes with streams comprised of liquids, gases, powders, slurries and melts). Sources of signal noise include:

- electrical interference
- jitter (clock related irregularities such as variations in sample spacing)
- quantizing of signal samples into overly-broad discrete "buckets" from low resolution or improperly specified instrumentation (*e.g.* too-large measurement span relative to operating range).

Process noise tends to be lower in frequency. This category borders on the philosophical as to what constitutes a disturbance to be controlled versus noise to be filtered. Bubbles and splashing that randomly corrupts liquid pressure drop measurements is an example of process noise that might benefit from filtering.

A less clear candidate for filtering is a temperature measurement in a poorly-mixed vessel. The mixing patterns can cause lower-frequency random variations in the temperature signal that are unrelated to changes in the bulk vessel temperature. It is important to emphasize that before we try to filter away a problem, we should first work to understand the source of the random error. Rather than "fix" the noise by hiding it with additional or modified algorithms, we should attempt to reduce or eliminate the problem through normal engineering and maintenance practices.

The Filtered Signal

The plot below shows the random behavior of a raw (unfiltered) PV signal and the smoother trace of a filtered PV signal.

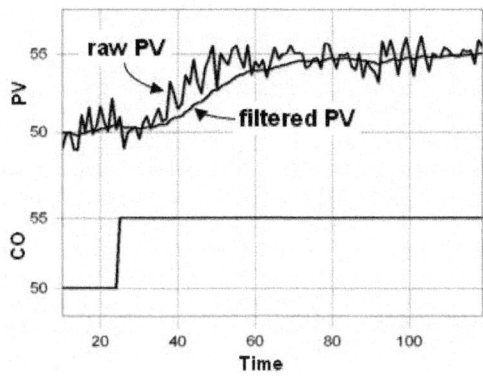

As a filter is able to receive a noisy signal and yield a signal with reduced random variation. A "better" filter design is one that decreases the random variation while retaining more of the true dynamic information of the original signal.

Filters can be analog (hardware) or digital (software); high, low or band pass; linear or nonlinear; designed in the time, Z-transform or frequency domain; and much more. Filters can collect and process data at a rate faster than the control loop sample time, so many data points can go into a single PV sample forwarded to the controller.

Clearly, filter design is a substantial topic. In these articles we offer only an introduction to the basic methods and ideas.

Filters Add Delay

The filtered signal in the plot above, though perhaps visually appealing, clearly lags behind the actual dynamic response of the unfiltered signal. More specifically, the filtered signal has an increased dead time and time constant relative to the behavior of the actual process.

Signal filters offer benefit in process control applications because they can temper the large CO moves caused by derivative action of noisy PV measurements.

Yet they add delay in sensing the true state of a process, and this has negative consequences in that as delay increases, the best achievable control performance decreases.

The design challenge is to find the careful balance between signal smoothing and information delay to achieve the controller performance we desire for our application.

External Filters in Control

There are three popular places to put external filters in the feedback loop. By "external," we mean that the filters are designed, installed and maintained separately from the controller.

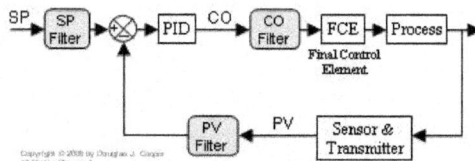

Internal filters that are part of the controller architecture itself are introduced later in this chpater.

Set Point Filters

Set point filters are not associated with the noisy PV problem, but are included here to make.

A set point filter takes a step change in SP, and as shown below, forwards a smooth transition signal to the controller.

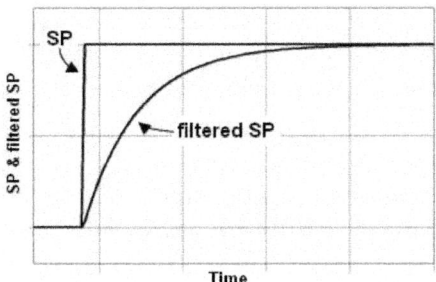

SP filters do not influence the disturbance rejection (regulatory) performance of a controller. Hence, these filters permit a controller to be tuned aggressively to reject disturbances, yet the smoothed SP transition results in a moderate set point tracking (servo) performance from this same aggressive controller.

SP filters are also used to limit overshoot at the top of a set point ramp or step change. If this is our design objective, an alternative is to eliminate the filter and employ a controller that uses proportional on PV rather than proportional on error. For example (compare to PI with proportional on error here):

$$CO = CO_{bias} - KcPV + \frac{Kc}{Ti} \int e(t)dt$$

Finally, and unfortunately, SP filters are occasionally used as a bandage to mask the fact that a controller is simply poorly designed and/or tuned. We all recognize that this is a practice to be avoided.

PV Filters

Signal filters are frequently placed between the sensor transmitter and the controller (or more likely, the multiplexer feeding the controller). In process control applications, these filters should be analog (hardware) devices designed specifically to minimize high frequency electrical interference.

While measurement noise does degrade derivative action in that it leads to chatter in the CO signal, this "noise leads to CO chatter" effect is very modest for proportional action. And interestingly, integral action is unaffected by noise because the constant summing of error literally averages the random variations in the signal.

Since filtering adds delay and this hurts best possible control performance, and because noise is not an issue for proportional and integral action, it is generally poor practice to filter the PV signal external to the controller for anything beyond electrical interference.

The preferred approach is to selectively filter only that signal destined for the derivative computation.

CO Filters

While PV filters smooth the signal feeding the controller, CO filters smooth the noise or "chatter" in the CO signal sent to the final control element. Even if PV signal noise does not appears to cause performance problems, a CO filter can offer potential benefits as it reduces fluctuations in the controller output and this reduces wear on the FCE.

If a noisy PV is an issue in our controller, our first attempts should be to locate and correct the problem. If after that exercise, our decision is to design and implement a filter in our feedback loop, CO filters are attractive alternatives.

Internal Filters in Control

For feedback control, filtering need only be applied to the signal feeding the derivative term. As stated before, noise does not present a problem for proportional and integral action. These elements will perform best without the delay introduced from a signal filter.

When we selectively filter just that signal feeding the derivative calculation, the filter becomes part of the controller architecture. Hence, we can still use our design recipe, though the correlations for tuning this four mode form are different from the four mode "PID with CO Filter" form mentioned above.

There are two common architectures that are identical in capability, though different in presentation.

We filter the PV signal before feeding it to the derivative mode of the PID algorithm for computation:

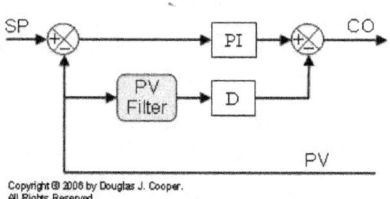

We can also compute the derivative action with the noisy signal and then filter the computed result:

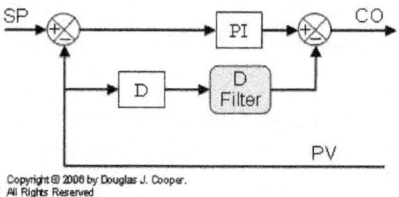

If the same filter form is used (*e.g.*, first-order), we can show mathematically that both options above are identical.

As it turns out, many commercial controllers use this internal derivative filtering form where they implement a first-order filter and fix the filter time at one-tenth of the derivative time value.

PID WITH CONTROLLER OUTPUT (CO) FILTER

The derivative action of a PID controller can cause noise in the measured process variable (PV) to be amplified and reflected as "chatter" in the controller output (CO) signal. Signal filters, implemented as either analog hardware or digital software, offer a popular solution to this problem.

If noise is impacting controller performance, our first attempts should be to locate and correct the underlying fault. Filters are poor cures for a bad design or failing equipment.

If we decide to employ a filter in our loop, an algorithm designed to smooth the controller output signal holds some allure.

The CO Filter Architecture

Below is a loop architecture with a PID controller followed by a controller output filter.

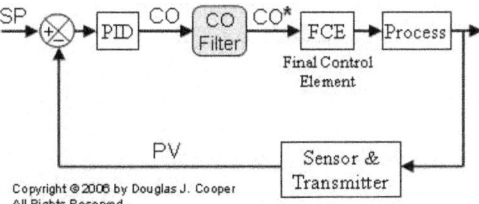

The benefits of this architecture include:

*   A CO filter works to limit large controller output moves regardless of the underlying cause. As a result, CO filters can reduce persistent controller output fluctuations that cause wear in a mechanical final control element (FCE).

*   A CO filter is a single solution that addresses both a noisy PV measurement problem and computational oddities that may exist in our vendor's particular PID algorithm,

*   Perhaps most important, the tuning recipe we have employed so successfully with PI and PID algorithms can be directly applied to a PID with CO filter architecture.

PID Plus External Filter

The filter computation is performed after the PID controller has computed the CO. Below, the PID controller output is computed using the non-interacting, dependent, ideal form, but any of the popular algorithms can be used with this "PID plus filter" architecture:

$$CO = CO_{bias} + Kce(t) + \frac{Kc}{Ti}\int e(t)dt + KcTd\frac{de(t)}{dt}$$

Where:

CO = controller output signal (the wire out)

CO_{bias} = controller bias; set by bumpless transfer

e(t) = current controller error, defined as SP – PV

SP = set point

PV = measured process variable (the wire in)

Kc = controller gain, a tuning parameter

Ti = reset time, a tuning parameter

Td = derivative time, a tuning parameter

The Filtering Algorithm

A first order filter yields a smoothed CO* value as:

$$T_f\frac{DCO^*}{dt} + CO^* = CO$$

Where:

CO = raw PID controller output signal

CO* = filtered CO signal sent to FCE (*e.g.*, valve)

T_f = filter time constant, a tuning parameter

A comparison of the first order filter above to a general first order plus dead time(FOPDT) model form reveals that:

- The gain (or scaling factor) of the filter is one. That is, the filtered CO* has the same zero, span and units as the CO value from the PID algorithm.

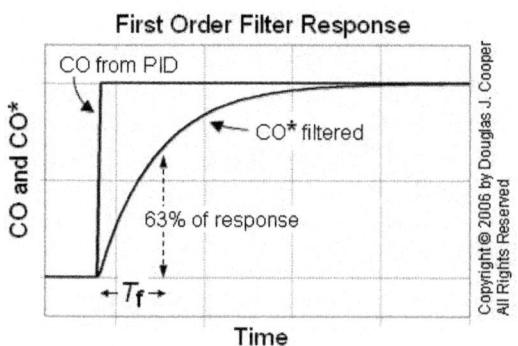

First Order Filter Response

- There is no dead time built into the filter. The CO* forwarded to the FCE is computed immediately after the PID algorithm yields the raw (unfiltered) CO signal.

- The degree of filtering, or how quickly CO* moves toward the unfiltered CO value, is set by T_f, the filter time constant.

Filtering Adds Delay

The degree of smoothing depends on the size of the filter time constant, T_f:

- A smaller T_f means CO* moves quickly and follows closer to changes in CO, so there is little filtering or smoothing of the signal.

- A larger T_f means CO* responds more slowly to changes in CO, so the filtering or smoothing is greater.

Shown below is a series of CO signals from a PID controller. Also shown is the filtered CO* trace using the same T_f as in the plot above.

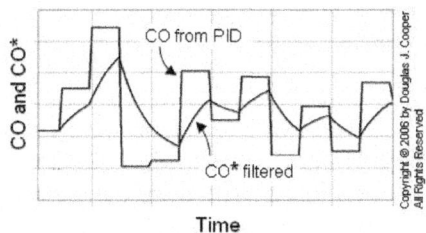

As smoothing (or filter time, T_f) increases, the filtered signal may become more visually appealing, but more filtering means additional information delay in the control loop computation.

As delay increases in a control loop, the best achievable control performance decreases. The design challenge is to find the careful balance between signal smoothing and information delay to achieve the controller performance we desire for our process.

Combining Controller Plus Filter

To use our tuning recipe, we must first combine the controller and filter into a single unified equation. Since both equations above have CO isolated on one side of the equal sign, we can set the two equations equal to yield:

$$T_f \frac{DCO^*}{dt} + CO^* = CO_{bias} + Kc\,e(t) + \frac{Kc}{Ti}\int e(t)\,dt + Kc\,Td\frac{de(t)}{dt}$$

Moving the left-most term to the right-hand side produces the unified PID with Filter equation:

$$CO^* = CO_{bias} + Kc\,e(t) + \frac{Kc}{Ti}\int e(t)\,dt + Kc\,Td\frac{de(t)}{dt} - T_f\frac{dCO^*}{dt}$$

Unified PID With Filter Form

For design and tuning purposes going forward, we will use the unified PID with Filter equation form, as if it were represented by the schematic:

As implied by this diagram, we will drop the CO versus CO* nomenclature and simply write the unified PID with Filter equation as:

$$CO = CO_{bias} + Kc\,e(t) + \frac{Kc}{Ti}\int e(t)\,dt + Kc\,Td\frac{de(t)}{dt} - T_f\frac{dCO}{dt}$$

Please be aware that this is identical in every way to the PID with external filter algorithm derived earlier in this chapter. We recast it only so the tuning correlations are consistent in appearance and application with those of the PI and PID forms presented in earlier articles.

Filter Time Constant

Many commercial controllers that include some form of a PID with Filter algorithm cast the filter time constant as a fraction of the controller derivative time, or:

$$T_f = \alpha Td$$

The unified PID with Filter algorithm then becomes:

$$CO = CO_{bias} + Kc\,e(t) + \frac{Kc}{Ti}\int e(t)\,dt + Kc\,Td\frac{de(t)}{dt} - \alpha Td\frac{dCO}{dt}$$

We will use this αTd form in the example that follows. If your controller output filter uses T_f, the conversion is computed: $T_f = \alpha Td$.

Discrete-Time Implementation

While a CO filter offers potential for benefit in loops with noise and/or delicate mechanical FCEs, if our vendor does not offer the option, we must program the filter ourselves. Fortunately, the code is straightforward.

The first order filtering algorithm is expressed:

$$T_f\frac{dCO^*}{dt} + CO^* = CO$$

In discrete-time form, we write:

$$T_f\frac{(CO^*new - CO^*old)}{T} + CO^*old = CO$$

or

$$CO^*new = CO^*old + (T/T_f)(CO - CO^*old)$$

Where T is the loop sample time and CO is the unfiltered PID signal.

In a computer program, the "new" CO* is computed directly from the "old" CO* value at each loop, so we do not need to keep track of these labels. The filter computation can be programmed in one line as:

$$COstar = COstar + (T/T_f)(CO - COstar)$$

Applying the CO Filter

We explore an example of PID control with CO filtering applied to the heat exchanger and jacketed stirred reactor process.

PID WITH CO FILTER CONTROL OF THE HEAT EXCHANGER

The same tuning recipe we successfully demonstrated for PI control and PID controlled sign and tuning can be used when a controller output (CO) filter is added to the heat exchanger process control loop.

Here we explore PID with CO Filter control using the unified (controller with internal filter) form. The unified form is identical to a PID with external first-order CO filter implementation. Hence, the methods we use and observations we make apply equally to both internal and external filter architectures.

We follow the same four-step design and tuning recipe used for all control implementations.

Steps 1-3 of the PID with CO Filter design are identical to our previous PI and PID control case studies.

Step 1: Design Level of Operation (DLO)

- Design PV and SP = 138 °C with operation ranging from 138 to 140 °C
- Expected warm liquid flow disturbance = 10 L/min

Step 2: Collect Data at the DLO

PI control chapter referenced above for a summary.

Step 3: Fit an FOPDT Model to the Dynamic Data

The first order plus dead time (FOPDT) model approximation of the heat exchanger data from step 2 is:

- Process gain (how far), Kp = –0.53 °C/%
- Time constant (how fast), Tp = 1.3 min
- Dead time (how much delay), θp = 0.8 min

Step 4: Use the Parameters to Complete the Design

Vendors market the PID algorithm in a number of different forms, creating a confusing array of choices for the practitioner. The addition of a CO filter makes a bad situation worse. A filter adds another adjustable (tuning) parameter and significantly increases the number of possible algorithm forms.

Trial and error tuning of a four mode (four tuning parameter) controller with filter while our process is making product is a sure path to waste and expense. With so many interacting variables, we will likely settle for an operation that "isn't horrible" rather than a performance that is near optimal.

The dilemma is real and our tuning recipe is the answer. Yet for success in this final step, it is critical that we match our algorithm with its proper tuning correlations.

The way to do this reliably is with loop tuning software. In fact, a good package will help with all of the steps, from data collection and model fitting through vendor algorithm selection and final performance analysis. When our task list includes maintaining and tuning loops during production, commercial software will pay for itself in days.

Sample Time and Bumpless Transfer

As explained in the PI control study, best practice is to set loop sample time $T \leq 0.1Tp$ (10 times per time constant or faster). For this example, $T = 1.0$ sec.

Also, like most commercial controllers, we employ bumpless transfer. Thus, when switching to automatic, SP is set equal to the current PV and CO_{bias} is set equal to the current CO.

Controller Action

Controller gain is negative for the heat exchanger, yet most commercial controllers require that a positive value of Kc be entered. The way we indicate a negative sign is to choose the direct acting option during implementation. If the wrong control action is entered, the controller will quickly drive the final control element to full on/open or full off/closed and remain there until a proper control action entry is made.

Specify Desired Performance

We use the industry-proven Internal Model Control (IMC) tuning correlations in this study. IMC correlations employ a closed loop time constant, Tc, that describes the desired speed or quickness of our controller in responding to a set point change or rejecting a disturbance. We must decide whether we seek:

- An *aggressive* controller with a rapid response and some overshoot:

$$Tc \text{ is the larger of } 0.1 \cdot Tp \text{ or } 0.8 \cdot \theta p$$

- A *moderate* controller that will move the PV reasonably fast yet produce little to no overshoot in a set point response:

$$Tc \text{ is the larger of } 1 \cdot Tp \text{ or } 8 \; \theta p$$

- A *conservative* controller that will move the PV in the proper direction, but quite slowly:

$$Tc \text{ is the larger of } 10 \cdot Tp \text{ or } 80 \; \theta p$$

Ideal PID With Filter Example

An external first-order filter into a unified dependent, ideal, non-interacting PID with internal filter form:

$$CO = CO_{bias} + Kc\,e(t) + \frac{Kc}{Ti}\int e(t)dt + Kc\,Td\frac{de(t)}{dt} - \alpha Td\frac{dCO}{dt}$$

If our vendor offers the option, the preferred algorithm in industrial practice is PID with derivative on measurement(derivative on PV).

While a CO filter can largely address derivative kick, the filter term must be made larger than otherwise necessary to do so. Thus, there remains a performance benefit to derivative on measurement even when using a CO filter.

$$CO = CO_{bias} + Kc\,e(t) + \frac{Kc}{Ti}\int e(t)dt - Kc\,Td\frac{de(t)}{dt} - \alpha Td\frac{dCO}{dt}$$

The IMC tuning correlations for either of the above PID with CO Filter forms are:

Kc	Ti	Td	α
$\dfrac{1}{Kp}\left(\dfrac{Tp + 0.5\,\theta p}{Tc + \theta p}\right)$	$Tp + 0.5\,\theta p$	$\dfrac{Tp\,\theta p}{2Tp + \theta p}$	$\dfrac{Tc(Tp + 0.5\,\theta p)}{Tp(Tc + \theta p)}$

We start our study by choosing an aggressive response tuning:

Aggressive Tc = the larger of $0.1 \cdot Tp$ or $0.8\;\theta p$

$$= \text{larger of } 0.1(1.3 \text{ min}) \text{ or } 0.8(0.8 \text{ min})$$

$$= 0.64 \text{ min}$$

Using this Tc and our Kp, Tp and θp from Step 3 in the tuning correlations above yields the aggressive PID w/ Filter tuning values below. Also listed are the PID Ideal and PI controller tuning values from earlier studies:

PID w/ Filter:

Kc = –2.2 %/°C   Ti = 1.7 min   Td = 0.31 min   a = 0.6

PID:

Kc = –3.1 %/°C   Ti = 1.7 min   Td = 0.31 min

PI:

Kc = –1.7 %/°C   Ti = 1.3 min

Below we compare the performance of these three aggressively tuned controllers side-by-side for the heat exchanger process simulation. Shown are three set point step pairs from 138 °C up to 140 °C and back again. Though not shown, the disturbance flow rate remains constant at 10 L/min throughout the study.

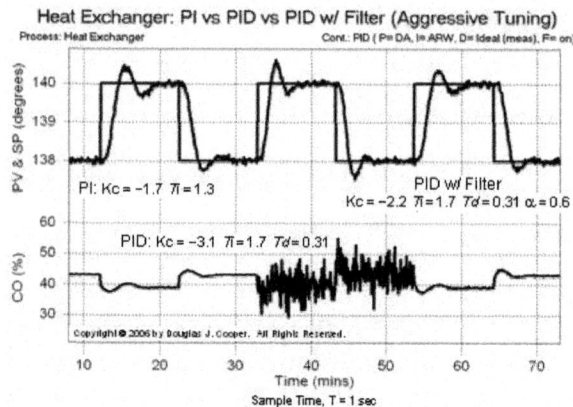

To the left is set point tracking performance for the PI controller. The middle set point steps show the performance of the PID controller. Derivative action enables a slightly faster rise time and settling time, but the derivative action causes the noise in the PV signal to be amplified and reflected as "chatter" in the control output signal.

To the right is the set point tracking performance of the PID w/ CO Filter controller. Indeed, the filter does an impressive job of cleaning up the chatter in the controller output signal without degrading performance.

In truth, however, the four tuning parameter PID w/ Filter performs similar to the two tuning parameter PI controller.

Tune One, Tune Them All

To complete this study, we compare the dependent, ideal, non-interacting form above to the performance of the dependent, interacting form:

$$CO = CO_{bias} + Kc\left(1 + \frac{Td}{Ti}\right)e(t) + \frac{Kc}{Ti}\int e(t)dt - KcTd\frac{dPV}{dt} - \alpha Td\frac{dCO}{dt}$$

The tuning correlations for the dependent, interacting form are:

Kc	Ti	Td	α
$\dfrac{1}{Kp}\left(\dfrac{Tp}{Tc + \theta p}\right)$	Tp	$0.5\,\theta p$	$\dfrac{Tc}{Tc + \theta p}$

For variety, we choose moderate response tuning for this comparison:

Moderate Tc = the larger of $1 \cdot Tp$ or $8 \cdot \theta p$

$\qquad$ = larger of $1(1.3 \text{ min})$ or $8(0.8 \text{ min})$

$\qquad$ = 6.4 min

Using this Tc and our model parameters in the proper tuning correlations, we arrive at these moderate tuning values:

Dependent, Interacting:

$\qquad$ $Kc = -0.34 \ \%/°C$ $Ti = 1.3$ min $Td = 0.40$ min $a = 0.9$

Dependent, Ideal:

$\qquad$ $Kc = -0.44 \ \%/°C$ $\quad Ti = 1.7$ min $Td = 0.31$ min $a = 1.2$

We see that moderate tuning provides a reasonably fast PV response while producing no overshoot. But more important, we establish that the interacting form and the ideal form provide identical performance when tuned with their own correlations.

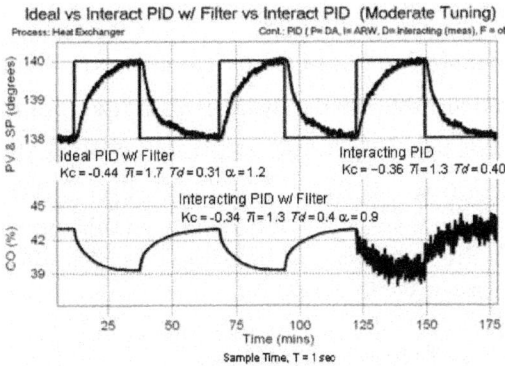

The third set point step shows the performance of a straight PID with no filter. This reinforces the benefits of a CO filter if derivative action is being contemplated.

Observations

Our study of the heat exchanger process has shown that PID controllers provide minor performance improvements over PI controllers. Yet derivative action causes noise in the PV to be reflected as chatter in the CO signal, and this counterbalances the small benefits of the derivative term.

We explore CO filters and learn that they can correct the chatter problem. But now we have elevated a difficult two tuning parameter PI controller design into an extremely challenging four parameter PID w/ Filter controller design. And at best, this extra effort still provides only modest performance benefits.

Unless the economic impact of a loop is substantial, many practitioners conclude that the PI controller is the best choice. It is faster to implement, easier to maintain, and provides performance approaching that of the PID w/ Filter controller.

PID WITH CO FILTER DISTURBANCE REJECTION IN THE JACKETED STIRRED REACTOR

The control objective for the jacketed stirred reactor process is to minimize the impact on reactor operation when the temperature of the liquid entering the cooling jacket changes. We have previously established the performance capabilities of a PI controller in rejecting the impact of this disturbance.

Here we explore the performance of a PID with controller output (CO) filter algorithm in meeting this same disturbance rejection objective. We use the unified PID with CO filter controller in this study. The unified form is identical to a PID with external first-order CO filter implementation. Thus, the methods and observations from this investigation apply equally to both controller architectures.

The important variables for the jacketed reactor are (view a process graphic):

CO = signal to valve that adjusts cooling jacket liquid flow rate (controller output, %)

PV = reactor exit stream temperature (measured process variable, °C)

SP = desired reactor exit stream temperature (set point, °C)

D = temperature of cooling liquid entering the jacket (major disturbance, °C)

We follow our industry proven recipe to design and tune our PID with CO filter controller. Recall that steps 1-3 of a design remain the same regardless of the controller used. For this process and objective, the results of steps 1-3 are summarized from previous investigations:

Step 1: Design Level of Operation (DLO)

DLO are summarized:

- Design PV and SP = 90 °C with approval for brief dynamic (bump) testing of ±2 °C
- Design D = 43 °C with occasional spikes up to 50 °C

Step 2: Collect Process Data around the DLO

When CO, PV and D are steady near the design level of operation, we bump the jacketed stirred reactor to generate CO-to-PV cause and effect process response data.

Step 3: Fit a FOPDT Model to the Dynamic Process Data

We approximate the dynamic behavior of the process by fitting a first order plus dead time (FOPDT) dynamic model to the test data from step 2. The results of the modeling study are summarized:

- process gain (direction and how far), $Kp = -0.5 \,°C/\%$
- time constant (how fast), $Tp = 2.2$ min
- dead time (how much delay), $\theta p = 0.8$ min

Step 4: Use the FOPDT Parameters to Complete the Design

Algorithm Form

A PID with CO filter controller, regardless of whether the filter is internal or external, presents us with a "four adjustable tuning parameter" challenge. As more parameters are included in our controller, the array of vendor algorithm forms increases.

The various controller forms are all capable of delivering a similar, predictable performance, as long as we match our algorithm with its proper tuning correlations. Certainly, a "guess and test" approach to tuning a four-mode controller while our process is making product is a sure path to wasting feedstock and utilities, creating safety and environmental concerns, and putting plant profitability at risk.

Modern loop tuning software will not only guide data analysis and model fitting, but will ensure our tuning matches our vendor's algorithm. Such software will even display expected final performance prior to implementation. If our task list includes maintaining and tuning control loops during production, such software will pay for itself in days.

Sample Time

Best practice is to set loop sample time to $T \le 0.1Tp$ (10 times per time constant or faster). We meet this design criterion with the widely-available vendor option of $T = 1.0$ sec.

Controller Action

Kp, and thus Kc, are negative for our jacketed stirred reactor process. Most commercial controllers have us specify a negative Kc by entering a positive value into the controller and then choosing the "direct acting" option. If the wrong control action is entered, the controller will drive the final control element to full on/open or full off/closed and remain there until a proper control action entry is made.

Specify Desired Performance

We use the industry-proven Internal Model Control (IMC) tuning correlations in this study. IMC correlations employ a closed loop time constant, Tc, that describes the desired speed or quickness of our controller in responding to a set point change or rejecting a disturbance.

Our PI control study describes what to expect from an aggressive, moderate or conservative controller. Once our desired performance is chosen, the closed loop time constant is computed:

- aggressive performance: Tc is the larger of $0.1 \cdot Tp$ or $0.8 \cdot \theta p$
- moderate performance: Tc is the larger of $1 \cdot Tp$ or $8 \cdot \theta p$
- conservative performance: Tc is the larger of $10 \cdot Tp$ or $80 \cdot \theta p$

The Tuning Correlations

How to combine an external first-order filter into the unified ideal PID with internal filter form:

$$CO = CO_{bias} + Kc\,e(t) + \frac{Kc}{Ti}\int e(t)\,dt + Kc\,Td\frac{de(t)}{dt} - \alpha Td\frac{dCO}{dt}$$

If our vendor offers the option, the preferred algorithm in industrial practice is PID with derivative on measurement, PV:

$$CO = CO_{bias} + Kc\,e(t) + \frac{Kc}{Ti}\int e(t)\,dt - Kc\,Td\frac{de(t)}{dt} - \alpha Td\frac{dCO}{dt}$$

The IMC tuning correlations for either of the above PID with CO filter forms are the same and listed in the chart below. The PI controller and ideal PID controller forms:

	Controller Gain Kc	Reset Time Ti	Deriv Time Td	Filter Const α
PI	$\dfrac{1}{Kp}\dfrac{Tp}{(\theta p + Tc)}$	Tp		
Ideal PID	$\dfrac{1}{Kp}\left(\dfrac{Tp + 0.5\theta p}{Tc + 0.5\theta p}\right)$	$Tp + 0.5\theta p$	$\dfrac{Tp\,\theta p}{2Tp + \theta p}$	
PID w/ CO Filter	$\dfrac{1}{Kp}\left(\dfrac{Tp + 0.5\theta p}{Tc + \theta p}\right)$	$Tp + 0.5\theta p$	$\dfrac{Tp\,\theta p}{2Tp + \theta p}$	$\dfrac{Tc(Tp + 0.5\theta p)}{Tp(Tc + \theta p)}$

PID With CO Filter Disturbance Rejection Study

In the plots below, we compare the performance of the PID with CO filter controller side-be-side with that of the PI controller and the ideal (unfiltered) PID controller.

Our objective is rejecting the impact on reactor operation when the temperature of cooling liquid entering the reactor jacket changes. We test both moderate and aggressive response tuning for the three controllers.

Moderate Response Tuning:

For a controller that will move the PV reasonably fast while producing little to no overshoot, choose:

Moderate Tc = the larger of 1·Tp or 8·θp
= larger of 1(2.2 min) or 8(0.8 min)
= 6.4 min

Using this Tc and the Kp, Tp and θp values listed in step 3 at the top of this chapter, the moderate IMC tuning values are:

PI: Kc = −0.61 %/°C   Ti = 2.2 min
PID: Kc = −0.77 %/°C   Ti = 2.6 min   Td = 0.34 min
PID w/ Filter: Kc = −0.72 %/°C   Ti = 2.6 min Td = 0.34 min a = 1.1

Aggressive Response Tuning:

For an active or quickly responding controller where we can tolerate some overshoot and oscillation as the PV settles out, specify:

Aggressive Tc = the larger of 0.1·Tp or 0.8·θp
= larger of 0.1(2.2 min) or 0.8(0.8 min)
= 0.64 min

and the aggressive IMC tuning values are:

PI: Kc = −3.1 %/°C   Ti = 2.2 min
PID: Kc = −5.0 %/°C   Ti = 2.6 min   Td = 0.34 min
PID w/ Filter: Kc = −3.6 %/°C   Ti = 2.6 min Td = 0.34 min a = 0.5

Implement and Test

A comparison of the three controllers in rejecting a disturbance change in the cooling jacket inlet temperature, D, is shown below. This plot shows controller performance when using the moderate tuning values computed above. Note that the set point remains constant at 90 °C throughout the study.

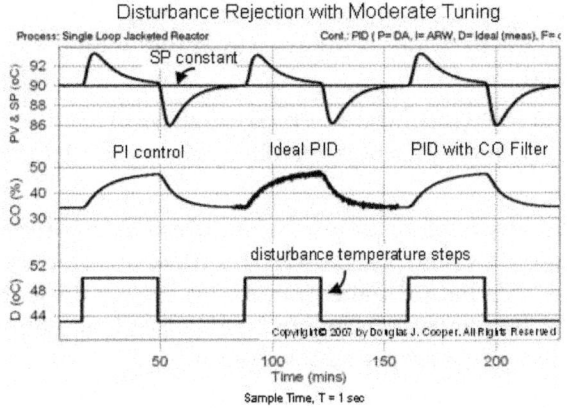

Disturbance Rejection with Moderate Tuning

The PI controller performance is shown to the left in the plot above. The ideal PID performance is in the middle. The plot reveals that the benefit of derivative action is marginal at best. There is a clear penalty, however, in that derivative action causes the modest noise in the PV signal to be amplified and reflected as "chatter" in the CO signal.

To the right in the plot above is the performance of the PID with CO filter controller. The filter is effective in reducing the controller output chatter caused by the derivative action without degrading performance. In truth, however, the four tuning parameter PID with filter performs similar to the two tuning parameter PI controller.

The disturbance rejection performance of the controllers when tuned for aggressive action is shown below. Note that the axis scales for the plots both above and below are the same to permit a visual comparison.

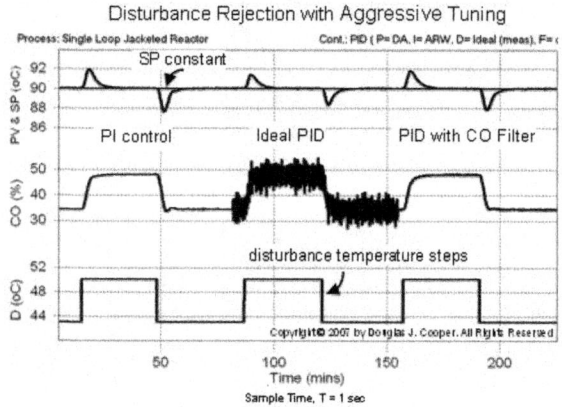

The aggressive tuning provides a smaller maximum deviation from set point and a faster settling time relative to the moderate tuning performance. The only obvious difference is that as a PID controller (middle of plot) becomes more aggressive in its actions, the CO chatter grows as a problem and filtering solutions become increasingly beneficial.

But ultimately, just as with the moderate tuning case, the two mode (or two tuning parameter) PI controller compares favorably with the four mode PID with CO filter controller.

While not our design objective, presented below is the set point tracking ability of the aggressively tuned controllers when the disturbance temperature is held constant:

The set point tracking response of the ideal PID controller is marginally better in that it shows a slightly shorter rise time, smaller overshoot and faster settling time. The CO chatter that comes as a price for these minor benefits will likely increase maintenance costs as our final control element (*e.g.*, valve, pump or compressor) wears from this excessive activity.

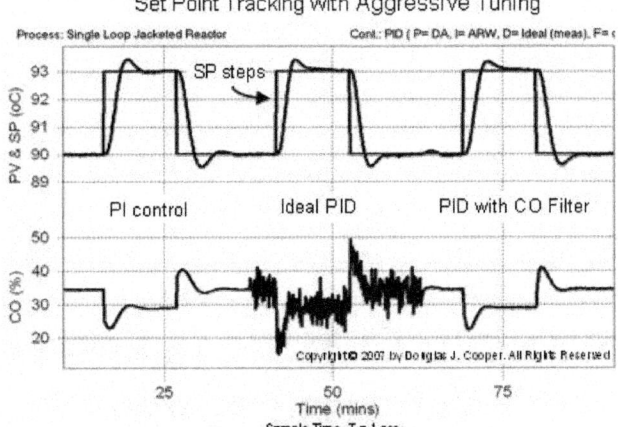

Set Point Tracking with Aggressive Tuning

The four mode PID with CO filter addresses the chatter, but it is not clear that the added complexity is worth the marginal performance benefits.

Thus, many practitioners conclude that the PI controller provides the best balance of complexity and performance. It is faster to implement, easier to maintain, and provides performance approaching that of the PID with CO filter controller.

Chapter 6

ADVANCED CLASSICAL CONTROL ARCHITECTURES

THE CASCADE CONTROL ARCHITECTURE

Two popular control strategies for improved disturbance rejection performance are cascade control and feed forward with feedback trim.

Improved performance comes at a price. Both strategies require that additional instrumentation be purchased, installed and maintained. Both also require additional engineering time for strategy design, tuning and implementation.

The cascade architecture offers alluring additional benefits such as the ability to address multiple disturbances to our process and to improve set point response performance. In contrast, the feed forward with feedback trim architecture is designed to address a single measured disturbance and does not impact set point response performance in any fashion.

The Inner Secondary Loop

The dashed line in the block diagram below circles a feedback control loop like we have discussed in dozens of articles on *controlguru.com*. The only difference is that the words "inner secondary" have been added to the block descriptions. The variable labels also have a "2" after them.

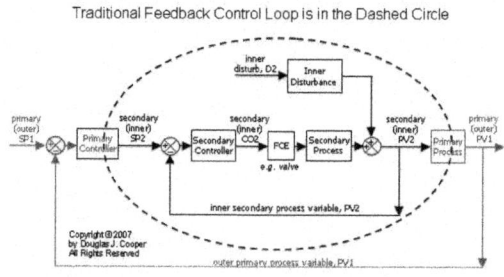

Traditional Feedback Control Loop is in the Dashed Circle

So,

SP2 = inner secondary set point

CO2 = inner secondary controller output signal

PV2 = inner secondary measured process variable signal

And

D2 = inner disturbance variable (often not measured or available as a signal)

FCE = final control element such as a valve, variable speed pump or compressor, *etc.*

The Nested Cascade Architecture

To construct a cascade architecture, we literally nest the secondary control loop inside a primary loop as shown in the block diagram below.

Note that outer primary PV1 is our process variable of interest in this implementation. PV1 is the variable we would be measuring and controlling if we had chosen a traditional single loop architecture instead of a cascade.

Cascade Structure is a Control Loop within a Control Loop

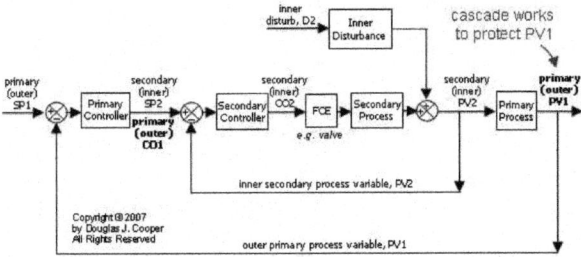

Because we are willing to invest the additional effort and expense to improve the performance response of PV1, it is reasonable to assume that it is a variable important to process safety and/or profitability. Otherwise, it does not make sense to add the complexity of a cascade structure.

Naming Conventions

Like many things in the PID control world, vendor documentation is not consistent. The most common naming conventions we see for cascade (also called nested) loops are:

- secondary and primary
- inner and outer
- slave and master.

In an attempt at clarity, we are somewhat repetitive in this chapter by using labels like "inner secondary" and "outer primary."

Two PVs, Two Controllers, One Valve

Notice from the block diagrams that the cascade architecture has:

- two controllers (an inner secondary and outer primary controller)
- two measured process variable sensors (an inner PV2 and outer PV1)
- only *one* final control element (FCE) such as a valve, pump or compressor.

How can we have two controllers but only one FCE? The controller output signal from the outer primary controller, CO1, becomes the set point of the inner secondary controller, SP2.

The outer loop literally commands the inner loop by adjusting its set point. Functionally, the controllers are wired such that SP2 = CO1.

This is actually good news from an implementation viewpoint. If we can install and maintain an inner secondary sensor at reasonable cost, and if we are using a <u>PLC</u> or<u>DCS</u> where adding a controller is largely a software selection, then the task of constructing a cascade control structure may be reasonably straightforward.

Early Warning is Basis for Success

An essential element for success in a cascade design is the measurement and control of an "early warning" process variable.

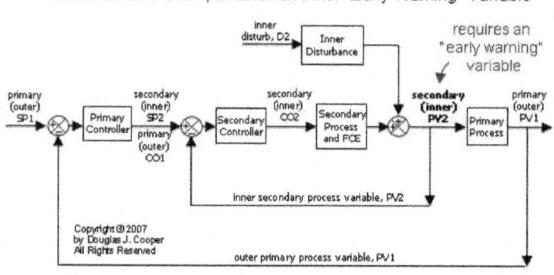

Cascade Control Depends on an Inner "Early Warning" Variable

In the cascade architecture, inner secondary PV2 serves as this early warning process variable. Given this, essential design characteristics for selecting PV2 include that:

- it be measurable with a sensor,
- the same FCE (*e.g.*, valve) used to manipulate PV1 also manipulates PV2,
- the same disturbances that are of concern for PV1 also disrupt PV2, and
- PV2 responds *before* PV1 to disturbances of concern and to FCE manipulations.

Since PV2 sees the disruption first, it provides our "early warning" that a disturbance has occurred and is heading toward PV1. The inner secondary controller can begin corrective action immediately. And since PV2 responds first to final control element (*e.g.*, valve) manipulations, disturbance rejection can be well underway even before primary variable PV1 has been substantially impacted by the disturbance.

With such a cascade architecture, the control of the outer primary process variable PV1 benefits from the corrective actions applied to the upstream early warning measurement PV2.

Disturbance Must Impact Early Warning Variable PV2

Even with a cascade structure, there will likely be disturbances that impact PV1 but do not impact early warning variable PV2.

Disturbance Must Hit the Inner PV for Cascade to Provide Benefit

The inner secondary controller offers no "early action" benefit for these outer disturbances. They are ultimately addressed by the outer primary controller as the disturbance moves PV1 from set point.

On a positive note, a proper cascade can improve rejection performance for any of a host of disturbances that directly impact PV2 before disrupting PV1.

An Illustrative Example

To illustrate the construction and value of a cascade architecture, consider the liquid level control process shown below. This is a variation on our gravity drained tanks, so hopefully, the behavior of the process below follows intuitively from our previous investigations.

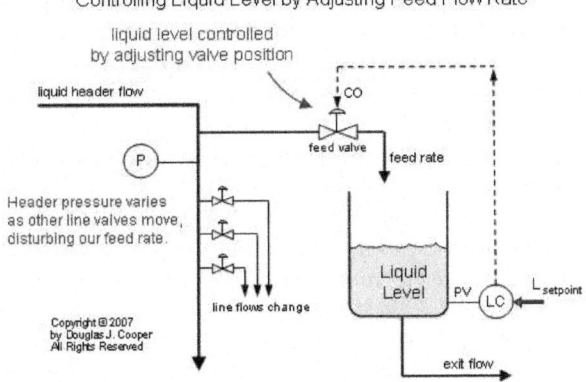

Controlling Liquid Level by Adjusting Feed Flow Rate

The tank is essentially a barrel with a hole punched in the bottom. Liquid enters through a feed valve at the top of the tank. The exit flow is liquid draining freely by the force of gravity out through the hole in the tank bottom.

The control objective is to maintain liquid level at set point (SP) in spite of unmeasured disturbances. Given this objective, our measured process variable (PV) is liquid level in the tank. We measure level with a sensor and transmit the signal to a level controller.

After comparing set point to measurement, the level controller (LC) computes and transmits a controller output (CO) signal to the feed valve. As the feed valve opens and closes, the liquid feed rate entering the top of the tank increases and decreases to raise and lower the liquid level in the tank.

This "measure, compute and act" procedure repeats every loop sample time, T, as the controller works to maintain tank level at set point.

The Disturbance

The disturbance of concern is the pressure in the main liquid header. The header supplies the liquid that feeds our tank. It also supplies liquid to several other lines flowing to different process units in the plant.

Whenever the flow rate of one of these other lines changes, the header pressure can be impacted. If several line valves from the main header open at about the same time, for example, the header pressure will drop until its own control system corrects the imbalance. If one of the line valves shuts in an emergency action, the header pressure will momentarily spike.

As the plant moves through the cycles and fluctuations of daily production, the header pressure rises and falls in an unpredictable fashion. And every time the header pressure changes, the feed rate to our tank is impacted.

Problem with Single Loop Control

The single loop architecture attempts to achieve our control objective by adjusting valve position in the liquid feed line. If the measured level is higher than set point, the controller signals the valve to close by an appropriate percentage with the expectation that this will decrease feed flow rate accordingly.

But feed flow rate is a function of two variables:

- feed valve position, and
- the header pressure pushing the liquid through the valve (a disturbance).

To explore this, we conduct some thought experiments:

Thought Experiment #1: Assume that the main header pressure is perfectly constant over time. As the feed valve opens and closes, the feed flow rate and thus tank level increases and decreases in a predictable fashion. In this case, a single loop structure provides acceptable level control performance.

Thought Experiment #2: Assume that our feed valve is set in a fixed position and the header pressure starts rising. Just like squeezing harder on a spray bottle, the valve position can remain constant yet the rising pressure will cause the flow rate through the fixed valve opening to increase.

Thought Experiment #3: Now assume that the header pressure starts to rise at the same moment that the controller determines that the liquid level in our tank is too high. The controller can be closing the feed valve, but because header pressure is rising, the flow rate through the valve can actually be increasing.

As presented in Thought Experiment #3, The changing header pressure (a disturbance) can cause a contradictory outcome that can confound the controller and degrade control performance.

A Cascade Control Solution

For high performance disturbance rejection, it is not valve position, but rather, feed flow rate that must be adjusted to control liquid level.

Because header pressure changes, increasing feed flow rate by a precise amount can sometimes mean opening the valve a lot, opening it a little, and because of the changing header pressure, perhaps even closing the valve a bit.

Below is a classic level-to-flow cascade architecture. An inner secondary sensor measures the feed flow rate. An inner secondary controller receives this flow measurement and adjusts the feed flow valve.

Level-to-Flow Cascade Control

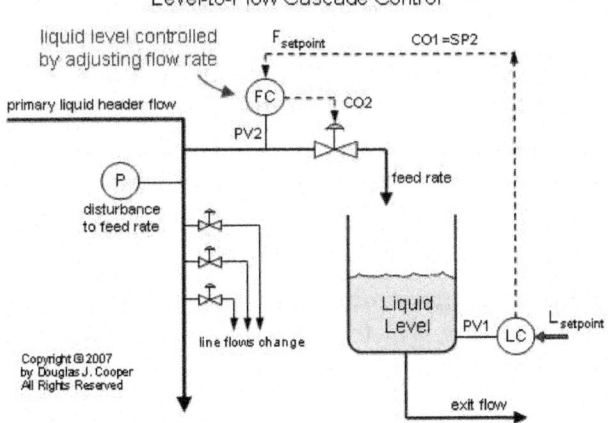

With this cascade structure, if liquid level is too high, the primary level controller now calls for a decreased liquid feed flow rate rather than simply a decrease in valve opening. The flow controller then decides whether this means opening or closing the valve and by how much.

Note in the diagram that, true to a cascade, the level controller output signal (CO1) becomes the set point for the flow controller (SP2).

Header pressure disturbances are quickly detected and addressed by the secondary flow controller. This minimizes any disruption caused by changing header pressure to the benefit of our primary level control process.

The Level-to-Flow Cascade Block Diagram

As shown in the block diagram below, our level-to-flow cascade fits into our block diagram structure. As required, there are:

- Two controllers – the outer primary level controller (LC) and inner secondary feed flow controller (FC)

- Two measured process variable sensors – the outer primary liquid level (PV1) and inner secondary feed flow rate (PV2)

- One final control element (FCE) – the valve in the liquid feed stream

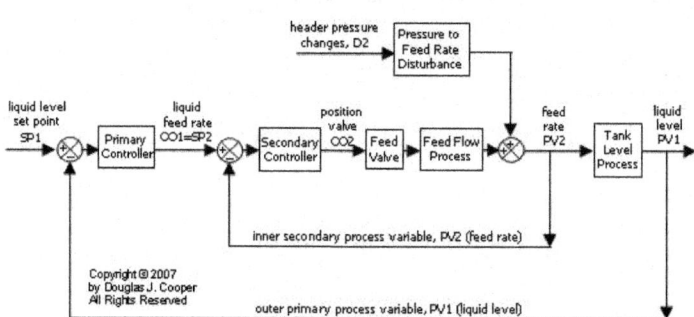

Formal Level-to-Flow Cascade Structure

As required for a successful design, the inner secondary flow control loop is nested inside the primary outer level control loop. That is:

- The feed flow rate (PV2) responds *before* the tank level (PV1) when the header pressure disturbs the process or when the feed valve moves.

- The output of the primary controller, CO1, is wired such that it becomes the set point of the secondary controller, SP2.

- Ultimately, level measurement, PV1, is our process variable of primary concern. Protecting PV1 from header pressure disturbances is the goal of the cascade.

Design and Tuning

The inner secondary and outer primary controllers are from the PID family of algorithms. We have explored the design and tuning of these controllers, implementing a cascade builds on many familiar tasks.

There are a number of issues to consider when selecting and tuning the controllers for a cascade. We explore next an implementation recipe for cascade control.

AN IMPLEMENTATION RECIPE FOR CASCADE CONTROL

When improved disturbance rejection performance is our goal, one benefit of a cascade control (nested loops) architecture over a feed forward strategy is that implementing a cascade builds upon our existing skills.

The cascade block diagram is presented. As shown, a cascade has two controllers. Implementation is a familiar task because the procedure is essentially to employ our controller design and tuning recipe twice in sequence.

The Cascade Structure is a Secondary Loop Inside a Primary Loop

The cascade architecture variables listed in the block diagram include:

CO2 = inner secondary controller output signal to the FCE

PV2 = the early warning inner secondary process variable

SP2 = CO1 = inner secondary set point equals the outer primary controller output

PV1 = outer primary process variable

SP1 = outer primary set point

D2 = disturbances that impact the early warning PV2 before they impact PV1

FCE = final control element (*e.g.*, a valve) is continuously adjustable between on/open and off/closed, and that impacts PV2 before it impacts PV1

Two Bump Tests Required

Two bump tests are required to generate the dynamic process response data needed to design and tune the two controllers in a cascade implementation.

- *First the Inner Secondary Controller*

A reliable procedure begins with the outer primary controller in manual mode (open loop) as we apply the design and tuning recipe to the inner secondary controller.

Thus, with our process steady at (or as near as practical to) its design level of operation (DLO), we generate dynamic CO2 to PV2 process response data with either a manual mode (open loop) bump test or a more sophisticated SP driven (closed loop) bump test.

The objective for the inner secondary controller is timely rejection of disturbances D2 based on the measurement of an "early warning" secondary process variable PV2. Good disturbance rejection performance is therefore of fundamental importance for the inner secondary controller.

The output signal of the outer primary controller becomes a continually updated set point for the inner secondary controller (SP2 = CO1). Since we expect the inner secondary controller to respond crisply to these rapidly changing set point commands, it must also be tuned to provide good SP tracking performance.

In the perfect world, we would balance disturbance rejection and set point tracking capability for the inner secondary controller. But we cannot shift our attention to the outer primary controller until we have tested and approved the inner secondary controller performance.

In production processes with streams comprised of gases, liquids, powders, slurries and melts, disturbances are often unmeasured and beyond our ability to manipulate at will. So while we desire to balance disturbance rejection and set point tracking performance for the inner secondary controller, in practice, SP tracking tests tend to provide the most direct route to validating inner secondary controller performance.

- *Then the Outer Primary Controller*

Once implemented, the inner secondary controller literally becomes part of the "process" from the outer primary controller's view.

As a result, *any* alteration to the inner secondary controller (*e.g.*, tuning parameter adjustments, algorithm modifications, sample time changes) can change the process gain, Kp, time constant, T_p, and/or dead time, θ_p, of the outer loop CO1 to PV1 dynamic response behavior. This, in turn, impacts the design and tuning of the outer primary controller.

Thus, we must design, tune, test, accept and then "lock down" the inner secondary controller, leaving it in automatic mode with a fixed configuration. Only then, with our process steady and at (or very near) its DLO, can we proceed with the second bump test to complete the design and tuning of the outer primary controller.

Software Provides Benefit

Given that the outer primary controller design and tuning is based on the specifics of the inner secondary loop, a guess and test approach to a cascaded implementation can prove remarkably wasteful, time consuming and expensive.

In production operations, a commercial software package that automates the controller design and tuning tasks will pay for itself as early as a first cascade tuning project.

Minimum Criteria for Success

A successful cascade implementation requires that early warning process variable PV2 respond *before* outer primary PV1 both to disturbances of concern (D2) *and* to final control element (FCE) manipulations (*e.g.*, a valve).

Responding first to disturbances means that the inner secondary D2 to PV2 dead time, θp, must be smaller than the overall D2 to PV1 dead time, or:

$$\theta p\ (D2 \rightarrow PV2) < \theta p\ (D2 \rightarrow PV1)$$

Responding first to FCE manipulations means that the inner secondary CO2 to PV2 dead time must be smaller than the overall CO2 to PV1 dead time, or:

$$p\ (CO2 \rightarrow PV2) < p\ (CO2 \rightarrow PV1)$$

If these minimum criteria are met, then a cascade control architecture can show benefit in improving disturbance rejection.

P-Only vs. PI for Inner Secondary Controller

A subtle design issue relates to the choice of control algorithm for the inner secondary controller. While perhaps not intuitive, an inner secondary P-Only controller will provide better performance than a PI controller in many cascade implementations.

Defining Performance

We focus all assessment of control performance on outer primary process variable PV1. Performance is "improved" if the cascade structure can more quickly and efficiently minimize the impact of disturbances D2 on PV1. Given the nature of the cascade structure, it is assumed that D2 first disrupts PV2 as it travels to PV1.

Since PV2 was selected because of its value as an early warning variable, our interest in PV2 control performance extends only to its ability to provide protection to outer primary process variable PV1. We are otherwise unconcerned if PV2 displays offset, shows a large response overshoot, or any other performance characteristic that might be considered undesirable in a traditional measured PV.

Is the Inner Loop "Fast" Relative to the Overall Process?

A cascade architecture with a P-Only controller on the inner secondary loop will provide improved disturbance rejection performance over that achievable with a traditional single loop controller if the minimum criteria for success.

A PI controller on the inner loop *may* provide even better performance than P-Only, but only if the dynamic character of the inner secondary loop is "fast" relative to that of the overall process.

If the inner secondary loop dynamic character is not sufficiently fast, then a PI controller on the inner loop, even if properly designed and tuned, will not perform as well as P-Only. It is even possible that a PI controller could degrade performance to an extent that the cascade architecture performs worse than a traditional (non-cascade) single loop controller.

Why PI controllers Need a "Fast" Inner Loop

At every sample time T, the outer primary controller computes a controller output signal that is fed as a new set point to the inner secondary controller (CO1 = SP2). The inner secondary controller continually makes CO2 moves as it works to keep PV2 equal to the ever-changing SP2.

The cascade will fail if the inner loop cannot keep pace with the rapid-fire stream of SP2 commands. If the inner secondary controller "falls behind" (or more specifically, if the CO2 actions induce dynamics in the inner loop that do not settle quickly relative to the dynamic behavior of the overall process), the benefit of the early warning PV2 measurement is lost.

A P-Only controller can provide energetic control action when tracking set points and rejecting disturbances. Its very simplicity can be a useful attribute in a cascade implementation because a P-Only controller quickly completes its response actions to any control error (E2 = SP2 – PV2). While P-Only controllers display offset when operation moves from the DLO, this is not considered a performance problem for inner secondary PV2.

With two tuning parameters, a PI controller has a greater ability to track set points and reject disturbances. However, this added sophistication yields a controller with a greater tendency to "roll" or oscillate. And the ability to eliminate offset, normally considered a positive attribute for PI controllers, can require a longer series of control actions that extends how quickly a loop settles.

Thus, a PI controller generally needs more time (a faster inner loop) to exploit its enhanced capability relative to that of a P-Only controller. If the dynamic nature of a particular cascade does not provide this time, then an inner-loop P-Only controller is the proper choice.

The Cascade Implementation Recipe

Below is a generally conservative approach for cascade control implementation. Note that the recipe assumes that:

- Early warning PV2 responds *before* PV1 both to D2 *and* CO2 changes as the minimum criteria for cascade success.
- A first order plus dead time (FOPDT) model of dynamic process data yields a process time constant, T_p, that is much larger than the dead time, -p. This permits us to focus on the time constant as the marker for "fast" dynamic process behavior.
  1. Starting from a steady operation, step, pulse or otherwise perturb CO2 around the design level of operation (DLO). Collect CO2, PV2 and PV1 dynamic data as the process variables respond. The inner secondary loop can be in automatic (closed loop) if the controller is <u>sufficiently active</u> to

force a clear dynamic response in PV2. The outer primary loop is normally in manual (open loop).

2. Fit a FOPDT model to the inner secondary process ($CO_2 \to PV2$) data and another to the overall process ($CO_2 \to PV1$) data, yielding:

	Inner Secondary	Overall Process
Process Gain (how far)	Kp ($CO_2 \to PV2$)	Kp ($CO_2 \to PV1$)
Time Constant (how fast)	Tp ($CO_2 \to PV2$)	Tp ($CO_2 \to PV1$)
Dead Time (how much delay)	-p ($CO_2 \to PV2$)	-p ($CO_2 \to PV1$)

Note from the block diagram at the top of the chapter that the inner secondary process dynamics are contained within the overall process dynamics. Therefore, the physics of a cascade implies that the time constant and dead time values for the inner process will not be greater than those of the overall process, or:

$$Tp \ (CO_2 \to PV2) \le Tp \ (CO_2 \to PV1) \text{ and}$$

$$p \ (CO_2 \to PV2) \le p \ (CO_2 \to PV1)$$

3. Use the time constants to decide whether the inner secondary dynamics are fast enough for a PI controller on the inner loop.

Case 1: If the inner secondary process is not at least 3 times faster than the overall process, it is not fast enough for a PI controller.

- If $3 \times Tp \ (CO_2 \to PV2) > Tp \ (CO_2 \to PV1)$ => Use P-Only controller

Case 2: If the inner process is 3 to 5 times faster than the overall process, then P-Only will perform similar to PI control. Use our own preference.

- If $3 \times Tp \ (CO_2 \to PV2) \le Tp \ (CO_2 \to PV1) \le 5 \times Tp \ (CO_2 \to PV2)$ => Use either P-Only or PI controller

Case 3: If the inner process is more than 5 times faster than the overall process, it is "fast" and a PI controller will provide improved performance.

- If $5 \times Tp \ (CO_2 \to PV2) < Tp \ (CO_2 \to PV1)$ => Use PI controller

4. When we have determine whether the inner secondary controller should be a P-Only or PI algorithm, we tune it and test it. Once acceptable performance has been achieved, leave the inner secondary controller in automatic; it now literally becomes part of the outer primary process.

5. Select an algorithm with integral action for the outer primary controller (PI, PID orPID with CO filter) to ensure that offset is eliminated. Tune the primary controller using the design and tuning recipe. Note that bumping the outer primary process requires stepping, pulse or otherwise perturbing the set point (SP2) of the inner secondary controller.

6. With both controllers in automatic, 3 tuning of the cascade is complete.

Cascade Control of the Jacketed Stirred Reactor

Once a cascade control architecture is put in service, we must remember that every time the inner secondary controller is changed in any way, the outer primary controller should be reevaluated for performance and retuned as necessary.

We should also be aware that the recipe presented above is a general procedure intended for broad application. Thus, there will be occasional exceptions to the rules.

A CASCADE CONTROL ARCHITECTURE FOR THE JACKETED STIRRED REACTOR

Our control objective for the jacketed stirred reactor process is to minimize the impact on reactor operation when the temperature of the liquid entering the cooling jacket changes. We have previously explored the modes of operation and dynamic CO-to-PV behavior of the reactor. We also have established the performance of a single loopPI controller and a PID with CO Filter controller in this disturbance rejection application.

Here we consider a cascade architecture as a means for improving the disturbance rejection performance in thejacketed stirred reactor.

The Single Loop Jacketed Stirred Reactor

The reactor exit stream temperature is controlled by adjusting the flow rate of cooling liquid through an outer shell (or cooling jacket) surrounding the main vessel.

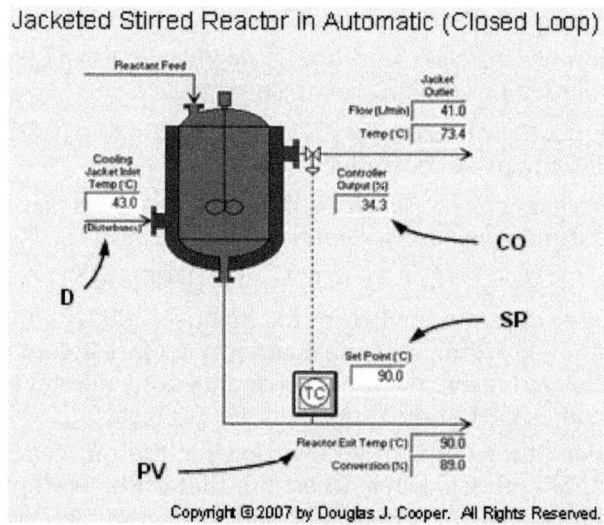

Jacketed Stirred Reactor in Automatic (Closed Loop)

As labeled above for the single loop case:

CO = signal to valve that adjusts cooling jacket liquid flow rate (controller output, %)

PV = reactor exit stream temperature (measured process variable, °C)

SP = desired reactor exit stream temperature (set point, °C)

D = temperature of cooling liquid entering the jacket (major disturbance, °C)

The control objective is to maintain the reactor exit stream temperature (PV) at set point (SP) in spite of unmeasured changes in the temperature of cooling liquid entering the jacket (D).

We measure exit stream temperature with a sensor and transmit the signal to a temperature controller. After comparing SP to PV, the temperature controller computes and transmits a CO signal to the cooling jacket liquid flow valve.

As the valve opens and closes, the flow rate of liquid through the jacket increases and decreases. Like holding a hot frying pan under a water faucet, higher flow rates of cooling liquid remove more heat. Thus, a higher flow rate of cooling liquid through the jacket cools the reactor vessel, lowering the reactor exit stream temperature.

Problems with Single Loop Control

The single loop architecture in the diagram above attempts to achieve our control objective by adjusting the flow rate of cooling liquid through the jacket.

If the measured temperature is higher than set point, the controller signals the valve to increase cooling liquid flow by an appropriate percentage with the expectation that this will decrease reactor exit stream temperature accordingly.

The temperature of the cooling liquid entering the jacket (D) can change, sometimes rather quickly. This can disrupt reactor operation as reflected in the measured reactor exit stream temperature PV.

So reactor exit stream temperature PV is a function of two variables:

- cooling liquid flow rate, and
- the temperature of the cooling liquid entering the cooling jacket (D).

To explore this, we conduct some thought experiments:

Thought Experiment #1: Assume that the temperature of the cooling liquid entering the jacket (D) is constant over time. If the cooling liquid flow rate increases by a certain amount, the reactor exit stream temperature will decrease in a predictable fashion (and vice versa). Thus, a single loop structure should provide good temperature control performance.

Thought Experiment #2: Assume that the temperature of cooling liquid entering the jacket (D) starts rising over time. A warmer cooling liquid can carry away less heat from the vessel. If the cooling liquid flow rate is constant through the jacket, the reactor will experience less cooling and the exit stream temperature will increase.

Thought Experiment #3: Now assume that the temperature of cooling liquid entering the jacket (D) starts to rise at the same moment that the reactor exit stream temperature moves above set point. The controller will signal for a cooling liquid flow rate increase, yet because the cooling liquid temperature is rising, the heat removed from the reactor vessel can actually decrease. Until further corrective action is taken, the reactor exit stream temperature can increase.

As presented in Thought Experiment #3, the changing temperature of cooling liquid entering the jacket (a disturbance) can cause a contradictory outcome that can confound the controller and degrade control performance.

Cascade Control Improves Disturbance Rejection

As we established in our study of the cascade control architecture, an essential element for success in a cascade (nested loops) design is the measurement and control of an "early warning" process variable, PV2, as illustrated in the block diagram below.

Cascade Control Requires an Inner Secondary "Early Warning" Variable

Since disruptions impact PV2 first, it provides our "early warning" that a disturbance is heading toward our outer primary process variable, PV1. The inner secondary controller can begin corrective action immediately. And since PV2 responds first to valve manipulations, disturbance rejection can begin before PV1 has been visibly impacted.

A Reactor Cascade Control Architecture

The thought experiments above highlight that it is problematic to control exit stream temperature by adjusting the cooling liquid flow rate.

An approach with potential for "tighter" control is to adjust the temperature of the cooling jacket itself. This provides a clear process relationship in that, if we seek a higher reactor exit stream temperature, we know we want a higher cooling jacket temperature. If we seek a lower reactor exit stream temperature, we want a lower cooling jacket temperature.

Because the temperature of cooling liquid entering the jacket changes, increasing cooling jacket temperature by a precise amount may mean decreasing the flow rate of cooling liquid a lot, decreasing it a little, and perhaps even increasing the flow rate a bit.

A "cheap and easy" proxy for the cooling jacket temperature is the temperature of cooling liquid exiting at the jacket outlet. Hence, we choose this as our inner secondary process variable, PV2, as we work toward the construction of a nested cascade control architecture.

Adding a temperature sensor that measures PV2 provides us the early warning that changes in D, the temperature of cooling liquid entering the jacket, are about to impact the reactor exit stream temperature, PV1.

The addition of a second temperature controller (TC2) completes construction of a jacketed reactor control cascade as shown in the graphic below.

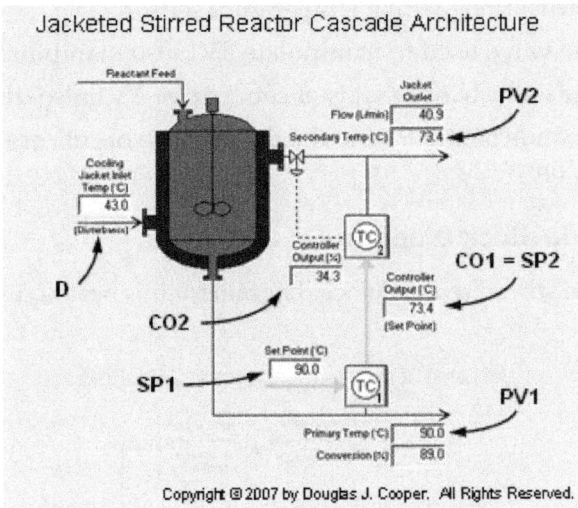

Now, our inner secondary control loop measures the temperature of cooling liquid exiting at the jacket outlet (PV2) and sends a signal (CO2) to the valve adjusting cooling jacket flow rate. The valve increases or decreases the flow rate of cooling liquid if the jacket temperature needs to fall or rise, respectively.

Our outer loop maintains reactor exit stream temperature (our process variable of primary interest and concern) as PV1. Note in the graphic above that the controller output of our primary controller, CO1, becomes the set point of our inner secondary controller, SP2.

If PV1 needs to rise, the primary controller signals a higher set point for the jacket temperature (CO1 = SP2). The inner secondary controller then decides if this means opening or closing the valve and by how much.

Thus, variations in the temperature of cooling liquid entering the jacket (D) are addressed quickly and directly by the inner secondary loop to the benefit of PV1.

The cascade architecture variables are identified on the above graphic and listed below:

PV2 = cooling jacket outlet temperature is our "early warning" process variable (°C)

CO2 = controller output to valve that adjusts cooling jacket liquid flow rate (%)

SP2 = **CO1** = desired cooling jacket outlet temperature (°C)

PV1 = reactor exit stream temperature (°C)

SP1 = desired reactor exit stream temperature (°C)

D = temperature of cooling liquid entering the jacket (°C)

The inner secondary PV2 (cooling jacket outlet temperature) is a proper early warning process variable because:

- PV2 is measurable with a temperature sensor.
- The same valve used to manipulate PV1 also manipulates PV2.
- The same disturbance that is of concern for PV1 also disrupts PV2.
- PV2 responds before PV1 to the disturbance of concern and to valve manipulations.

Reactor Cascade Block Diagram

The jacketed stirred reactor block diagram for this nested cascade architecture is shown below.

Jacketed Stirred Reactor Cascade Block Diagram

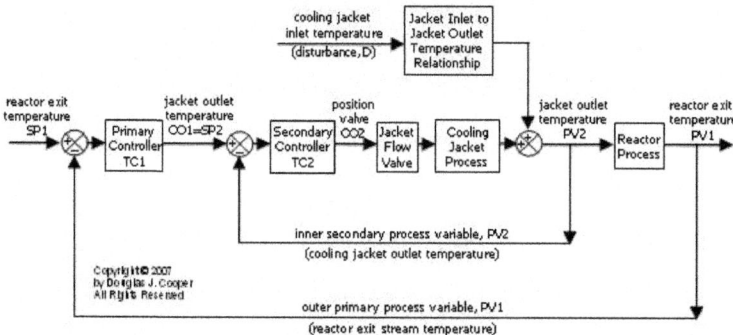

As expected for a nested cascade, this architecture has:

- two controllers (an inner secondary and outer primary controller)
- two measured process variable sensors (an inner PV2 and outer PV1)
- only one valve (to adjust cooling liquid flow rate)

Tuning a Cascade

With a cascade architecture established, we apply our implementation recipe for cascade control and explore the disturbance rejection capabilities of this structure.

CASCADE DISTURBANCE REJECTION IN THE
JACKETED STIRRED REACTOR

Our control objective for the jacketed stirred reactor is to maintain reactor exit stream temperature at set point in spite of disturbances caused by a changing cooling liquid temperature entering the vessel jacket. In previous articles, we have established the design level of operation for the reactor and explored the performance of a single loop PI controller and a PID with CO Filter controller in meeting our control objective.

We also have proposed a cascade control architecture for the reactor that offers potential for improving disturbance rejection performance. We now apply our proposed architecture following the implementation recipe for cascade control. Our goal is to demonstrate the implementation procedure, understand the benefits and drawbacks of the method, and explore cascade disturbance rejection performance for this process.

The Reactor Cascade Control Architecture

The reactor cascade architecture shown in the graphic below:

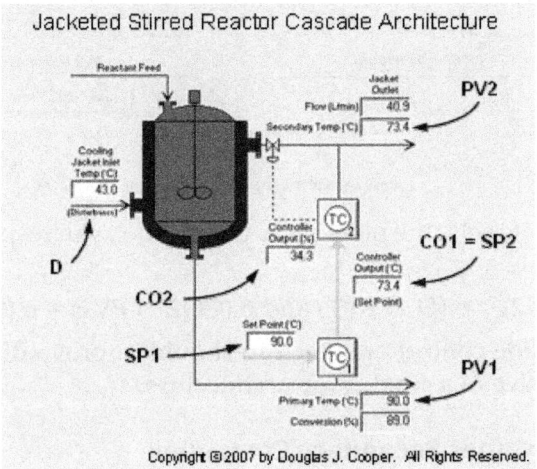

Jacketed Stirred Reactor Cascade Architecture

where:

CO2 = controller output to valve that adjusts cooling jacket liquid flow rate (%)

PV2 = cooling jacket outlet temperature is our "early warning" process variable (°C)

SP2 = CO1 = desired cooling jacket outlet temperature (°C)

PV1 = reactor exit stream temperature (°C)

SP1 = desired reactor exit stream temperature (°C)

D = temperature of cooling liquid entering the jacket (°C)

Design Level of Operation (DLO)

For our DLO are presented in this chapter and are summarized:

- Design PV1 and SP1 = 90 °C with approval for brief dynamic testing of ±2 °C
- Design D = 43 °C with occasional spikes up to 50 °C

Minimum Criteria for Success

A successful cascade implementation requires that early warning process variable PV2 respond *before* outer primary PV1 both to changes in the jacket cooling liquid temperature disturbance (D) *and* to changes in inner secondary controller output signal CO2. The plots below verify that both of these criteria are met with the architecture shown in the graphic above.

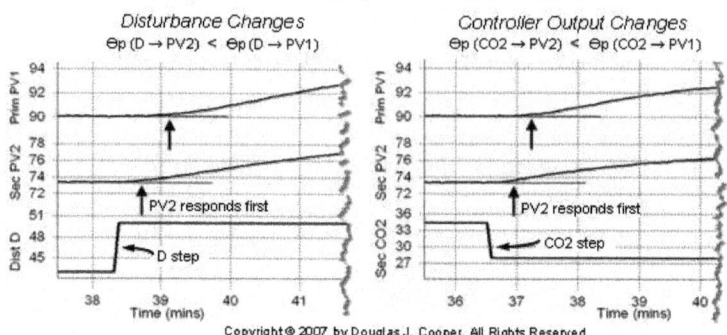

Verify That PV2 Responds Before PV1 To:

Expressed concisely, the plots show that the delay in response (or dead time, θp) follows the rule:

$$p\ (D \rightarrow PV2) < p\ (D \rightarrow PV1)\ \text{and}\ p\ (CO2 \rightarrow PV2) < p\ (CO2 \rightarrow PV1)$$

Thus, a cascade control architecture should improve disturbance rejection performance relative to a single loop architecture.

P-Only vs. PI for Inner Secondary Controller

The cascade implementation recipe first helps us decide if the inner secondary controller is fast enough for a PI algorithm or if it is better suited for a P-Only algorithm. The decision is based on the size of the inner secondary time constant relative to that of the overall process time constant.

To compute these time constants, we need to bump the process and analyze the dynamic process response data. In this example we choose to place both inner secondary and outer primary controllers in manual mode (open loop) during the bump test. An alternative not explored here is to have the inner secondary controller in automatic mode with tuning <u>sufficiently active</u> to force a clear dynamic response in PV2.

Following the steps of the cascade implementation recipe:

1. With the process steady at the design level of operation (DLO), we perform adoublet test, choosing here to move CO2 from 34.3% → 39.3 % → 29.3% → 34.3%. We record both PV2 and PV1 dynamic data as the process responds. We use this "step 1 data set" as we proceed with the implementation recipe.

2. As shown in the plots below, we use commercial software to fit a first order plus dead time (FOPDT) model to the inner secondary process (CO2 −› PV2) dynamic data and another to the overall process (CO2−› PV1) dynamic data:

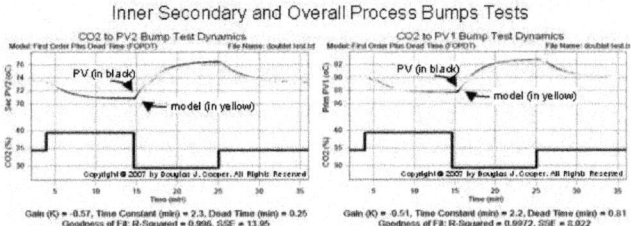

Inner Secondary and Overall Process Bumps Tests

The FOPDT model parameters from these bump tests are summarized:

	Inner Secondary	Overall Process
	(CO2 −› PV2)	(CO2 −› PV1)
Process gain, Kp =	− 0.57 °C/%	− 0.51 °C/%
Time constant, Tp =	2.3 min	2.2 min
Dead time, p =	0.25 min	0.81 min

The cascade block diagram implies that the time constant and dead time values for the inner process are contained within and contribute to the dynamics of the overall process. Thus, we can surmise that:

Tp (CO2 −› PV2) ≤ Tp (CO2 −› PV1) and p (CO2 −› PV2) ≤ p (CO2 −› PV1)

Yet the values in the table above indicate that this seemingly fundamental relationship does not hold true for the jacketed stirred reactor. In fact, when using a FOPDT model approximation of the dynamic response data, the time constant of the inner secondary process is slightly larger (longer) than that of the overall process.

The process graphic at the top of this chapter shows a main reactor vessel with a large volume of liquid relative to that of the cooling jacket. The function of the cooling jacket is to remove heat energy to regulate reactor vessel temperature. It seems logical that because of the volume differences, as long as the liquid temperature in the large reactor is changing, the temperature of the liquid in the small cooling jacket must follow. And the liquid in the reactor acts as a heat source because the <u>chemical reaction inside the vessel generates heat energy faster</u> as reactor temperature rises (and vice versa).

This intertwined relationship of heat generation and removal combined with relative sizes of the reactor and jacket offers one physical rationalization as to why the jacket (inner secondary) Tp might reasonably be longer than that of the vessel (overall process) Tp.

A simpler explanation is that the sensor used to measure temperature in the cooling jacket outlet flow was improperly specified at the time of purchase, and unfortunately, it responds slowly to actual cooling liquid changes. This additional response time alone could account for the observed behavior.

In any case, the dead time of the overall process is three times that of the inner secondary controller, and thus, PV2 provides a clear early warning that we can exploit for a cascade design. So while the simple cascade recipe has limitations that require our judgment, it provides valuable insights that enable us to proceed.

3. The cascade implementation recipe uses a time constant comparison to decide whether the inner secondary loop is fast enough for a PI controller. We reach a true statement with case 1 of the decision tree in step 3 of the recipe. That decision is:

if $3 \times Tp$ (CO2 $\rightarrow$ PV2) > Tp (CO2 $\rightarrow$ PV1) => Use P-Only controller

or using the parameters in the table above:

if 3(2.3 min) > 2.2 min => Use a P-Only controller

Inner Secondary Controller

4. Our cascade implementation recipe states that a P-Only algorithm is the best choice for the inner secondary controller. To explore this decision, we run four trials and compare P-Only and PI algorithms side-by-side. We are able to use the same "step 1 data set" for the design and tuning of all four of these inner secondary controllers:

TRIAL 1: moderate P-Only

TRIAL 2: aggressive P-Only

TRIAL 3: moderate PI

TRIAL 4: aggressive PI

As listed in the table in step 2 above, a first order plus dead time (FOPDT) model fit of the "step 1 data set" collected around our design level of operation (DLO) yields inner secondary FOPDT model parameters:

CO2 $\rightarrow$ PV2 model: Kp = -0.57 °C/%; Tp = 2.3 min; θp = 0.25 min

Following our controller design and tuning recipe, we use these FOPDT model parameters in rules and correlations to complete the secondary controller design.

P-Only Controller design and tuning, including the use of the ITAE tuning correlation for computing controller gain, Kc, from FOPDT model parameters. Following those details:

P-Only algorithm:

CO = CObias + Kc·e(t)

TRIAL 1: Moderate P-Only

$$Kc = \frac{0.2}{Kp}\left(\frac{Tp}{\theta p}\right)^{1.22} = \frac{0.2}{-0.57}\left(\frac{2.3}{0.25}\right)^{1.22} = -5.3\,\%/°C$$

TRIAL 2: Aggressive P-Only

Kc = 2.5 (Moderate Kc) = 2.5 (-5.3) = - 13.3 %/°C

PI Controller design and tuning, including computing controller gain, Kc, and reset time, Ti, from FOPDT model parameters. Following those details:

PI algorithm:

$$CO = CO_{bias} + Kc \cdot e(t) + \frac{Kc}{Ti} \int e(t)dt$$

TRIAL 3: Moderate PI

Moderate Tc = the larger of $1 \cdot Tp$ or $8 \cdot \theta p$

$\qquad$ = larger of 1(2.3 min) or 8(0.25 min)

$\qquad$ = 2.3 min

$$Kc = \frac{1}{Kp} \frac{Tp}{(\theta p + Tc)} = \frac{1}{-0.57} \frac{2.3}{(0.25 + 2.3)} = -1.6\%/°C$$

$Ti = Tp = 2.3$ min

TRIAL 4: Aggressive PI

Aggressive Tc = the larger of $0.1 \cdot Tp$ or $0.8 \, \theta p$

$\qquad$ = larger of 0.1(2.3 min) or 0.8(0.25 min)

$\qquad$ = 0.23 min

$$Kc = \frac{1}{Kp} \frac{Tp}{(\theta p + Tc)} = \frac{1}{-0.57} \frac{2.3}{(0.25 + 2.3)} = -8.4\%/°C$$

$Ti = Tp = 2.3$ min

The "step 1 data set" model parameters and the four inner secondary controllers designed from this data are summarized in the upper half of the table below:

Inner Secondary CO2 Bump Test: from 34.3% → 39.3% → 29.3% → 34.3%			
FOPDT CO2→ PV2 Model: Kp = -0.57 °C/%; Tp = 2.3 min; Θp = 0.25 min			

	TRIAL 1	TRIAL 2	TRIAL 3	TRIAL 4
Inner Secondary P-Only and PI Controllers from above inner FOPDT model				
Kc [=] %/°C	Moderate P-Only	Aggressive P-Only	Moderate PI	Aggressive PI
Ti [=] min	Kc = -5.3	Kc = -13.3	Kc = -1.6; Ti = 2.3	Kc = -8.4; Ti = 2.3
Outer Primary FOPDT CO1=SP2 → PV1 Models using above inner controllers				
Kp (°C/°C)	0.70	0.82	0.90	0.93
Tp (min)	0.61	0.43	2.2	0.50
Θp (min)	0.63	0.48	0.82	0.62
Outer Primary PI Controllers with moderate tuning using above FOPDT models				
Kc [=]°C/°C	Moderate PI	Moderate PI	Moderate PI	Moderate PI
Ti [=] min	Kc =0.15 Ti =0.61	Kc =0.12 Ti =0.43	Kc =0.33 Ti =2.2	Kc =0.10 Ti =0.50

Outer Primary Controller

5. We normally would implement one inner secondary controller and test it for acceptable performance. Here, we "accept" each of the four trial controllers and turn our attention to the outer primary loop.

The four inner secondary controllers are thus implemented one after the other in a series of studies. When in automatic mode, each literally becomes part of the overall process. And because each is different, we must perform four separate bump tests and compute four sets of FOPDT model parameters to describe the four different outer primary (overall) dynamic process behaviors.

Recall that the output of the outer primary controller output, CO1, becomes the set point of the inner secondary controller, SP2. Thus, bumping CO1 is the same as bumping SP2 (*i.e.*, CO1=SP2).

With each of the four inner secondary trial controllers in automatic, we choose to bump CO1=SP2 from 73.4 °C −>76.4 °C −> 70.4 °C −> 73.4 °C. We again use commercial software to fit a FOPDT dynamic model to the CO1=SP2 −>PV2 dynamic response data sets as shown in the plots below. The FOPDT model parameters (Kp, Tp, θp) from each bump test fit are summarized in the table above.

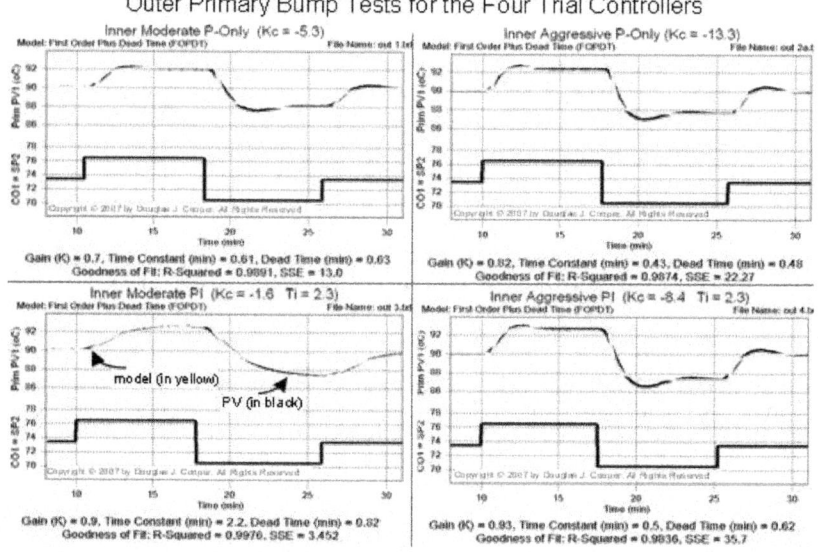

Outer Primary Bump Tests for the Four Trial Controllers

We find that industry practitioners, when designing a strategy for an application as challenging as reactor temperature control, generally seek a moderate response performance. Integral action will always be included to eliminate offset issues. Thus, we pair each inner secondary trial controller with a moderately tuned outer primary PI controller.

Following the identical procedure detailed in trial 3 of step 4 above, we compute four sets of moderate PI controller tuning parameters (one for each inner secondary controller). These are listed in the last row of the above table.

Compare Disturbance Rejection Performance

With both controllers in automatic, design and implementation of the cascade is complete. The objective of this cascade is to minimize the disruption to primary process variable PV1 when disturbance D changes.

The specific D of concern in this study is that the temperature of cooling liquid entering the jacket, normally at 43 °C, is known to spike occasionally to 50 °C.

The lower trace in the plot below shows disturbance steps (temperature changes) from 43 °C up to 50 °C and back. There are four trials shown, one for each of the inner secondary and outer primary controller pairs listed in the table above.

The middle portion of the plot shows the constantly moving inner secondary SP2=C01 in gold and the ability of early warning PV2 to track this ever-changing set point in black.

Recall that offset for early warning variable PV2 is not a concern in many cascade implementations. Here, for example, PV2 is cooling liquid temperature and cooling liquid is not a product destined for market. So our central focus is on how control actions based on this early warning variable help us minimize disturbance disruptions to PV1 and not on how precisely PV2 tracks set point.

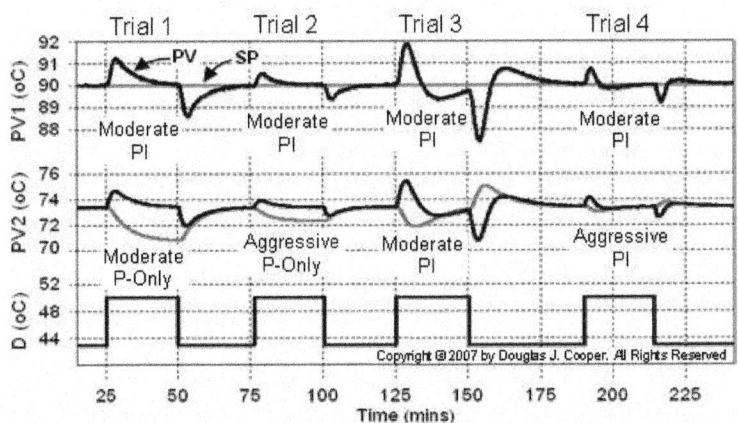

Disturbance Rejection Performance of Reactor Cascade

Our success is shown on the upper portion of the plot. The outer primary set point, SP1 (in gold) remains constant throughout the study. Our interest is in the ability of each of the cascade implementations to maintain reactor exit temperature PV1 at SP1 in spite of the abrupt disturbance changes.

Some Observations

In general, a more successful (or "better") cascade performance is one where:

- There is a smaller maximum deviation from set point during the disturbance, and

- PV1 most rapidly returns to and settles out at set point after a disturbance.

- Trials 2 and 4 both have aggressively tuned inner secondary controllers, and these two implementations both have the smallest deviations from set point and settle most rapidly back to SP. This supports the notion that *inner secondary controllers should energetically attack early warning PV2 disruptions* for best cascade performance.

- While an aggressively tuned inner secondary P-Only and PI controller (trials 2 and 4) performed with similar success, the moderately tuned inner secondary PI controller (trial 3) displayed markedly degraded performance. This high sensitivity to inner loop PI tuning strengthens the "use P-Only" conclusion made in step 3 of our cascade implementation recipe.

Comparing Cascade to Single Loop

The central question is whether the extra effort associated with cascade control provides sufficient payoff in the form of improved disturbance rejection performance.

To the left in the plot below is the performance of our trial 2 cascade implementation. The performance of an aggressively tuned single loop PI controller in rejecting the same disturbance is shown to the right.

Cascade vs Single Loop Disturbance Rejection Performance

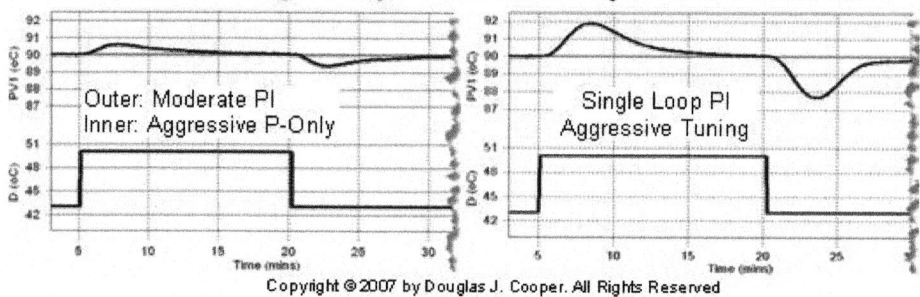

The cascade architecture reduces the maximum deviation from SP during the disturbance from ±2 °C for the single loop controller down to ±0.5 °C for the cascade. Settling time is shortened from about 10 minutes for the single loop controller down to about 8 minutes for the cascade.

If the financial return from such improved performance is greater than the cost to install and maintain the cascade, then choose cascade control.

Set Point Tracking Performance

While not our design objective, presented below is the set point tracking performance of the four cascade implementations:

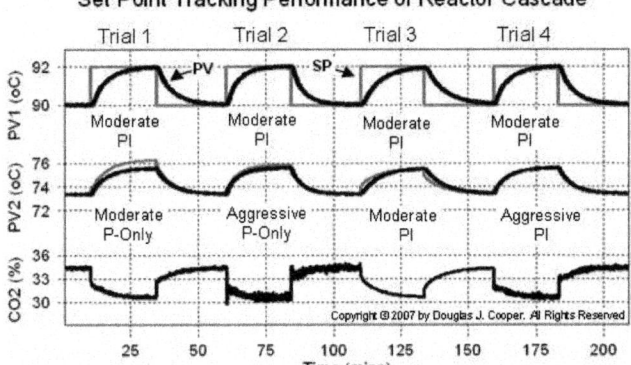

Cascade control is best suited for improved disturbance rejection. As shown above, its impact on set point tracking performance is minimal. While one might argue that our "best" cascade design for disturbance rejection (trial 2) also provides the most rapid SP tracking response, this same improvement can be obtained with more aggressive tuning of a single loop PI controller.

THE FEED FORWARD CONTROLLER

The most popular architectures for improved disturbance rejection performance are cascade control and the "feedfor ward with feedback trim" architecture introduced below.

Like cascade, feed forward requires that additional instrumentation be purchased, installed and maintained. Both architectures also require additional engineering time for strategy design, implementation and tuning.

Cascade control will have a small impact on set point tracking performance when compared to a traditional single-loop feedback design and this may or may not be considered beneficial depending on the process application. The feed forward element of a "feed forward with feedback trim" architecture does not impact set point tracking performance in any way.

Feed Forward Involves a Measurement, Prediction and Action

Consider that a process change can occur in another part of our plant and an identifiable series of events then leads that "distant" change to disturb or disrupt our measured process variable, PV.

The traditional PID controller takes action only when the PV has been moved from set point, SP, to produce a controller error, $e(t) = SP - PV$. Thus, disruption to stable operation is already in progress before a feedback controller first begins to respond. From this view, a feedback strategy simply starts too late and at best can only work to minimize the upset as events unfold.

In contrast, a feed forward controller measures the disturbance, D, while it is still distant. As a feed forward element receives the measured D, uses it to predict an impact on PV, and then computes preemptive control actions, CO feed forward, that counteract the predicted impact as the disturbance arrives. The goal is to maintain the process variable at set point (PV = SP) throughout the disturbance event.

where:

CO = controller output signal

D = measured disturbance variable

e(t) = controller error, SP – PV

FCE = final control element (*e.g.*, valve, variable speed pump or compressor)

PV = measured process variable

SP = set point

When to Consider Cascade Control

The cascade architecture requires that an "early warning" secondary measured process variable, PV2, be identified that is inside (responds before) the primary measured process variable, PV1. Essential elements for success include that:

- PV2 is measurable with a sensor.
- The same final control element (FCE) used to manipulate PV1 also manipulates PV2.
- The same disturbances that are of concern for PV1 also disrupt PV2.
- PV2 responds *before* PV1 to disturbances of concern and to FCE manipulations.

One benefit of a cascade architecture is that it uses two traditional controllers from the PID family, so implementation is a familiar task that builds upon our existing skills. Also, cascade control will help improve the rejection of *any* disturbance that first disrupts the early warning variable, PV2, prior to impacting the primary process variable, PV1.

When to Consider Feed Forward with Feedback Trim

Feed forward anticipates the impact of a measured disturbance on the PV and deploys control actions to counteract the impending disruption in a timely fashion. This can significantly improve disturbance rejection performance, but only for the particular disturbance variable being measured.

Feed forward with feedback trim offers a solution for improved disturbance rejection if no practical secondary process variable, PV2, can be established (*i.e.*, a process variable cannot be located that is measureable, provides an early warning of impending disruption, and responds first to FCE manipulations).

Feed forward also has value if our concern is focused on one specific disturbance that is responsible for repeated, costly disruptions to stable operation. To provide benefit, the additional measurement must reveal process disturbances *before* they arrive at our PV so we have time to compute and deploy preemptive control actions.

The Feed-Forward-Only Controller

Pure feed-forward-only controllers are rarely found in industrial applications where the process flow streams are composed of gases, liquids, powders, slurries or melts. Nevertheless, we explore this idea using a thought experiment on the shell-and-tube heat exchanger simulation and available for exploration and study in commercial software.

The PV to be controlled is the exit temperature on the tube side of the exchanger. To regulate this exit temperature, the CO signal adjusts a valve to manipulate the flow rate of cooling liquid on the shell side. A side stream of warm liquid combines with the hot liquid entering the exchanger and acts as a measured disturbance, D, to our process.

Because there is no feedback of a PV measurement in our controller architecture, feed-forward-only presents the interesting notion of open loop control. As such, it does not have a tendency to induce oscillations in the PV as can a poorly tuned feedback controller.

If we could mathematically describe how each change in D impacts PV (D $\rightarrow$ PV) and how each change in CO impacts PV (CO $\rightarrow$ PV), then we could develop a math model that predicts what manipulations to make in CO to maintain PV at set point whenever D changes.

But this would only be true if:

- we have perfect understanding of the D $\rightarrow$ PV and CO $\rightarrow$ PV dynamic relationships,
- we can describe these perfect dynamic relationships mathematically,
- these relationships never change,
- there are no other unmeasured disturbances impacting PV, and
- set point, SP, is always held constant.

The reality, however, is that with only a single measured D, a predictive model cannot account for many phenomena that impact the D $\rightarrow$ PV and CO $\rightarrow$ PV behavior. These may include changes in:

- the temperature and flow rate of the hot liquid feed that mixes with our warm disturbance stream on the tube side,
- the temperature of the cooling liquid on the shell side,
- the ambient temperature surrounding the exchanger that drives heat loss to the environment,
- the shell/tube heat transfer coefficient due to corrosion or fouling, and
- valve performance and capability due to wear and component failure.

Since all of the above are unmeasured, a fixed or stationary model cannot account for them when it computes control action predictions. Installing additional sensors and enhancing the feed forward model to account for each would improve performance but would lead to an expensive and complex architecture. And since there are more potential disturbances and external influences then those listed above, that still would not be sufficient.

This highlights that feed-forward-only control is problematic and should only be considered in rare instances. One situation where it may offer value is if a PV critical to process operation simply cannot be measured or inferred using currently available technology. Feed-forward-only control, in spite of its weaknesses and pitfalls, then offers some potential for improved operation.

Feed Forward with Feedback Trim

The "feed forward with feedback trim" control architecture is the solution widely employed in industrial practice. It balances the capability of a feed forward element to take preemptive control actions for *one* particularly disruptive disturbance while permitting a traditional feedback control loop to:

- reject all other disturbances and external influences that are not measured,
- provide set point tracking capability, and
- correct for the inevitable simplifying approximations in the predictive model of the feed forward element that make preemptive disturbance rejection imperfect.

To construct the architecture, a feedback controller is first implemented and tested following our controller design and tuning recipe as if it were a stand-alone entity. The feed forward controller is then designed based on our understanding of the relationship between the D −› PV and CO −› PV variables. This is generally expressed as a math function that can range in complexity from a simple multiplier to complex differential equations.

With the architecture completed, the disturbance flow is measured and passed to a feed forward element that is essentially a combination disturbance/process model. The model uses changes in D to predict an impact on PV, and then computes control actions, CO feed forward to compensate for the predicted impact.

Conceptual Feed Forward with Feedback Trim Diagram

The COfeed forward control actions are combined with CO feedback to create an overall control action, CO total, to send to the final control element (FCE).

To illustrate the control strategy in a more tangible fashion, we present the feed forward with feedback trim architecture.

The basis for the feed forward element math function, $f(D) = -(GD/Gp)$.

The more accurate the feed forward math function is in computing control actions that will counteract changes in the measured disturbance in a timely fashion, the less impact those disturbances will have on our measured process variable.

Practitioner's note: a potential application of feed forward control exists if we hear an operator say something like, "Every time event A happens, process variable X is upset. I can usually help the X controller by switching to manual mode and moving the controller output." If the variable associated with event A is already being measured and logged in the process control system, sufficient data is likely available to allow the implementation of a feed forward element to our feedback controller.

Improved disturbance rejection performance comes at a price in terms of process engineering time for model development and testing, and instrument engineering time for control logic programming. Like all projects, such investment decisions are made on the basis of cost and benefit.

FEED FORWARD USES MODELS WITHIN THE CONTROLLER ARCHITECTURE

Both "feed forward with feedback trim" and cascade control can provide improved disturbance rejection performance. They have different architectures, however, and choosing between the two depends on our specific control objective and the ability to obtain certain process measurements.

Feed Forward and Feedback Trim are Largely Independent

The feed forward with feedback trim architecture is constructed by coupling a feed-forward-only controller to a traditional feedback controller.

The feed forward controller seeks to reject the impact of one specific disturbance, D, that is measured *before* it reaches our primary process variable, PV, and starts its disruption to stable operation. Typically, this D is one that has been identified as causing repeated and costly upsets, thus justifying the expense of both installing a sensor to measure it, and developing and implementing the feed forward computation element to counteract it.

where:

CO = controller output signal

D = measured disturbance variable

e(t) = controller error, SP – PV

FCE = final control element (*e.g.*, valve, variable speed pump or compressor)

PV = measured process variable

SP = set point

The feedback controller is designed and tuned like any stand-alone controller from the PID family of algorithms. The only difference is that it must allow for its controller output signal, CO feedback, to be combined with a feed forward controller output signal, CO feed forward, to arrive at a total control action, CO total.

Feed Forward Element Uses a Process and Disturbance Model

The function block element that computes the feed forward controller output signal, CO feed forward, is constructed by combining a process model and disturbance model.

The blocks circled with dotted lines below show where data is collected when developing the process and disturbance models. The feed forward controller *does not* include these circled blocks as separate elements in its architecture.

The process model (CO –› PV) in the feed forward element describes or predicts how each change in CO will impact PV. The disturbance model (D –› PV) describes or predicts how each change in D will impact PV.

In practice, these models can range from simple scaling multipliers (*static* feed forward) through sophisticated differential equations (*dynamic* feed forward). Sophisticated dynamic models can better describe actual process and disturbance behaviors, often resulting in improved disturbance rejection performance. Such models can also be challenging to derive and implement, increasing the time and expense of a project.

Dynamic Feed Forward Based on the FOPDT Model

We first develop a general feed forward element using dynamic models (differential equations). Later, we will explore how we can simplify this general construction into a static feed forward element. Static feed forward is widely employed in industrial practice, in part because it can be implemented with an ordinary multiplying relay that scales the disturbance signal.

A dynamic feed forward element accounts for the "how far" gain, the "how fast" time constant and the "how much delay" dead time behavior of both the process (CO –› PV) and disturbance (D –› PV) relationships.

The simplest differential equation that describes such "how far, how fast, and with how much delay" behavior for either the process or disturbance dynamics is the familiar first order plus dead time (FOPDT) model.

- *The* CO –› PV *Process Model*
- Describing the CO –› PV process behavior with a FOPDT model is not a new challenge. For example, we presented all details as we developed the FOPDT dynamic CO –› PV model for the gravity drained tanks process from step test data as:

Which matches the general FOPDT (first order plus dead time) dynamic model form:

where for a change in CO, the FOPDT model parameters are:

- K_p = process gain (the direction and how far PV will travel)
- T_p = process time constant (how fast PV moves after it begins its response)
- θ_p = process dead time (how much delay before PV first begins to respond)

This equation describes how each change in CO causes PV to respond (CO -› PV) as time, t, passes. Our past modeling experience also includes developing and documenting FOPDT CO -› PV models for the heat exchanger and jacketed reactor processes.

- *The D -› PV Disturbance Model*
- The procedure used to develop the FOPDT (first order plus dead time) process models can be used in an identical fashion to develop a dynamic D -› PV disturbance model. While we do not show the graphical calculations at this point, consider the plot below from the gravity drained tanks process.

Instead of analyzing a plot where a CO step forces a PV response while D remains constant, here we would analyze a D step forcing a PV response while CO remains constant.

We presume that an analogous graphical modeling procedure can be followed to determine the "how far, how fast, and with how much delay" dynamic D -› PV disturbance model:

where for a step change in D:

- K_D = disturbance gain (the direction and how far PV will travel)
- T_D = disturbance time constant (how fast PV moves after it begins its response)
- θ_D = disturbance dead time (how much delay before PV first begins to respond)

Dynamic Feed Forward as a Thought Experiment

A feed forward element typically performs a model computation every sample time, T, to address any changes in measured disturbance, D. To help us visualize events, we discuss this as if it occurs as a two step "prediction and corrective action" procedure for a single disturbance:

1. The D -› PV disturbance model receives a change in the measured value of D and predicts an open-loop or uncontrolled "impact profile" on PV. This includes a prediction of how much delay will pass before the disruption first arrives at PV, the direction PV will travel for this particular D once it begins to respond, and how fast and how far PV will travel before it settles out at a predicted new steady state.

2. The CO -› PV process model then uses this PV impact profile to back-calculate a series of corrective control actions, CO feed forward. These are CO moves sent to the final control element (FCE) to cause an "equal but opposite" response in PV such that it remains at set point, SP, throughout the event. Thus, the CO model seeks to exactly counteract the disruption profile predicted in step 1. The first CO actions are delayed as needed so

they meet D upon arrival at PV. A series of CO actions are then deployed to counteract the predicted disruption over the life of the event.

Even sophisticated dynamic models are too simple to precisely describe the behavior of real processes. So although a feed forward element can dramatically reduce the impact of a disturbance on our PV, it will not provide a perfect "prediction and corrective action" disturbance rejection.

To account for model inaccuracies, the feed forward signal is combined with traditional feedback control action, COfeedback, to create a total controller output, COtotal. Whether it be a P-Only, PI, PID or PID w/ CO Filter algorithm, the feedback controller plays the important role of:

- minimizing the impact of disturbance variables other than D that can disrupt the PV,

- providing set point tracking capability to the overall strategy, and

- correcting for the simplifying approximations used in constructing the feed forward computation element that ultimately makes it imperfect in its actions.

The Sign of CO feed forward Requires Careful Attention

CO feed forward is added as:

$$COtotal = COfeedback + COfeedforward$$

We write the equation this way because it is consistent with typical vendor documentation and standard piping/process & instrumentation diagrams (P&IDs).

The "plus" sign in the equation above requires our careful attention. To understand the caution, consider a case where the D $\to$ PV disturbance model predicts that D will cause the PV to *move up* in a certain fashion or pattern over a period of time. According to our thought experiment above, the CO $\to$ PV process model must compute feed forward CO actions that cause the PV to *move down* in an identical pattern.

But the sign of both the process gain, Kp, and disturbance gain, KD, together determine whether we need to send an increasing or decreasing signal to the FCE to compensate for a particular D. If Kp and KD are both positive or both negative, for example, then as a disturbance D moves up, the COfeedforward signal must move down to compensate. If Kp and KD are of opposite sign, then D and COfeedforward move in the same direction to counteract each other.

We show a standard "plus" sign. But the computed feed forward signal, COfeedforward, will be positive or negative depending on the signs of the process and disturbance gains as just described. This ensures that the impact of the disturbance and the compensation from the feed forward element move in opposite directions to provide improved disturbance rejection performance.

Dynamic Feed Forward in Math

Suppose we define a generic process model, Gp, and generic disturbance model, GD, as:

Gp = generic CO –› PV process model (*describing how a CO change will impact PV*)

GD = generic D –› PV disturbance model (*describing how a D change will impact PV*)

We allow Gp and GD to have forms that can range from simple to the sophisticated.

For example, they can both be pure gain values Kp and KD and nothing more. They can be full FOPDT differential equations that include Kp, T_p and θ_p, and KD, T_D and θ_D. One or the other (or both) can be non-self regulating (integrating), or perhaps self regulating but second or third order.

Leaving the exact form of the models undefined for now, we develop our feed forward element with the following steps:

1) Our generic CO –› PV process model, Gp, allows us to compute a PV response to changes in CO as:

$$PV = Gp \cdot CO$$

With the generic model approach, we can rearrange the above to compute controller output actions that would reproduce a known or specified PV as:

$$CO = (1/Gp) \cdot PV$$

2) Our generic D –› PV disturbance model, GD, lets us compute a PV response to changes in D as:

$$PV = GD \cdot D$$

3) Following the logic in the above thought experiment, we use the D –› PV model of step 2 to predict an impact profile on PV for any measured disturbance D:

$$PVimpact = GD \cdot D$$

4) We then use our rearranged equation of step 1 to back-calculate a series of corrective feed forward control signals that will move PV in a pattern that is opposite (and thus negative in sign) to the predicted PV impact profile from Step 3:

$$COfeedforward = -(1/Gp) \cdot PVimpact$$

5) We finish by substituting the "PVimpact = GD·D" equation of step 3 into the COfeedforward equation of step 4:

$$COfeedforward = -(1/Gp) \cdot (GD \cdot D)$$

and rearrange to arrive at our final feed forward computational element composed of a disturbance model divided by a process model:

$$COfeedforward = -(GD/Gp) \cdot D$$

Note: Above is a math argument that we hope seems reasonable and easy to follow. But please be aware that for such manipulations to be proper, all variables and equations must first be mapped into the Laplace domain using Laplace transforms. The ease with which complex equations can be manipulated in the Laplace domain is a major reason control theorists use it for their derivations.So, to be mathematically correct, we must first cast all variables and models into the Laplace domain as:

$$PV(s) = Gp(s) \cdot CO(s)$$ when the disturbance is constant

and

$$PV(s) = GD(s) \cdot D(s)$$ when the controller output signal is constant

where the Laplace domain models Gp(s) and GD(s) are called transfer functions.

At the end of our derivation, our final feed forward computational element should be expressed as:

$$COfeedforward(s) = - [GD(s)/Gp(s)] \cdot D(s)$$

Thus, while we had omitted important details, our feed forward equation is indeed correct and we will use it going forward. We will continue to downplay the complexities of the math for now as we focus on methods of use to industry practitioners.

Conceptual Feed Forward with Feedback Trim Diagram

As promised in our introductory chapter on the feed forward architecture, we now have the basis for why we express the feed forward element math function, $f(D) = -(GD/Gp)$ as shown in our generalize feed forward with feedback trim.

Implementation and Testing of Feed Forward with Feedback Trim

We next explore the widely used and surprisingly powerful static feed forward controller. We will discover the ease with which we can develop such an architecture, and also explore some of the limitations of this simplified approach.

STATIC FEED FORWARD AND DISTURBANCE REJECTION IN THE JACKETED REACTOR

The purpose of the feed forward controller of the feed forward with feedback trim architecture is to reduce the impact of one specific disturbance, D, on our primary process variable, PV. An additional sensor must be located upstream in our process so we have a disturbance measurement that provides warning of impending disruption. The feed forward element uses this D measurement signal to compute and implement corrective control actions so the disturbance has minimal impact on stable operation.

Here we build on the mathematical foundation of this previous material as we explore the popular and surprisingly powerful static feed forward computational element for this disturbance rejection architecture.

Static Feed Forward Uses the Simplest Model Form

If we define a generic process model, Gp, and generic disturbance model, GD, as:

Gp = generic CO −› PV process model (*describing how a CO change will impact PV*)

GD = generic D −› PV disturbance model (*describing how a D change will impact PV*)

then we can show details to derive a general feed forward computational element as a ratio of the disturbance model divided by the process model:

$$COfeedforward = -(GD/Gp) \cdot D$$

Models Gp and GD can range from the simple to the sophisticated. With static feed forward, we limit Gp and GD to their constant "which direction and how far" gain values:

$$Gp = Kp \text{ (the CO −› PV process gain)}$$

$$GD = KD \text{ (the D −› PV disturbance gain)}$$

And the static feed forward element is thus a simple gain ratio multiplier:

$$COfeedforward = -(KD/Kp) \cdot D \text{ (static feed forward element)}$$

The static feed forward controller does not consider how the controller output to process variable (CO −› PV) dynamic behavior differs from the disturbance to process variable (D −› PV) dynamic behavior.

We do not account for the size of the process time constant, Tp, relative to the disturbance time constant, TD. As a consequence, we cannot compute and deploy a series of corrective control actions over time to match how fast the disturbance event is causing the PV to move up or down.

We do not consider the size of the process dead time, θp, relative to the disturbance dead time, θD. Thus, we cannot delay the implementation of corrective actions to coordinate their arrival with the start of the disturbance disruption on PV.

For processes where the CO −› PV dynamic behavior is very similar to the D −› PV dynamic behavior, like many liquid level processes for example, static feed forward will perform virtually the same as a fully dynamic feed forwardcontroller in rejecting our measured disturbance.

Visualizing the action of this static feed forward element as a two step "prediction and corrective action" procedure for a single disturbance:

1. The D −› PV disturbance gain, KD, receives a change in D and predicts the total final impact on PV. The computation can only account for information contained in KD, which includes the direction and how far PV will ultimately travel in response to the measured D before it settles out at a new steady state.

2. The CO −› PV process gain, Kp, then uses this disturbance impact prediction of "which direction and how far" to back-calculate one CO move as

a corrective control action, COfeedforward. This COfeedforward move is sent immediately to the final control element (FCE) to cause an "equal but opposite" response in PV.

Limited in Capability but (Reasonably) Easy to Implement

The static feed forward element makes one complete and final corrective action for every measured change in D. It does not delay the feed forward signal so it will meet the D impact when it arrives at PV. It does not compute and deploy a series of CO actions to try and counteract a predicted disruption pattern over an event life.

Even with this limited performance capability, the benefit of the static form that makes it popular with industry practitioners is that it is reasonably straightforward to implement in a production environment. We can construct a static feed forward element with:

- a sensor/transmitter to measure disturbance D
- a scaling relay that multiplies signal D by our static feed forward ratio, $(- KD/Kp)$
- a summing junction that adds COfeedforward to COfeedback to produce COtotal

COfeedforward is Normally Zero

An important implementation issue is that COfeedforward should equal zero when D is at its design level of operation (DLO) value. Thus, the D used in our calculations is actually the disturbance signal from the sensor/transmitter (Dmeasured) that has been shifted or biased by the design level of operation disturbance value (DDLO), or:

$$D = Dmeasured - DDLO$$

With this definition, both D and COfeedforward will be zero when the disturbance is at its normal or expected value. Such a biasing capability is included with most all commercial scaling relay function blocks.

Static Feed Forward and the Jacketed Reactor Process

We have previously explored the modes of operation and dynamic CO $\rightarrow$ PV behavior of the jacketed stirred reactor process. We also have established the performance of a single loop PI controller, a PID with CO Filter controller and a cascade control implementation when our control objective is to minimize the impact of a disturbance caused by a change in the temperature of the liquid entering the cooling jacket.

Here we explore the design, implementation and performance of a static feed forward with feedback trim architecture for this same disturbance rejection objective.

Limitations of the Single Loop Architecture

The control objective is to maintain the reactor exit stream temperature (PV) at set point (SP) in spite of changes in the temperature of cooling liquid entering the jacket (D) by adjusting controller output (CO) signals to the cooling liquid flow valve.

A Feed Forward with Feedback Trim Reactor Architecture

Below is the jacketed stirred reactor process with a feed forward with feedback trim controller architecture.

The loop architecture from this commercial software simulation shows that D is measured, scaled and transmitted as COfeedforward to the controller. There it is combined with the traditional feedback signal to produce the COtotal sent to the cooling jacket flow valve. This is a simplified representation of the same feed forward with feedback trim conceptual diagram shown earlier in this chapter.

Design Level of Operation (DLO)

The DLO used in our previous disturbance rejection studies for the jacketed stirred reactor are presented and are summarized:

- Design PV and SP = 90 °C with approval for brief dynamic testing of ±2 °C
- Design D = 43 °C with occasional spikes up to 50 °C

We note that D moves between 43 °C and 50 °C. We seek a single DLO value that lets us conveniently compare results from two different design methods explored below.

We choose here a DDLO as the average value of (43+50)/2 = 46.5 °C and acknowledge that other choices (such as simply using 43 °C) are reasonable. As long as we are consistent in our methodology, the conclusions we draw when comparing the two methods will remain unchanged.

When D = 46.5 °C and CO = 40%, then our measured process variable settles at the design value of PV = 90 °C. This relationship between the three variables explains the DLO values indicated on the plots that follow.

Design Method 1: Compute KD/Kp From Historic Data

Below is a trend from our data historian showing approximately three hours of operation from the jacketed reactor process under PI control. No feed forward controller is active. The set point (SP) is constant and a number of disturbance events force the PV from SP. All variables are near our DLO as described above.

If we recall the definition that $Kp = \Delta PV/\Delta CO$ and $KD = \Delta PV/\Delta D$, then for our static feed forward design:

$$
\begin{aligned}
COfeedforward &= -\,(KD/Kp)\cdot D \\
&= -\,[(\Delta PV/\Delta D)/(\Delta PV/\Delta CO)]\cdot D \\
&= -\,[(\Delta CO/\Delta D)]\cdot D
\end{aligned}
$$

With this last equation, our design challenge is reduced to finding a disturbance change that lasts long enough for the controller output response to settle. If this event occurs reasonably close to our DLO, then we can directly compute our gain ratio feed forward element by measuring the disturbance and controller output changes from a plot.

About the Feedback Controller

The plot shows disturbance rejection performance when the process is using a PI controller tuned for an aggressive response action. The details of the design, tuning and testing of this process and controller combination.

Note that with this "measure from a plot" approach of Method 1, the process can be controlled by any algorithm from the PID family, though integral action must be included to return the PV to SP (eliminate offset) after a disturbance. Also, our feedback controller tuning can range from conservative (sluggish) through an aggressive (active) response without affecting the applicability of the method.

Our interest is limited to finding a ΔD disturbance event with a corresponding ΔCO controller output signal response that lasts long enough for the PV to be returned to SP. Then as shown above, the ΔD and ΔCO relationship can be measured directly from the plot data.

Accounting for Negative Feedback

If we are using automatic mode (closed loop) data as shown in the plot above, we must account for the negative feedback of our controller in our calculations. A controller always takes action that moves the CO signal in a direction that counteracts the developing controller error. Thus, when using automatic mode (closed loop) data as above, we must consider that a negative sign has been introduced into the signal relationship.

On the plot above, we have labeled a ΔD disturbance change with corresponding ΔCO controller output signal response. We introduce the sign change from negative feedback of our controller and compute:

$$COfeedforward = - [(\Delta CO/\Delta D)] \cdot D \cdot (-1 \text{ for negative feedback})$$
$$= [(14\%)/(7 \text{ °C})] \cdot D$$
$$= [2 \text{ %}/\text{°C}] \cdot D$$

Design Method 2: Perform Two Independent Bump Tests

To validate that a feed forward gain ratio of 2 %/°C is a reasonable number as determined from automatic mode (closed loop) data, here we perform two independent step tests, compute individual values for KD and Kp, and then compare the ratio results to Method 1.

This approach is largely academic because the challenges of steadying a real process and then stepping individual parameters in such a perfect fashion is unrealistic in the chaotic world of most production environments. This exercise holds value, however, because it provides an alternate route that confirms the results presented in Method 1.

We follow the established procedure for computing a process gain, Kp, from a manual mode (open loop) step test response plot. That is, we set our disturbance parameter at DDLO, set the controller output at a constant CO value and wait until the PV is steady. We then step CO to force a PV response that is centered around our DLO.

Such a step response plot is shown below for the jacketed stirred reactor. We measure and compute $Kp = \Delta PV/\Delta CO = -0.4$ °C/% as indicated on the plot.

We repeat the procedure to compute a disturbance gain, KD, from an open loop step response plot. Here, we set our CO signal at the DLO value of 40%, set the disturbance parameter at a constant D value and wait until the PV is steady. We then step D to force a PV response that is again centered around our DLO.

Such a disturbance step response plot. We measure and compute $KD = \Delta PV/\Delta D = 0.8$ °C/°C as labeled on the plot.

With values for Kp and KD, we compute our gain ratio feed forward multiplier. Since we are in manual mode (open loop), we need not account for any sign change due to negative feedback in our calculation.

$$COfeedforward = -(KD/Kp) \cdot D$$
$$= -[(0.8 °C/°C)/(-0.4 °C/\%)] \cdot D$$
$$= [2 \%/°C] \cdot D$$

Thus, with careful testing using a consistent and repeatable commercial process simulation, we observe that the practical approach of Method 1 provides the same results as the academic approach of Method 2.

Implement and Test

The all-important question we now consider is whether the extra effort associated with designing and implementing a "static feed forward with feedback trim" architecture provides sufficient payoff in the form of improved disturbance rejection performance.

To the left in the plot below is the performance of a dependent, ideal PI controller with aggressive tuning values for controller gain $Kc = -3.1$ %/°C and reset time $Ti = 2.2$ min.

To the right in the plot above is the disturbance rejection performance of our static feed forward with feedback trim architecture. The feed forward gain ratio multiplier used is the 2 %/°C as determined by two different methods described earlier in this chapter. The feedback controller remains the aggressively tuned PI algorithm.

The static feed forward controller makes one complete preemptive corrective CO action whenever a change in D is detected as noted in the plot above. There is no delay of the feed forward signal based on relative dead time considerations, and there is no series of CO actions computed and deployed based on relative time constant considerations.

Nevertheless, the static feed forward controller is able to reduce the maximum deviation from SP during a disturbance event to half of its original value. The settling time is also reduced, though less dramatically. Like any control project, the operations staff must determine if this represents a sufficient payback for the effort and expenses required.

Practitioner's note: Our decision to add feed forward to a feedback control loop is driven by the character of the disturbance and its effect on our PV. If the controller can react more quickly than the D can change, feed forward is not likely to significantly improve control. However, if the disturbance changes rapidly and fairly often, feed forward control can be a powerful tool to stabilize our process.

No Impact on Set Point Tracking Performance

While not our design objective, presented below is the set point tracking performance of the single loop PI controller compared to that of the static feed forward with feedback trim architecture. The same aggressive PI tuning values used above are maintained for this study.

Feed forward with feedback trim is designed for the improved rejection of one measured disturbance. A feed forward controller has no impact on set point tracking performance when the disturbance is constant. This makes sense since the computed CO feed forward signal does not change unless D changes. Indeed, with no change in the measured disturbance, both architectures provide identical performance.

THE RATIO CONTROL ARCHITECTURE

The ratio control architecture is used to maintain the flow rate of one stream in a process at a defined or specified proportion relative to that of another. A common application for ratio control is to combine or blend two feed streams to produce a mixed flow with a desired composition or physical property. Consistent with other articles in this e-book, applications of interest are processes with streams comprised of gases, liquids, powders, slurries or melts.

The conceptual the flow rate of one of the streams feeding the mixed flow, designated as the *wild feed*, can change freely. Its flow rate might change based on product demand, maintenance limitations, feedstock variations, energy availability, the actions of another controller in the plant, or it may simply be that this is the stream we are least willing to manipulate during normal operation.

The other stream shown feeding the mixed flow is designated as the *controlled feed*. A final control element (FCE) in the controlled feed stream receives and reacts to the controller output signal, COc, from the ratio control architecture. While the conceptual diagrams in this chapter show a valve as the FCE, we note that other flow manipulation devices such as variable speed pumps or compressors may also be used in ratio control implementations.

Ratio Control Conceptual Diagram

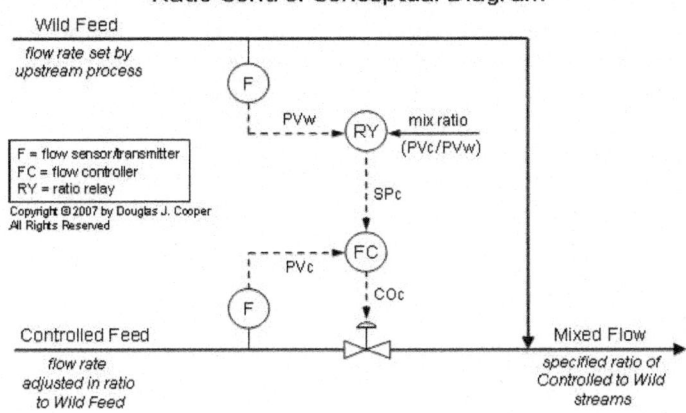

Wild Feed

flow rate set by upstream process

PVw   mix ratio (PVc/PVw)

F = flow sensor/transmitter
FC = flow controller
RY = ratio relay

SPc

PVc

COc

Controlled Feed

flow rate adjusted in ratio to Wild Feed

Mixed Flow

specified ratio of Controlled to Wild streams

Relays in the Ratio Architecture

We measure the flow rate of the wild feed and pass the signal to a relay, designated as RY in the diagram. The relay is typically one of two types:

A *ratio relay*, where the mix ratio is entered once during configuration and is generally not available to operations staff during normal operation.

A *multiplying relay* , where the mix ratio is presented as an adjustable parameter on the operations display and is thus more readily accessible for change.

In either case, the relay multiplies the measured flow rate of the wild feed stream, PVw, by the entered mix ratio to arrive at a desired or set point value, SPc, for the controlled feed stream. A flow controller then regulates the controlled feed flow rate to this SPc, resulting in a mixed flow stream of specified proportions between the controlled and wild streams.

Linear Flow Signals Required

A ratio controller architecture as requires that the signal from each flow sensor/transmitter change linearly with flow rate. Thus, the signals from the wild stream process variable, PVw, and the controlled stream process variable, PVc, should increase and decrease in a straight-line fashion as the individual flow rates increase and decrease.

Turbine flow meters and certain other sensors can provide a signal that changes linearly with flow rate. Unfortunately, a host of popular flow sensors, including inferential head flow elements such as orifice meters, do not. Additional computations (function blocks) must then be included between the sensor and the ratio relay to transform the nonlinear signal into the required linear flow-to-signal relationship.

Flow Fraction (Ratio) Controller

A classic example of ratio control is the blending of an additive into a process stream. An octane booster is blended with straight-run gasoline stream being produced by an atmospheric distillation column. For any number of reasons, the production rate of straight-run gasoline will vary over time in a refinery. Therefore, the amount of octane booster required to produce the desired octane rating in the mixed product flow must also vary in a coordinated fashion.

Flow Fraction (Ratio) Controller

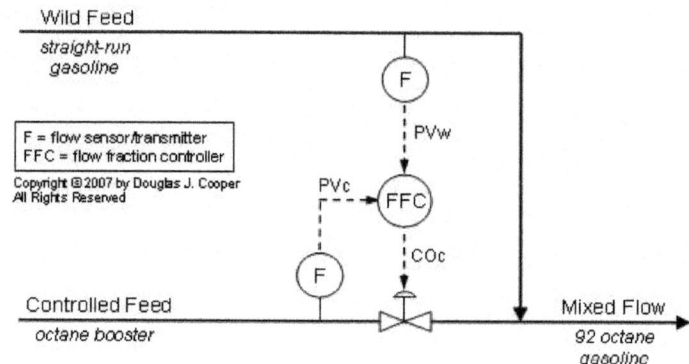

Rather than using a relay, we present an alternative ratio control architecture based on a flow fraction controller (FFC). The FFC is essentially a "pure" ratio controller in that it receives the wild feed and controlled feed signals directly as inputs. A ratio set point value is entered into the FCC, along with tuning parameters and other values required for any controller implementation.

Ratio Relay or Flow Fraction Controller

The flow fraction (ratio) controller is a preconfigured option in many modern computer based DCS or advanced PLCcontrol systems. It provides exactly the same functionality as the ratio relay combined with a single-input single-output controller.

The choice of using a relay or an FFC is a practical matter. The entered ratio multiplier value in a relay is not a readily accessible parameter. It therefore requires a greater level of permission and access to adjust. Consequently, the use of the ratio relay has the advantage (or disadvantage depending on the application) of requiring a higher level of authorization before a change can be made to the ratio multiplier.

Multiplying Relay With Remote Input

The ratio controller presents an additional level of complexity in that, like the cascade architecture, our ratio controller is contained within and is thus part of a larger control strategy.

In the example below, an analyzer sensor measures the composition or property we seek to maintain in the mixed flow stream. The measured value is compared to a set point value, SPA, and a mix ratio controller output signal, COA, is generated based on the difference. Thus, like a cascade, the outer analyzer controller continually sends mix ratio updates to the inner ratio control architecture.

Ratio Relay with Remote Input

The updated mix ratio COA value enters the multiplying relay as an external set point. The objective of this additional complexity is to correct for any unmeasured changes in the wild feed or controlled feed, thus maintaining the mixed flow composition or property at the set point value, SPA.

The term "analyzer" is used broadly here. Hopefully, we can indentify a fast, inexpensive and reliable sensor that allows us to infer the mixed flow composition or property of interest. Examples might include a capacitance probe, an in-line viscometer, or a pH meter.

If we are required to use a chromatograph, spectrometer or other such instrument, we must allow for the increased maintenance and attention such devices often demand. Perhaps more important, the time to complete a sample and analysis cycle for these devices can introduce a long dead time into our feedback loop. As dead time increases, best attainable control performance decreases.

RATIO CONTROL AND METERED-AIR COMBUSTION PROCESSES

A ratio control strategy can play a fundamental role in the safe and profitable operation of fired heaters, boilers, furnaces and similar fuel burning processes. This is because the air-to-fuel ratio in the combustion zone of these processes directly impacts fuel combustion efficiency and environmental emissions.

A requirement for ratio control implementation is that both the fuel feed rate and combustion air feed rate are measured and available as process variable (PV) signals.

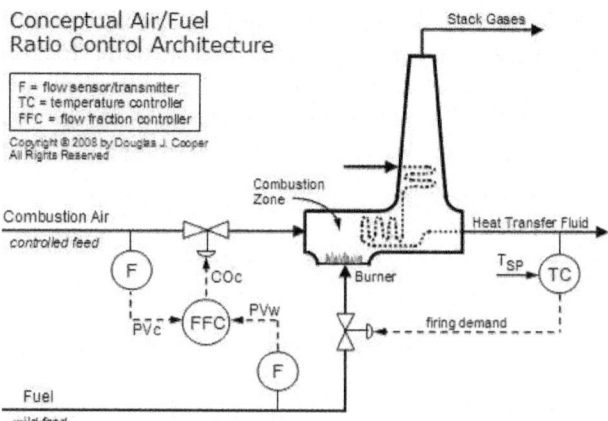

In this representative architecture, the fuel flow rate is adjusted to maintain the temperature of a heat transfer fluid exiting a furnace. On other processes, fuel flow rate might be adjusted to maintain the pressure in a steam header, the duct temperature downstream of the burner, or similar variable that must be regulated for efficient operation.

The combustion air feed rate is then adjusted by a flow fraction (ratio) controller to maintain a desired air/fuel ratio. While a simple sensor and valve, we will expand and modify this conceptual architecture as we progress:

The final control element (FCE) for the combustion air stream, rather than being a valve, is more commonly a variable speed blower, perhaps with adjustable dampers or louvers.

Measuring combustion air flow rate is challenging and can involve measuring a pressure drop across a portion of the combustion gas exhaust flow path.

In different applications, the air flow rate can be the wild feed while fuel flow rate is the controlled feed.

Stack gas analyzers add value and sophistication as they monitor the chemistry associated with combustion efficiency and environmental emissions.

Why Air/Fuel Ratio is Important

In combustion processes, air/fuel ratio is normally expressed on a mass basis. We get maximum useful heat energy if we provide air to the combustion zone at a mass flow rate (*e.g.*, lb/min, Kg/hr) that is properly matched to the mass flow rate of fuel to the burner.

Consider this generic equation for fuel combustion chemistry:

$$\text{Fuel} + \text{Air} = \frac{\text{Useful}}{\text{Heat}} + CO_2 + H_2O + CO + \frac{\text{Unburned}}{\text{Fuel}} + \frac{\text{Waste Heat}}{\text{Up the Stack}}$$

Increases as combustion air Decreases

Increases as combustion air Increases

Where:

CO_2 = carbon dioxide

CO = carbon monoxide

H_2O = water

Air = 21% oxygen (O_2) and 79% nitrogen (N_2)

Fuel = hydrocarbon such as natural gas or liquid fuel oil

Air is largely composed of oxygen and nitrogen. It is the oxygen in the air that combines with the carbon in the fuel in a highly energetic reaction called combustion. When burning hydrocarbons, nature strongly prefers the carbon-oxygen double bonds of carbon dioxide and will yield significant heat energy in an exothermic reaction to achieve this CO_2 form.

Thus, carbon dioxide is the common green house gas produced from the complete combustion of hydrocarbon fuel. Water vapor (H_2O) is also a normal product of hydrocarbon combustion.

Aside: nitrogen oxide (NOx) and sulfur oxide (SOx) pollutants are not included in our combustion chemistry equation. They are produced in industrial combustion processes principally from the nitrogen and sulfur originating in the fuel. As the temperature in the combustion zone increases, a portion of the nitrogen in the air can also convert to NOx.

Too Little Air Increases Pollution and Wastes Fuel

The oxygen needed to burn fuel comes from the air we feed to the process. If the air/fuel ratio is too small in our heater, boiler or furnace, there will not be enough oxygen available to completely convert the hydrocarbon fuel to carbon dioxide and water.

A too-small air/fuel ratio leads to incomplete combustion of our fuel. As the availability of oxygen decreases, noxious exhaust gases including carbon monoxide will form first. As the air/fuel ratio decreases further, partially burned and unburned fuel can appear in the exhaust stack, often revealing itself as smoke and soot. Carbon monoxide, partially burned and unburned fuel are all poisons whose release is regulated by the government.

Incomplete combustion also means that we are wasting expensive fuel. Fuel that does not burn to provide useful heat energy, including carbon monoxide that could yield energy as it converts to carbon dioxide, literally flows up our exhaust stack as lost profit.

Too Much Air Wastes Fuel

The issue that makes the operation of a combustion process so interesting is that if we feed too much air to the combustion zone (if the air/fuel ratio is too high), we also waste fuel, though in a wholly different manner.

Once we have enough oxygen available in the burn zone to complete combustion of the hydrocarbon fuel to carbon dioxide and water, we have addressed the pollution portion of our combustion chemistry equation. Any air fed to the process above and beyond that amount becomes an additional process load to be heated.

As the air/fuel ratio increases above that needed for complete combustion, the extra nitrogen and unneeded oxygen absorb heat energy, decreasing the temperature of the flame and gases in the combustion zone. As the operating temperature drops, we are less able to extract useful heat energy for our intended application.

So when the air/fuel ratio is too high, we produce a surplus of hot air. And this hot air simply carries its heat energy up and out the exhaust stack as lost profit.

Theoretical (Stoichiometric) Air

The relationship between the air/fuel ratio, pollution formation and wasted heat energy provides a basis for control system design. In a meticulous laboratory experiment with exacting measurements, perfect mixing and unlimited time, we could determine the precise amount of air required to just complete the conversion of a hydrocarbon fuel to carbon dioxide and water. This minimum amount is called the "theoretical" or "stoichiometric" air.

Unfortunately, real combustion processes have imperfect mixing of the air with the fuel. Also, the gases tend to flow so quickly that the air and fuel mix have limited contact time in the combustion zone. As such, if we feed air in the exact theoretical or stoichiometric proportion to the fuel, we will still have incomplete combustion and lost profit.

Real burners generally perform in a manner similar to the graph below. The cost associated with operating at increased air/fuel ratios is the energy wasted in heating extra oxygen and nitrogen. Yet as the air/fuel ratio is decreased, losses due to incomplete combustion and pollution generation increase rapidly.

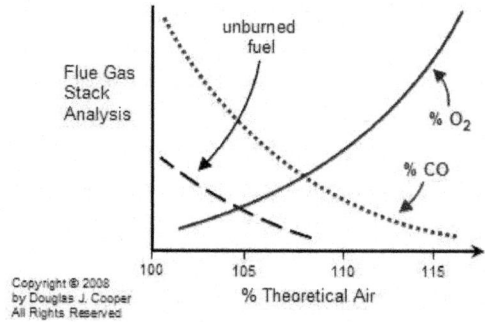

For any particular burner design, there is a target air/fuel ratio that balances the competing effects to minimize the total losses and thus maximize profit. As a gas or liquid fuel burner generally balances losses by operating somewhere between 105% to 120% of theoretical air. This is commonly referred to as operating with 5% to 20% excess air.

Sensors Should be Fast, Cheap and Easy

Fired heaters, boilers and furnaces in processes with streams composed of gases, liquids, powders, slurries and melts are found in a broad range of manufacturing, production and development operations. Knowing that the composition of the fuel, the design of the burners, the configuration of the combustion zone, and the purpose of the process can differ for each implementation hints at a dizzying array of control strategy design and tuning possibilities.

To develop a standard control strategy, we require a flexible method of measuring excess air so we can control to a target air/fuel ratio. We normally seek sensors that are reliable, inexpensive, easy to install and maintain, and quick to respond. If we cannot get these qualities with a direct measurement of the process variable (PV) of interest, then an effective alternative is to measure a related variable if it can be done with a "fast, cheap and easy" sensor option.

Excess air is an example of a PV that is very challenging to directly measure in the combustion zone, yet oxygen and energy content in the stack gases is an appropriate alternative. As it turns out, operating with 5% to 20% excess air equates to having about 1% to 3% oxygen by volume in the stack gases.

Measuring the Stack Gases

By measuring exhaust stack gas composition, we obtain information we need to properly monitor and control air/fuel ratio in the combustion zone. Stack analyzers fall into two broad categories:

- *Dry Basis Extractive Analyzers* pull a gas sample from the stack and cool it to condense the water out of the sample. Analysis is then made on the dry stack gas.

- *Wet Basis In Situ Analyzers* are placed in very close proximity to the stack. The hot sample being measured still contains the water vapor produced by combustion, thus providing a wet stack gas analysis.

A host of stack gas (or flue gas) analyzers can be purchased that measure O_2. The wet basis analyzers yield a lower oxygen value than dry basis analyzers by perhaps 0.3% – 0.5% by volume.

Instruments are widely available that also include a carbon monoxide measurement along with the oxygen measurement. A common approach is to pass the stack gas through a catalyst chamber and measure the energy released as the carbon monoxide and unburned fuel converts to carbon dioxide. The analyzer results are expressed as an equivalent percent CO in the sample. The single number, expressed as a CO measurement but representing fuel wasted because of insufficient air, simplifies control strategy design and process operation.

With a measurement of O_2 and CO (representing all lost fuel) in the stack of our combustion process, we have critical PV measurements needed to implement an air/fuel ratio control strategy. Note that it is the responsibility of the burner manufacturer and/or process design staff to specify the target set point values for a particular combustion system prior to controller tuning.

Air Flow Metering

Combustion processes generally have combustion air delivered in one of three ways:

- A forced draft process uses a blower to feed air into the combustion zone.
- An induced draft process has a blower downstream of the burner that pulls or draws air through the combustion zone.
- A natural draft process relies on the void left as hot exhaust gases naturally rise up the stack to draw air into the combustion zone.

Even with a blower, measuring the air feed rate delivered at low pressure through the twists and turns of irregular ductwork and firebrick is not cheap or easy. A popular alternative is to measure the pressure drop across some part of the exhaust gas stream. The bulk of the exhaust gas is nitrogen that enters with the combustion air. As long as the air/fuel ratio adjustments are modest, the exhaust gas flow rate will track the combustion air feed rate quite closely.

Thus, a properly implemented differential pressure measurement is a "fast, cheap and easy" method for inferring combustion air feed rate. The figure below illustrates such a measurement across a heat transfer section and up the exhaust stack.

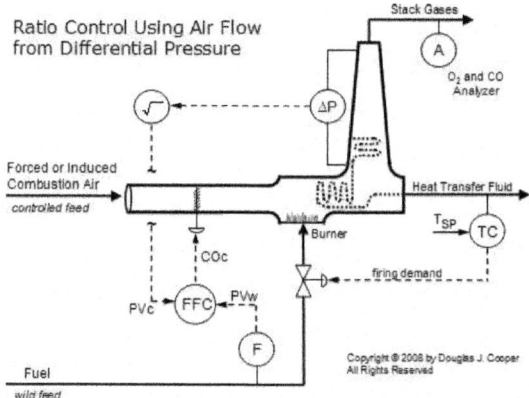

Also shown is that the controller output signal from the flow fraction (ratio) controller, COC, adjusts louvers to modulate the flow through the combustion zone. As the louvers open and close to permit more or less flow, the differential pressure measurement will increase or decrease, respectively.

The signal from the wild and controlled flow sensors must change linearly with flow rate. The differential pressure transmitter connected across a portion of the exhaust gas path becomes a linear gas flow sensor by recognizing that total gas flow, F, is proportional to the square root of the pressure differential (ΔP), or

$F = \alpha\sqrt{\Delta P}$. Thus, the controlled feed process variable signal, PVC, is linear with flow when the square root of the differential pressure signal is extracted as shown in the diagram.

Practitioner's Note: The differential pressure measurement must not be connected across the portion of the gas flow path that includes the adjustable louvers. Each change in louver position changes the F vs. ΔP relationship. Success would require that during calibration, we somehow determine a different coefficient α for each louver position. This unrealistic task is easily avoided by proper location of the differential pressure taps.Calibrating the differential pressure signal to a particular air feed rate is normally achieved while the fired heater, boiler or furnace is operating with the air/fuel ratio controller in manual mode. The maximum or full scale differential pressure calibration is determined by bringing the fuel flow firing rate to maximum (or as close as practical) and then adjusting the air feed flow rate until the design O_2 level is being measured in the stack gas.The differential pressure being measured by these sensors is very small and the exhaust gas contains water vapor that can condense in sensing lines. Even one or two inches of condensate in one side of the differential pressure transmitter can dramatically corrupt the measurement signal.

Choosing Air or Fuel for Firing Rate Control

With a means of measuring both the combustion air flow and the fuel flow, and with a stack analyzer to permit calibration and monitoring, we can implement the simple air/fuel ratio control.

Because it can be measured accurately, fuel feed rate is a popular choice for the firing rate control variable. Yet in certain applications it is more desirable to employ the combustion air flow rate in this capacity.

If fuel is the firing rate control variable, a rapid increase in firing rate with air following behind in time will lead to incomplete combustion. On the other hand, if air is made the firing rate control variable, a rapid decrease in firing rate will lead to the same situation.

OVERRIDE (SELECT) ELEMENTS AND THEIR USE IN RATIO CONTROL

A select element receives two input signals and forwards one of them onward in the signal path. A *low select*, shown below to the left, passes the lowest of the two signals, while a *high select*, shown to the right, passes the larger value onward.

A select element can be implemented as a DCS or PLC function block, as a few lines of computer code, or as a simple hardware circuit. And while the elements above are using electrical current, they can also be designed to select between high and low voltage or digital (discrete) counts.

The above pictures are not meant to imply that the selected output value has anything to do with signal location. If the 12 mA signal shown entering on the lower input were to drop down to 5 mA while the 7 mA input entering from the left side remained constant, then the low select output would be 5 mA while the high select output would be 7 mA.

Logic Permits Increased Sophistication

The simple select element enables decision-making logic to be included in a control strategy, which in turn provides a means for increasing strategy sophistication. One such example is to use a select element to construct an architecture designed to control to a maximum or minimum limit or constraint.

Another popular application, and the one explored here, is to employ a select as an *override element* in a ratio control architecture. In particular, we explore how a select override might be included in an air/fuel ratio combustion control strategy to enhance safety, limit emissions and maximize useful energy from fuel.

Ratio Strategy Without Override

Before demonstrating the use of a select override, we consider a variation. The ratio architecture much like that used in the referenced chapter except here we choose to employ a ratio relay with remote input rather than a flow fraction controller.

In this design, the fuel mass flow rate is regulated by a flow controller whose set point, SPw, arrives as a firing demand from elsewhere in the plant. SPw might be generated, for example, by a controller adjusting the duct temperature downstream of a burner, the temperature of a heat transfer fluid exiting a furnace or the pressure in a steam header.

There is an implicit assumption in this architecture that the fuel mass flow rate closely tracks the firing demand set point, that is, PVw ≈ SPw. Thus, an integral term (*e.g.*, PI control) is required in the wild feed flow control algorithm.

Since SPw is set elsewhere, we are not free to adjust fuel flow rate separately to maintain a desired air/fuel ratio. It is thus appropriately designated as the "wild feed" in this construction.

As SPw (and thus PVw) increases and decreases, a ratio relay shown in the control diagram multiplies the incoming signal by a design air/fuel ratio value (or in the general case, a controlled/wild feed ratio value) to compute the combustion air set point, SPc, for the controlled feed flow loop.

Practitioner's Note: A ratio controller architecture requires that the signal from each mass flow sensor/transmitter change linearly with flow rate. Thus, the signals from the wild stream process variable, PVw, and the controlled stream process variable, PVc, should increase and decrease in a straight-line fashion as the individual mass flow rates increase and decrease. If the flow sensor is not linear, additional computations (function blocks) must be included between the sensor and the ratio relay to transform the nonlinear signal into the required linear flow-to-signal relationship.

If the fuel flow control loop and the combustion air control loop both respond quickly to flow commands COw and COc respectively, then the architecture above should maintain the desired air/fuel ratio even if the demand set point signal, SPw, moves rapidly and often.

Problem if the Combustion Air Loop is Slow

The diagram shows a valve as the final control element (FCE) adjusting the fuel mass flow rate, and a variable frequency drive (VFD) and blower assembly as the FCE adjusting the combustion air mass flow rate. Valves generally respond quickly to controller output signal commands, so we expect the fuel mass flow rate to closely track changes in COw.

In contrast, air blower assemblies vary in capability. Here we consider a blower that responds slowly to control commands, COc, relative to the valve (the time constant of the blower "process" is much larger than that of the valve).

While we desire that the mass flow rates of the two streams move together to remain in ratio, the different response times of the FCEs means that during a firing demand change (a change in SPw), the feed streams may not be matched at the desired air/fuel ratio for a period of time.

To illustrate, consider a case where the firing demand, SPw, suddenly increases. The fuel flow valve responds quickly, increasing fuel feed to the burner. The ratio relay will receive SPw and raise the set point of the combustion air mass flow rate, SPc, so the two streams can remain in ratio.

If the air blower response is slow, however, a fuel rich environment can temporarily develop. That is, there will be a period of time when we are below the desired 5% to 20% of excess air (below the 105% to 120% of theoretical or stoichiometric air) as we wait for the blower to ramp up and deliver more air to the burner.

If there is insufficient air for complete combustion, then carbon monoxide and partially burned fuel will appear in the exhaust stack. As such, we have a situation where we are wasting expensive fuel and violating environmental regulations.

Solution 1: Detune the Fuel Feed Controller

One solution is to enter conservative or sluggish tuning values into the fuel feed controller. By detuning (slowing down) the wild feed control loop so it moves as slowly as the combustion air blower, the two feed streams will be able to track together and stay in ratio. We thus avoid creating the fuel rich environment as just described.

Unfortunately, however, we also have made the larger firing demand control system less responsive, and this diminishes overall plant performance. In some process applications, a slow or sluggish ratio control performance may be acceptable. In the particular case of combustion control, it likely is not.

Solution 2: Use a Low Select Override

The addition of an override to our control architecture. The same as that above except a second ratio relay feeding a low select element has been included in the design.

The second ratio relay receives the *actual* measured combustion air mass flow rate, PVc, and computes a matching fuel flow rate based on the design air/fuel ratio. This "fuel flow matched to the actual air flow" value is transmitted to the low select element. The low select element also receives the firing demand fuel flow rate, SPw, set elsewhere in the plant.

A low select element passes the lowest of the two input signals forward. In this case, if SPw is a fuel rate that exceeds the availability of combustion air required to burn it, the select element will *override* the demand signal and forward the lower "fuel flow matched to the actual air flow" signal.

The override strategy thus ensures that the feed streams remain in ratio for a rapid increase in firing demand, SPw, but it has no effect when there is a rapid decrease in firing demand.

When SPw rapidly decreases, the fuel flow rate will respond quickly and we will be in a "lean" environment (too much combustion air) until the blower slows to match the decreased fuel rate. When there is more air than that needed for complete combustion, the extra nitrogen and unneeded oxygen absorb heat energy, decreasing the temperature of the flame and gases in the combustion zone.

So while the select override element has eliminated pollution concerns when firing demand rapidly *increases*, we produce a surplus of hot air that simply leaves the exhaust stack as lost profit when firing demand rapidly *decreases*. In effect, we have solved only half the air/fuel balance problem with a single select override element.

A Simulated Furnace Air/Fuel Ratio Challenge

To further our understanding of the select override in an air/fuel ratio strategy, we consider a furnace simulation available in commercial software. The furnace burns natural gas to heat a process liquid flowing through tubes in the fire box. Firing demand is determined by a temperature controller located on the process liquid as it exits the furnace.

Because the output of the firing demand temperature controller becomes the set point to the wild feed of the air/fuel ratio strategy, it is in fact a primary (or outer) controller in a cascade control architecture. If the temperature of the process liquid is too hot (greater than set point), the firing demand controller seeks to reduce energy input. If the temperature is below set point, it seeks to add energy.

Unlike the example above, combustion air is the wild feed in this furnace simulation. Thus, when the firing demand temperature controller is in automatic (the cascade is enabled) set point changes are transmitted to the air flow controller. If the temperature controller is in manual, the set point of the combustion air flow controller must be entered manually by operations staff.

Firing Demand in Manual

- We first consider furnace operation when the firing demand temperature controller is in manual mode.

Firing Demand in Automatic

- With operation steady, we switch the firing demand temperature controller to automatic.

Firing Demand Override

- A process upset requires that the high select element override the firing demand controller.

Cross-Limiting Ratio Control Strategy

If we were to increase the process liquid flow rate through the furnace, the firing demand controller would quickly ramp up the combustion air feed rate to provide more energy. Temporarily, there would be more air than that needed for complete combustion. That temporary surplus of hot air will carry its heat energy up and out the exhaust stack as lost profit.

So similar to the first example, a single select override element provides only half the solution depending on the direction that the upstream demand is moving.

The use of two select override elements in a cross-limiting ratio control strategy. While fairly complex, the cross limiting structure offers benefit in that it provides protection in an air/fuel ratio strategy both when firing demand is increasing *and* decreasing.

By careful sensor selection and scaling, we can maintain the "ratio with low select override" strategy as presented earlier in this chapter while eliminating the multiplying relays from our design. As long as we use control algorithms with an integrating term (PI or PID), the upstream demand signal becomes the set point for both controllers and the desired ratio will be maintained.

RATIO WITH CROSS-LIMITING OVERRIDE CONTROL OF A COMBUSTION PROCESS

We explored override control using select elements and learned that environmental and energy efficiency concerns for metered-air combustion processes can be partially addressed with a single select override element. Examples illustrated how a select override can either prevent having too much fuel *or* too much air in the air/fuel mixture fed to the burner of a combustion process, but one override element alone is not capable of preventing both scenarios.

We explore the addition of a second select override element to create a cross-limiting architecture that prevents the air/fuel ratio fed to the burner from becoming overly rich (too much fuel) or lean (too much air) as operating conditions change. Variations on this cross-limiting architecture are widely employed within the air/fuel ratio logic of a broad range of industrial combustion control systems.

Steam Boiler Process Example

To provide a larger context for this topic, we begin by considering a multi-boiler steam generation process.

Steam generation processes often have multiple boilers that feed a common steam header. When steam is needed anywhere in the plant, the load is drawn from this common header. Steam turbines, for example, drive generators, pumps and compressors. Steam is widely used for process heating, can be injected into production vessels to serve as a reactant or diluent, and may even be used to draw a vacuum in a vessel via jet ejectors.

With so many uses, steam loads can vary significantly and unpredictably over time in a plant. The individual boilers must generate and feed steam to the common header at a rate that matches these steam load draws. Controlling the steam header to a constant pressure provides an important stabilizing influence to plant-wide operation.

Plant Master Controller

* A popular multi-boiler architecture for maintaining header pressure is to use a single pressure controller on the common header that outputs a firing demand signal for *all* of the boilers in the steam plant. This steam header pressure controller is widely referred to as the *Plant Master*.

Based on the difference between the set point (SPP) and measured pressure in the header, the Plant Master controller computes a firing demand output that signals all of the boilers in the plant to increase or decrease firing, and thus, steam production.

Boiler Master Controller

* The Boiler Masters in the above multi-boiler process diagram are Auto/Manual selector stations with biasing (+/–) values. If all three of the Boiler Masters are in automatic, any change in the Plant Master output signal will pass through and create an associated change in the firing demand for the three boilers.

If a Boiler Master is in automatic, that boiler is said to be operating as a *swing boiler*. As such, its firing demand signal will vary (or swing) directly as the Plant Master signal varies. If each of the fuel flow meters are scaled so that 100% of fuel flow produces maximum rated steam output, then each boiler will swing the same amount as the Plant Master calls for variations in steam production.

But suppose Boiler B has cracked refractory brick in the fire box or some other mechanical issue that, until repaired, requires that it be operated no higher than, for example, 85% of its design steam production rate. That is, Boiler B has been *derated* and its maximum permissible steam generating capacity has been lowered from the original design rating. Two options we can consider include:

1. When a Boiler Master is in automatic, then: signal out = signal in + bias where the bias value is set by the operator. If the bias value of Boiler Master B is set in this example to –15%, then no matter what output is received from the Plant Master (0% to 100%), the firing demand signal will never exceed 85% (100% plus the negative 15% bias). In this mode of operation, Boiler B will still swing with Boiler A and Boiler C in response to the Plant Master, but it will operate at a firing rate 15% below the level of the other two boilers (assuming their bias values are zero).

2. If a boiler is suffering from refractory problems, then allowing the firing rate to swing can accelerate refractory degradation. Thus, Boiler Master B might alternatively be switched to manual mode where the output firing demand signal is set to a constant value. In manual mode, Boiler B is said to provide a *base load* of steam production. With the firing rate of Boiler B set manually from the Boiler Master, it is unresponsive to firing demand signal variations from the Plant Master. We then would have two swing boilers (Boiler A and Boiler C) and one base loaded boiler (Boiler B).

Combustion Control Process

Each furnace and steam boiler has its own control system. Of particular interest here is the maintenance of a specified air/fuel mass ratio for efficient combustion at the burners.

The air/fuel ratio control strategy receives a firing demand from the Boiler Master. Air mass flow rate may be measured downstream of the combustion zone and is thus shown as an input to the ratio control strategy.

Ratio with Cross-Limiting Override Control

Certain assumption are used in the presentation that follows:

1. Air/fuel ratio is normally expressed as a mass flow ratio of air to fuel.

2. The air and fuel flow transmitter signals are linear with respect to the mass flow rate and have been scaled to range from 0-100%.

3. The flow transmitters have been carefully calibrated so that both signals at the design air/fuel ratio are one to one. That is, if the fuel flow transmitter signal, PV_f, is 80%, then an air flow signal, PV_a, of 80% will produce an air flow rate that meets the design air/fuel mass ratio. This enables us to implement the ratio strategy without using multiplying relays.

We rearrange the loop layout to make the symmetry of the design more apparent. Specifically, we reverse the fuel flow direction and show the air mass flow rate transmitter as a generic measurement within the control architecture.

The flow transmitters have been calibrated so that when both signals match, we are at the design air/fuel mass flow ratio. Thus, because of the high select override, SPa is always the greater of the the firing demand signal or the value that matches the current fuel flow signal. And because of the low select override, SPf is always the lesser of the firing demand signal or the value that matches the current air flow signal.

The result is that if firing demand moves up, the high select will pass the firing demand signal through as SPa, causing the air flow to increase. Because of the low select override, the fuel set point, SPf, will not match the firing demand signal increase, but rather, will follow the increasing air flow rate as it responds upward.

And if the firing demand moves down, the low select will pass the firing demand signal through as SPf, causing the fuel flow to decrease. Because of the high select override, the air set point, SPa, will not match the firing demand signal decrease, but rather, will track the decreasing fuel flow rate as it moves downward.

In short, the control system ensures that during sudden operational changes that move us in either direction from the design air/fuel ratio, the burner will temporarily receive extra air until balance is restored. While a lean air/fuel ratio means we are heating extra air that then goes up and out the stack, it avoids the environmentally harmful emission of carbon monoxide and unburned fuel.

Variable Air/Fuel Ratio

The basic cross-limiting strategy we have described to this point provides no means for adjusting the air/fuel ratio. This may be necessary, for example, if the composition of our fuel changes, if the burner performance changes due to corrosion or fouling, or if the operating characteristics of the burner change as firing level changes.

A cross-limiting override control strategy that also automatically adjusts the air/fuel ratio based on the oxygen level measured in the exhaust stack.

As shown in the diagram, the signal from the air flow transmitter, PVraw, is multiplied by the output of the analyzer controller, COO2, and the product is forwarded as the measured air flow rate process variable, PVa.

With this construction, if the measured exhaust oxygen, PVO2, matches the oxygen set point, SPO2, then the analyzer controller (AC) output, COO2 will equal one and PVa will equal PVraw.

But if the oxygen level in the stack is too high, COO2 will become greater than one. By multiplying the raw air flow signal, PVraw, by a number greater than one, PVa appears to read high. And if the oxygen level in the stack is too low, we multiply PVraw with a number smaller than one so that PVa appears to read low.

The ratio strategy reacts based on the artificial PV values, adjusting the air/fuel ratio until the measured oxygen level, PVO2, is at set point SP02.

This manipulation to the air/fuel ratio based on measured exhaust oxygen is commonly called oxygen trim control. By essentially changing the effective calibration of the air flow transmitter to a new range, the signal ratio of the carefully scaled air and fuel transmitters can remain 1:1.

CASCADE, FEED FORWARD AND BOILER LEVEL CONTROL

One common application of cascade control combined with feed forward control is in level control systems for boiler steam drums.

The control strategies now used in modern industrial boiler systems had their beginnings on shipboard steam propulsion boilers. When boilers operated at low pressure, it was reasonably inexpensive to make the steam drum large. In a large drum, liquid level moves relatively slowly in response to disturbances (it has a long time constant). Therefore, manual or automatic adjustment of the feedwater valve in response to liquid level variations was an effective control strategy.

But as boiler operating pressures have increased over the years, the cost of building and installing large steam drums forced the reduction of the drum size for a given steam production capacity.

The consequence of smaller drum size is an attendant reduction in process time constants, or the speed with which important process variables can change. Smaller time constants mean upsets must be addressed more quickly, and this has led to the development of increasingly sophisticated control strategies.

Element Strategy

Most boilers of medium to high pressure today use a "3-element" boiler control strategy. The term "3-element control" refers to the number of process variables (PVs) that are measured to effect control of the boiler feedwater control valve. These measured PVs are:

- liquid level in the boiler drum,
- flow of feedwater to the boiler drum, and
- flow of steam leaving the boiler drum.

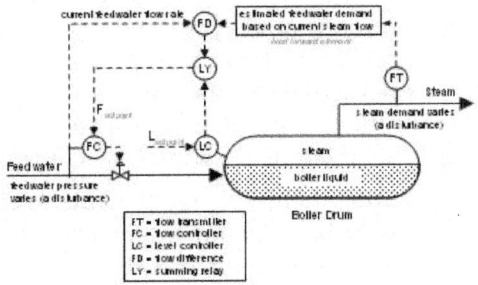

Maintaining liquid level in the boiler steam drum is the highest priority. It is critical that the liquid level remain low enough to guarantee that there is adequate disengaging volume above the liquid, and high enough to assure that there is water present in every steam generating tube in the boiler. These requirements typically result in a narrow range in which the liquid level must be maintained.

The feedwater used to maintain liquid level in industrial boilers often comes from multiple sources and is brought up to steam drum pressure by pumps operating in parallel. With multiple sources and multiple pumps, the supply pressure of the feedwater will change over time. Every time supply pressure changes, the flow rate through the valve, even if it remains fixed in position, is immediately affected.

So, for example, if the boiler drum liquid level is low, the level controller will call for an increase in feedwater flow. But consider that if at this moment, the feedwater supply pressure were to drop. The level controller could be opening the valve, yet the falling supply pressure could actually cause a decreased flow through the valve and into the drum.

Thus, it is not enough for the level controller to directly open or close the valve. Rather, it must decide whether it needs more or less feed flow to the boiler drum. The level controller transmits its target flow as a set point to a flow controller. The flow controller then decides how much to open or close the valve as supply pressure swings to meet the set point target.

This is a "2-element" (boiler liquid level to feedwater flow rate) cascade control strategy. By placing this feedwater flow rate in a fast flow control loop, the flow controller will immediately sense any variations in the supply conditions which produce a change in feedwater flow. The flow controller will adjust the boiler feedwater valve position to restore the flow to its set point before the boiler drum liquid level is even affected. The level controller is the primary controller (sometimes referred to as the master controller) in this cascade, adjusting the set point of the flow controller, which is the secondary controller (sometimes identified as the slave controller).

The third element in a "3-element control" system is the flow of steam leaving the steam drum. The variation in demand from the steam header is the most common disturbance to the boiler level control system in an industrial steam system.

By measuring the steam flow, the magnitude of demand changes can be used as a feed forward signal to the level control system. The feed forward signal can be added into the output of the level controller to adjust the flow control loop set point, or can be added into the output of the flow control loop to directly manipulate the boiler feedwater control valve. The majority of boiler level control systems add the feed forward signal into the level controller output to the secondary (feedwater flow) controller set point. This approach eliminates the need for characterizing the feed forward signal to match the control valve characteristic.

Actual boiler level control schemes do not feed the steam flow signal forward directly. Instead, the difference between the outlet steam flow and the inlet water flow is calculated. The difference value is directly added to the set point signal to the feedwater flow controller. Therefore, if the steam flow out of the boiler is suddenly increased by the start up of a turbine, for example, the set point to the feedwater flow controller is increased by exactly the amount of the measured steam flow increase.

Simple material balance considerations suggest that if the two flow meters are exactly accurate, the flow change produced by the flow control loop will make up exactly enough water to maintain the level without producing a significant upset to the level control loop. Similarly, a sudden drop in steam demand caused by the trip of a significant turbine load will produce an exactly matching drop in feedwater flow to the steam drum without producing any significant disturbance to the boiler steam drum level control.

Of course, there are losses from the boiler that are not measured by the steam production meter. The most common of these are boiler blow down and steam vents (including relief valves) ahead of the steam production meter. In addition, boiler operating conditions that alter the total volume of water in the boiler cannot be corrected by the feed forward control strategy. For example, forced circulation boilers may have steam generating sections that are placed out of service or in service intermittently. The level controller itself must correct for these unmeasured disturbances using the normal feedback control algorithm.

Notes on Firing Control Systems

In general, firing control is accomplished with a Plant Master that monitors the pressure of the main steam header and modulates the firing rate (and hence, the steam production rate) of one or more boilers delivering steam to the steam header. The firing demand signal is sent to all boilers in parallel, but each boiler is provided with a Boiler Master to allow the Plant Master demand signal to be overridden or biased. When the signal is overridden, the steam production rate of the boiler is set manually by the operator, and the boiler is said to be base-loaded. Most boilers on a given header must be allowed to be driven by the Plant Master to maintain pressure control. Boilers that have the Boiler Master set in automatic mode (passing the steam demand from the Plant Master to the boiler firing control system) are said to be swing boilers as opposed to base-loaded boilers.

The presence of heat recovery steam boilers on a steam header raises new control issues because the steam production rate is primarily controlled by the horsepower demand placed on the gas turbine providing the heat to the boiler. If the heat recovery boiler operates at a pressure above the header pressure, a separate pressure control system can be used to blow off excess steam from the heat recovery boiler when production is above the steam header demand. Note that for maximum efficiency, most heat recovery boilers are fitted with duct burners to provide additional heat to the boiler. The duct burner is controlled with a Boiler Master like any other swing boiler. As long as there are other large swing boilers connected to the steam header, the other fired boilers can reduce firing as required when output increases from the heat recovery boiler.

DYNAMIC SHRINK/SWELL AND BOILER LEVEL CONTROL

Boiler Start-up

As high pressure boilers ramp up to operating temperature and pressure, the volume of a given amount of saturated water in the drum can expand by as much as 30%. This natural expansion of the water volume during start-up is not dynamic shrink/swell, though it does provide its own unique control challenges.

The expansion (or more precisely, decrease in density) of water during start-up of the boiler poses a problem if a differential pressure or displacer instrument is used for level measurement. Such a level transmitter calibrated for saturated water service at say, 600 psig, will indicate higher than the true level when the drum is filled with relatively cool boiler feedwater at a low start-up pressure.

If left uncompensated at low pressure conditions, the "higher than true level" indication will cause the controller to maintain a lower than desired liquid level in the drum during the start-up period. If the low level trip device is actually sensitive to the interface (*e.g.* conductance probes or float switches), troublesome low level trip events become very likely during start-up.

This variation in the sensitivity of the level transmitter with operating conditions can be corrected by using the drum pressure to compensate for the output of the level transmitter. The compensation can be accomplished with great accuracy using steam table data. The compensation has no dynamic significance and can be used independent of boiler load or operating pressure.

Dynamic Shrink/Swell

Dynamic shrink/swell is a phenomenon that produces variations in the level of the liquid surface in the steam drum whenever boiler load (changes in steam demand) occur. This behavior is strongly influenced by the actual arrangement of steam generating tubes in the boiler.

We have significant experience with "water wall" boilers that have radiant tubes on three sides of the firebox. There is a steam drum located above the combustion chamber and a mud drum located below the combustion chamber.

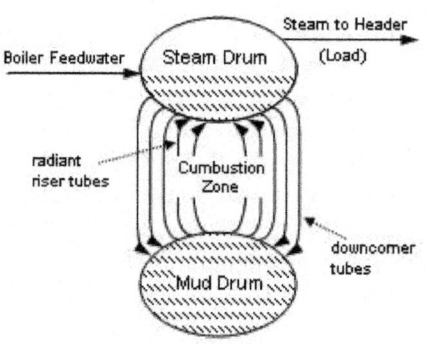

During operation, the tubes exposed to the radiant heat from the flame are always producing steam. As the steam rises in the tubes, boiler water is also carried upward and discharged into the steam drum. Tubes that are not producing significant steam flow have a net downward flow of boiler water from the steam drum to the mud drum.

The tubes producing large quantities of steam are termed risers and those principally carrying water down to the mud drum from the steam drum are termed downcomers. Excluding the tubes subject to radiant heat input from the firebox flame, a given tube will serve as a riser at some firing rates and a downcomer at other firing rates.

The mechanics of the natural convection circulation of boiler water within the steam generator is the origin of the dynamic shrink/swell phenomenon. Consider what happens to a boiler operating at steady state at 600 psig when it is subjected to a sudden increase in load (or steam demand).

A sudden steam load increase will naturally produce a drop in the pressure in the steam drum, because, initially at least, the firing rate cannot increase fast enough to match the steam production rate at the new demand level. When the pressure in the drum drops, it has a dramatic effect on the natural convection within the boiler. The drop in pressure causes a small fraction of the saturated water in the boiler to immediately vaporize, producing a large amount of boil-up from most of the tubes in the boiler. During the transient, most of the tubes temporarily become risers. The result is that the level in the steam drum above the combustion chamber rises.

However, this rise in level is actually an inverse response to the load change. Since, the net steam draw rate has gone up, the net flow of water to the boiler needs to increase, because the total mass of water in the boiler is falling. However, the level controller senses a rise in the level of the steam drum and calls for a reduction in the flow of feedwater to the boiler.

This inverse response to a sudden load increase is dynamic swell. Dynamic shrink is also observed when a sudden load decrease occurs. However, the dynamic shrink phenomenon does not disrupt the natural convection circulation of the boiler as completely as the dynamic swell effect. Consequently, the reduction in level produced by a sudden decrease in load is typically much smaller and of shorter duration than the effect produced by dynamic swell.

Control Strategy for Shrink/Swell

When a sudden load (steam demand) increase occurs, the feed forward portion of the strategy will produce an increase in the set point for the feedwater flow controller. This increase in feedwater flow controller set point will be countered to varying degrees by the level controller response to the temporary rise in level produced by the dynamic swell.

The standard tool used to minimize the impact of the swell phenomenon on the level in a three-element level control system is the lead-lag relay in the feed forward signal from the flow difference relay. This is the traditional means of dealing with mismatched disturbance and manipulated variable dynamics in feed forward systems, and is certainly applicable in this control strategy. When used in the three-element level control strategy, the lead-lag relay is commonly termed the "shrink/swell relay."

There are two significant limitations to the use of the lead-lag relay for shrink/swell compensation. To begin with, the response of most boilers to a load increase (swell event) is much more dramatic than the response to a load decrease (shrink event). In other words, the system response is very asymmetric. The lead-lag relay is perfectly symmetrical in responding to load changes in each direction and cannot be well matched to both directions.

Furthermore, the standard method of establishing the magnitudes of the lead time constant and lag time constant involves open loop tests of the process response to the disturbance (steam load) and to the manipulated variable (feedwater flow). A step test of the manipulated variable is generally not too difficult to conduct. However, changing the firing rate upward fast enough to actually produce significant swell is difficult without seriously upsetting the steam system, an event that is to be avoided in most operating plants. Therefore, the practitioner's only choice is to gather accurate data continuously and wait for a disturbance event that will exercise the shrink/swell relay's function.

When a lead-lag relay is to be added to an existing three-element boiler control scheme, operator knowledge of the boiler behavior in sudden load increase situations can guide the initial settings. For example, if the operators indicate that they must manually lead the feedwater valve by opening it faster than the control system will open it automatically, it is clear that a lead time constant larger than the lag time is required. Similarly, if the operator must retard the valve response to prevent excessively high level, the lead time constant must be less than the lag time. The lag time constant will typically fall in the range of one minute to three minutes. The ratio of the lead time constant to the lag time constant determines the magnitude of the initial response to the disturbance. If the ratio is one to one, the system behaves the same as a system with no lead-lag relay.

Ultimately, the system must be adjusted by watching the response to actual steam system upsets that require sudden firing increases. If the initial observed response of level to an upset is a rising level, the ratio of lead time to lag time should be decreased. The inverse is similarly true. If the recovery from an initial rise in level is followed by significant overshoot below the level target, the lag time should be reduced. If the recovery from an initial level drop is followed by a large overshoot above the level target, the lag time should be increased.

Chapter 7

A COMBINED FEED-FORWARD/FEED-BACK CONTROL SYSTEM FOR A QbD-BASED CONTINUOUS TABLET MANUFACTURING PROCESS

Ravendra Singh*, Fernando J. Muzzio, Marianthi Ierapetritou and Rohit Ramachandran*

Engineering Research Center for Structured Organic Particulate Systems (C-SOPS), Department of Chemical and Biochemical Engineering, Rutgers, The State University of New Jersey, Piscataway, NJ 08854, USA; E-Mails: fjmuzzio@yahoo.com (F.J.M.); marianth@soemail.rutgers.edu (M.I.)

* Authors to whom correspondence should be addressed; E-Mails: ravendra.singh@rutgers.edu (R.S.); rohit.r@rutgers.edu (R.R.); Tel.: +1-848-445-4944 (R.S.); +1-732-445-6278 (R.R.); Fax: +1-732-445-2581 (R.S. & R.R.).

Academic Editor: Michael Henson

ABSTRACT

Continuous pharmaceutical manufacturing together with PAT (Process Analytical Technology) provides a suitable platform for automatic control of the end product quality as desired by QbD (quality by design)-based efficient manufacturing. The precise control of the quality of the pharmaceutical product requires corrective actions in the process/raw material variability before product quality can be influenced. In this manuscript, a combined feed-forward/feed-back control system has been developed for a direct compaction continuous tablet manufacturing process. The feed-forward controller takes into account the effect of process disturbances proactively while the feed-back control system ensures the end product quality consistently. The coupled feed-forward/feed-back control system ensures the minimum variability in the final product quality irrespective of process and raw material variations. The performance of the combined control strategy has been evaluated through process simulation and is found to be more effective in comparison with a feed-back only control strategy and, therefore, demonstrates potential to further improve pharmaceutical tablet manufacturing operations.

Keywords

Feed-forward control; feed-back control; pharmaceutical; manufacturing; QbD; PAT.

1. INTRODUCTION

Pharmaceutical companies are one of the most strictly regulated companies where precise control of the end product quality is highly desired to satisfy the high standard set by regulatory authorities to ensure the efficiency and efficacy of the drug products [1,2]. Continuous pharmaceutical manufacturing enables the implementation of efficient automatic real time monitoring and control of the critical process parameters (CPP's) and critical quality attributes (CQA's) as desired for quality by design (QbD) — rather than quality by testing (QbT) — based manufacturing [3,4]. There are many factors that can affect the end product quality of the pharmaceutical products involving solid dosage forms. For example, the raw materials and process variability and any other measurable and non-measurable process disturbances can upset the process and thereby can affect the product quality. An automatic feed-back control system, though essential to ensure the end product quality in real time, can take action only after the disturbances affect the product quality. Therefore, a feed-back standalone strategy is insufficient to provide near perfect control of the process which is often needed for the manufacture of pharmaceutical products. A feed-forward control strategy takes actions before process disturbances can affect the product quality. However, a feed-forward only control strategy does not take into account the feed-back signal of the control variables and, as a result of other unmeasured process disturbances, the control variables or end product quality can deviate from the desired values. Therefore, a combined feed-forward/feed-back control strategy is needed in which the measured process disturbances (feed-forward signals) can be taken into account before it affects the end product quality and the feed-back signal of the control variables can be utilized to ensure the consistent predefined end product quality.

In the last few years, extensive work has been done to study the continuous tablet manufacturing process [3–6]. The main focus, however, has been on powder blend uniformity, drug concentration and final end product qualities. For example, Vanarase *et al.* (2010) have applied NIR for inline monitoring of powder blend homogeneity [7]. Singh *et al.* have demonstrated the advanced model predictive control (MPC) as well as classical PID (proportional integral derivative) based real time automatic feed-back control of the drug concentration [3]. Much less attention, however, has been paid to the development of PAT and control methods for physical properties of blends. Some physical properties (*e.g.*, blend bulk density, shear) have significant effects on the end product quality (*e.g.*, weight, hardness, dissolution) of the pharmaceutical tablets. Blend physical properties such as bulk density can vary during processing because of many factors. For example, variations in raw material properties (*e.g.*, particle size), feeder hopper level, amounts of lubrication, milling and blending action, applied shear in different processing stages can affect the blend bulk density significantly and thereby tablet weight, hardness and dissolution. Therefore, the inline real time monitoring of the physical properties of the blend (*e.g.*, bulk density) and its incorporation into the control system so that it does not affect the end product quality is highly desired [8]. However, due to the different levels of complexity associated with powder handling and the unavailability of the PAT and control methods, the integration of the physical properties to the control system is still a challenging task.

Feed-forward control systems have been very successful in various manufacturing industries to take proactive actions on process variations/disturbances. Using both feed-forward and feed-back control to respond in real time to disturbances throughout the multiple unit operations is a hallmark of continuous manufacturing [9]. However, to date, no attempts have been made to design a feed-forward control system for a pharmaceutical manufacturing process. Moreover, feed-back control strategy of presently available tablet press also needs to be improved because these control logics are based on empirical law rather than real control algorithm (*e.g.*, PID, MPC); all three control loops in these tablet press can not be activated at the same time meaning that tablet weight and hardness are not really controlled in real time, and the tablet weight and hardness control-loops are not decoupled. Therefore, new feedback control logic has been developed in this manuscript to address these issues as well.

In the last few years, very few attempts have been made toward the control of a tablet manufacturing process utilizing feed-back control algorithm. Hsu *et al.* (2010) have suggested a feed-back control system for a roller compactor, an important unit operation used for a dry granulated continuous tablet manufacturing process [10,11]. A detailed review on the feed-back control of a fluid bed granulation process has been performed by Burggraeve *et al.* (2012), and discussion has been provided by Bardin *et al.* (2004) on the control aspects for efficient operation of a high shear mixer [12,13]. Sanders *et al.* (2009) have performed extensive feed-back control studies using PID and model predictive control (MPC) methods on an experimentally validated fluidized bed granulation model [14]. Recently, Singh *et al.* (2014) have implemented a MPC based feed-back control system into the direct compaction tablet manufacturing process with focus on drug concentration control at blending unit operation [3]. However, no attempt has been made to design a combined feed-forward/feed-back control system for an integrated continuous tablet manufacturing process. Moreover, the simultaneous control of tablet weight and hardness, which are two important control variables in tablet press, is still a challenging task.

In this study, for the first time, a combined feed-forward/feed-back control system has been developed for an integrated continuous direct compaction pharmaceutical tablet manufacturing process. The tablet weight and hardness control loops have been decoupled so that these variables can be controlled simultaneously. The powder blend bulk density has been used as the feed-forward variable of a tablet press control system consisting of loops to control the tablet weight and hardness. The performance of the proposed control system has been evaluated for set point tracking and disturbances rejection ability through process simulation. The systematic application of the feed-forward/feed-back control can take proactive action on a process disturbance and therefore will enable the industrial practitioners to achieve a predefined end product quality more consistently.

2. CONTINUOUS TABLET MANUFACTURING PROCESS

2.1. Process Description

A continuous direct compaction tablet manufacturing process has been previously reported in Singh *et al.*, 2014 [15]. The process includes feeders, co-mill, blender and tablet press. There are three gravimetric feeders with the capability of adding more that feed the various formulation components (active pharmaceutical ingredient (API), excipient *etc.*). A co-mill is also integrated after the feeder hopper, primarily for de-lumping the powders and creating contact between components. The co-mill eliminates any large, soft lumps within the powder. The lubricant feeder is added after the co-mill to prevent over lubrication of the formulation in the co-mill. These feed streams are then connected to a continuous blender within which a homogeneous powder mixture of all the ingredients is generated. Subsequently, the outlet from the blender is fed to the tablet press via a rotary feed frame. The powder blend fills a die and is subsequently compressed to create a tablet. The NIR sensor for inline monitoring of blend uniformity, blend composition and blend bulk density has been integrated through a chute placed in between tablet press and blender.

2.2. Control Relevant Process Model

The control relevant process model involving transfer functions has been developed through step change response analysis in the system identification toolbox of Matlab (Mathworks, Natick, MA, USA). The integrated flowsheet model for direct compaction continuous tablet manufacturing process has been previously reported [6,16]. The mathematical model for powder blending, an important but complex unit operation, has been previously developed as described in [5]. The model for the tablet compression process is previously reported in Singh *et al.* (2013) [16]. This model is based on the Kawakita powder compression model and Kuentz-Leuenberger tablet hardness model [17,18]. The dissolution model was adapted from Kimber *et al.* (2011) [19]. The models for the different unit operations have been developed and included in gPROMS (Process Systems Enterprise, PSE, London, UK) to facilitate the integrated flowsheet modeling.

Input and output variables considered in the process identification stage are listed in Table 1. The output variables are main compression force, tablet weight and hardness. The process inputs with which these control variables have been related using transfer function modelling methodology are fill depth, punch displacement, and powder blend bulk density. The tablet weight and hardness have been also related with the main compression force using transfer function model. The transfer function model has been simulated in Simulink (Mathworks). The transfer function model has been widely used to design and evaluate the control system and it has been observed that the considered system can be modelled through the transfer function approach.

The developed transfer function model along with inputs and outputs variables are given in Table 1. Fill depth has been related with the main compression force through a second order transfer function represented by one zero and two poles. The poles and zeros plot is shown in Figure 1a. Figure shows that the process is stable and non-oscillatory. The values of one pole and zero are very small indicating that the process behavior is closer to a first order system. The phase diagram is shown in Figure 1b. Phase lag and phase lead reasons have been highlighted in the figure. From the phase diagram, the stability margin can be obtained. For a marginal stability, the gain margin should be more than 2.5 and phase margin (180° − Phase angle) should be more than 30 [20]. Note that, both a good gain margin and a good phase margin are needed but neither is sufficient by itself. The delay time for this process depends on tablet press operating conditions.

Table 1. Process transfer function model.

Transfer Functions	Models	Inputs	Outputs
$G_{p1}(S)$	$\dfrac{0.3782(77.3204S+1)}{43.6110S^2+25.5342S+1}$	Fill depth	Main compression force
$G_{p2}(S)$	$\dfrac{1.4459(19.4813S+1)}{13.1562S+1}$	Main compression force	Tablet hardness
$G_{p3}(S)$	$\dfrac{10.1868(0.2595S+1)}{10.9218S^2+0.07510S+1}$	Main compression force	Tablet weight
$G_{p4}(S)$	$\dfrac{1.1701}{106.6667S^2+4.8672S+1}$	Punch displacement	Main compression force
$G_d(S)$	$\dfrac{34.6667(247.1219S+1)}{453.5147S^2+86.4399S+1}$	Powder bulk density	Main compression force

The main compression force has been related with the tablet hardness through a first order transfer function represented by one pole and one zero. The pole and zero plots are shown in Figure 2 (see Gp2-P and Gp2-Z). As shown in the figure, the process is stable and non-oscillatory. The phase diagram is shown in Figure 3 (see Gp2(S)). Main compression force has been also related with the tablet weight through a second order transfer function model with one zero and two poles (see Table 1). Poles and zeros plots shown in Figure 2 (see Gp3-P and Gp3-Z) indicate that the process is stable and oscillatory. Phase diagram is shown in Figure 3 (see Gp3(S)). The dead time for this process is 2 s. Punch displacement has been related with the main compression force through a second order system with one zero and two imaginary poles. The poles and zeros plot shown in Figure 2 (Gp4-P)

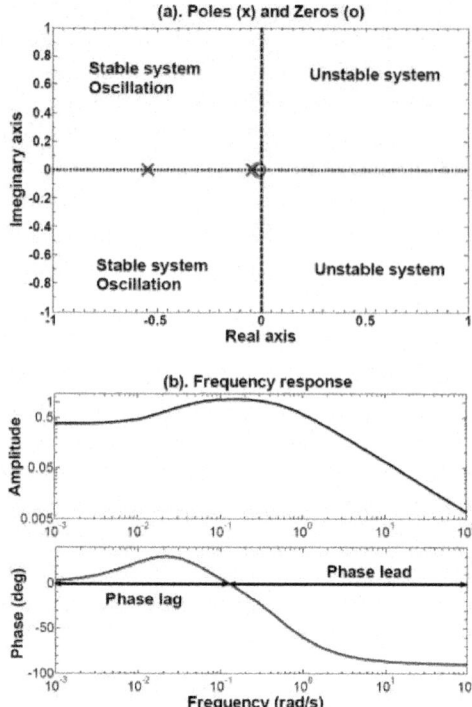

Figure 1. Poles-zeros plot and phase diagram of transfer function model relating fill depth with main compression force. (**a**) Poles-zeros plot; (**b**) Frequency response.

indicate that the response is stable but oscillatory. The phase diagram for this transfer function is shown in Figure 3 (see Gp4(S)). Similarly, the relationship between powder bulk density and main compression force has been modelled using second order transfer function. One zero and two real poles are involved in the model. Figure 2 shows that the system is stable and non-oscillatory (see Gd-P and Gd-Z). Figure 3 shows the phase diagram and marginal stability for this process (see Gd(s)). Note that, for each system identification case, the transfer function models have been identified for a range of numbers of zeros, and poles and the best fitted models are given in Table 2.

Table 2. Controller inputs and outputs.

Control Loops	Controllers	Control Variables	Inputs	Outputs
1: Feed-back	$G_{c1}(S)$	Main compression force (y_1)	Deviation of main compression force from set point	Fill depth (u_{11})
2: Feed-forward	$G_{c2}(S)$	-	Powder bulk density	Fill depth (u_{12})
3: Feed-back	$G_{c3}(S)$	Tablet weight (y_2)	Deviation of tablet weight from set point	Main compression force set point
4: Feed-back	$G_{c4}(S)$	Tablet hardness (y_3)	Deviation of tablet hardness from set point	Punch displacement (u_2)

"-" means there is no control variable associated with feed-forward controller.

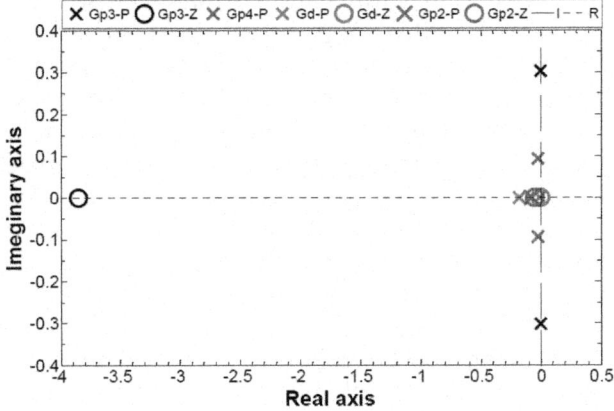

Figure 2. Poles (P)-zeros (Z) plot of transfer function models.

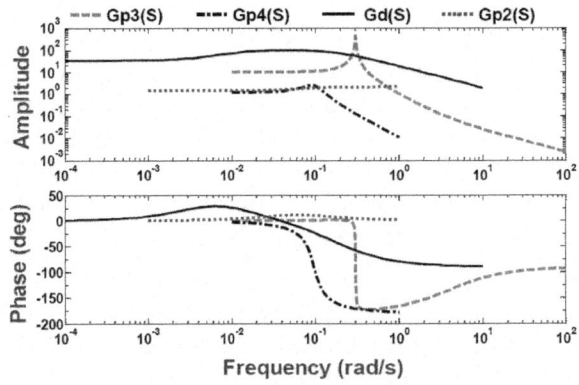

Figure 3. Phase diagram (frequency response) of process models.

3. COMBINED FEED-FORWARD/FEED-BACK CONTROL OF TABLET PRESS

A combined feed-forward/feed-back control system has been designed for a tablet press used in continuous tablet manufacturing process, using a standard approach. The final product specifications include tablet weight, tablet hardness, and tablet dissolution. The unit operations for considered process are feeders, co-mill, blender and tablet press. API, excipient and lubricant are important feeds to the process.

3.1. Architecture of the Combined Feed-Forward/Feed-Back Control System

The architecture of the combined feed-forward/feed-back control system for the tablet press is shown in Figure 4 and consists of four control loops. All controllers and corresponding inputs and outputs are listed in Table 2. Control loop 1 is a slave feed-back control loop for real time control of main compression force. This loop is based on real time measurement of the main compression force and maintains the main compression force at the specified set point. As shown in Figure 4, the powder bulk density, fill cam level and the distance between punch tips under maximum compression can affect the main compression force. The slave loop 1 is significantly faster than the master loop. Control loop 2 is a feed-forward loop. This loop is based on the real time measurement of the powder blend density. This loop compensates for disturbances in powder bulk density before it can affect the final end product quality and therefore minimizes the rejections. Similar to powder blend density, any other process variables can be fed-forward to the control system. Loop 3 is a master feed-back loop for tablet weight control. It provides the set point for control loop 1 (*i.e.*, set point of main compression force). This loop ensures consistent tablet weight throughput the operation. This loop should only be activated if the achieved tablet weight violates the specific tolerance limit. Loop 4 is a feed-back loop specifically designed for tablet hardness control. This loop is also only activated when the achieved tablet hardness violates the specified control limits. This loop also affects the set point of main compression force. This loop controls the hardness by adjusting the punch displacement. This loop ensures the consistent tablet hardness throughout the continuous plant operation. The control variables have been paired with appropriate actuators using knowledge of the physics of the system combined with relative gain array (RGA) analysis and dynamic sensitivity analysis. The studies on control variable and actuator pairing have been previously reported [16].

It can be seen in Figure 4 that the tablet weight and hardness control loops have been decoupled meaning that both variables can be controlled simultaneously. For example, on changing the tablet weight set point, the main compression force set point will change and control loop 1 will then track the new set point by manipulating the fill cam level. Similarly, if tablet hardness set point has been changed then control loop 4 will adjust the distance between punch tips and that will lead to a change in the main compression force. Loop 4 then will try to maintain the consistent tablet weight by not allowing the change in the fill cam level and, thereby, the set point of the main compression force will change. In this control mechanism, the loop 1 (designed to control the main compression force) is a shared slave loop for tablet weight and hardness control loops. Therefore, changing the tablet weight and/or hardness set point will lead to a change in the set point of the main compression force. A single slave loop has been shared by two master control loops in such a way that both control variables can be controlled simultaneously.

The spectroscopic technique is the most commonly used method to monitor the pharmaceutical tablet manufacturing process. NIR sensor has been proposed to measure the powder bulk density in real time. Some other methods based on load cell are available as well for real time measurement of powder bulk density [21]. The most desired measurement point for powder bulk density is the feed frame of the tableting press if that is possible for a given tablet press design and configuration.

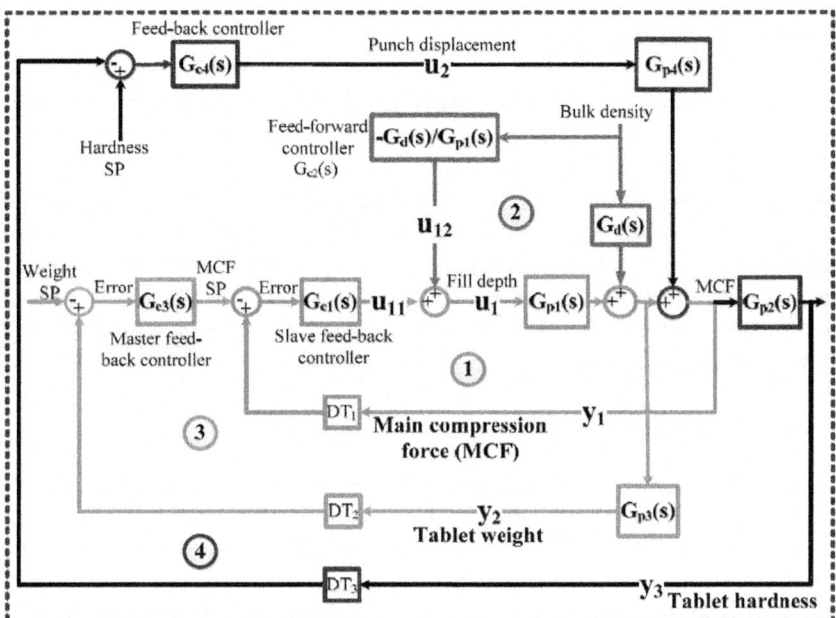

Figure 4. Architecture of the feed-forward/feed-back control system and decoupled tablet weight and hardness control loops. SP: Set point; MCF: Main compression force.

Previously, we have experimentally verified that the NIR sensor is fast enough to be used for real time process control [3]. It has been observed that the sensing delay for NIR sensor (micro NIR from JDSU, Milpitas, CA, USA) is around 1 s which is considered to be reasonable for real time control. Strain gauge or load cell are commonly used for measurement of the main compression force. The measurement time for strain gauge or load cell is in milliseconds and it has been used for real time control of the tablet press. Very limited number of methods is currently available for the inline/online measurement of tablet weight and hardness. For example, Checkmaster (Fette, Schwarzenbek, Germany) can be used for inline measurement of tablet weight and hardness. Within our research center (C-SOPS), the efforts are being made to develop the feed-back control relevant faster sensors for hardness and weight measurement that can be used. The sensing delay has been added into the process model to analyze its effect. The NIR sensor has been integrated within the plant and the control platform (DeltaV, Round Rock, TX, USA) through OPC (object linked and embedding [OLE] for process control) communication protocol. OPC is the standard system for the real-time plant data communication between control devices from different manufactures. The NIR sensor has been used for the real time monitoring of the API composition and blend RSD (relative standard deviation) along with monitoring of powder bulk density. The application of NIR based API composition measurement system for feed-back control has been experimentally previously demonstrated [3,15]. This system has been adapted to demonstrate the application of a NIR based powder blend density measurement system for feed-forward/feed-back control system. The process and disturbances time delays are reasonably small making the feed-forward controller realizable.

The process and controller models are represented through a transfer function. The process transfer functions involved in Figure 4 are given in Table 1. In order to obtain the transfer function for feed-forward controller, first a process model (Gd(s)) relating the input variable (or disturbances) to the control variable has been developed. Then, a process model (Gp1(s)) relating the actuator with the control variable has been identified. Subsequently, the characteristic equation for the feed-forward control loop has been derived. For a perfect controller, the characteristic equation can be equated to zero and thereby the model (-Gd(s)/Gp1(s)) for feed-forward controller can be generated.

3.2. Controller Parameters Tuning

Classical PID based feed-back controllers which are easier to implement into the manufacturing plant have been used to control the main compression force, tablet weight and hardness. There are three controllers involved in the process that consist of maximum nine tuning parameters. There are several methods and rules for PID controller parameter tuning [20,22]. In this work, these tuning parameters (gain, integral time constant, derivative time constant) have been tuned using the inbuilt tuning methodology in Simulink (Mathworks, Natick, MA, USA). Simulink has been widely used to design and implement the controller for manufacturing process and therefore it has been used for this study. MathWorks® (Natick, MA, USA) algorithm for tuning PID controllers is based on insuring the closed-loop stability, adequate performance and adequate robustness. This algorithm meets these objectives by tuning the PID parameters to achieve a good balance between performance and robustness. The algorithm designs an initial controller by choosing a bandwidth to achieve that balance, based upon the open-loop frequency response of linearized model. Upon interactively changing the response time, bandwidth, transient response, or phase margin using the PID Tuner interface, the algorithm computes new PID gains. The controller tuning ensures that the closed-loop system tracks reference changes and suppresses disturbances as rapidly as possible. The tuning method uses the rise time, settling time, overshoot, gain margin phase margin and closed-loop stability as an index to assess the performance of the controller parameters tuning. The process flowsheet model simulated in Simulink has been utilized to tune the controller parameters. The anti-windup reset algorithm, which ensures that the controller output lies within the specified upper and lower bounds, has been included [23]. Because of anti-windup, if the bounds are violated, the time derivative of the integral error is set to zero, and the controller output is clipped to the bounds. Once the controller output is back in the range of the bounds, the integral error will change according to the current error. The controllers need to be re-tuned if there are any changes in the control architecture and/or process. The tuned controller parameters are given in Table 3. The parameters reported in Table 3 correspond to the PID controller form given in Equation 1. A standard PID form, inbuilt in Simulink (Mathworks) has been adapted to represent in terms of controller gain, integral and derivative time constants, as given in Equation 1. Controller gain (KC), Integral time constant (τ_I), derivative time constant (τ_D) and filter coefficient (N) are the parameters of Equation 1. The filter coefficient sets the location of the poles in the derivative filter. Loop 1 is slave controller and loop 3 is master controller. As given in Table 3, the reset time of loops 3 is 3.52 times that of loop 1 meaning that loop 1 (slave) is faster than loop 3 (master). Note that in a cascade arrangement, the slave loop dynamics needs to be faster than the master loop dynamics. The derivative action has been found to be zero for all loops; therefore, the controller actions are based on proportional and integral actions only.

Table 3. Controller tuning parameters.

Controllers	Control Variables	Controller Gain (Kc)	Integral Time Constant (τ_I)	Derivative Time Constant (τ_D)
Gc1	Main compression force	0.06224 cm/KN	0.361027 s	0 s
Gc3	Tablet weight	0.010000 KN/g	1.272265 s	0 s
Gc4	Tablet hardness	0.000010 cm/Kp	0.029412 s	0 s

For this study, it was assumed that the process model is perfect and, therefore, the feed-forward controller did not require any tuning. However, if process/model mismatch is present, then the feed-forward controller would also require tuning (*e.g.*, tuning the lead time and the lag time, if the controller is implemented as a lead/lag element). More details of the feed-forward control law, tuning and the impact of process/model mismatch can be found in scientific literatures (*e.g.*, Marlin, 2000 [24]). It should be noted that for implementation of the control system, whole process model is not needed and only the models involved in feed-forward control algorithm are required. For a best feed-forward controller performance, the models involved in feed-forward control logic are desired to be as accurate as possible. The pharmaceutical manufacturing processes are normally very well designed (as required by regulator); therefore, through a set of experiments it is possible to obtain the accurate models to be used in the feed-forward controller. Note that for implementation of proposed control system into plant, only feed-forward controller model is needed and complete process models are not required. Feed-forward controller model needs to be re-developed if there are any changes in plant and materials involved.

$$G_c(S) = K_C \left[1 + \frac{1}{\tau_i} \left(\frac{1}{S} \right) + \tau_D \left(\frac{NS}{S+N} \right) \right] \tag{1}$$

Prior to the implementation of the control system to the continuous tablet manufacturing pilot-plant, the performance of the designed feed-forward/feed-back control system is evaluated using process simulation as described in this section. The model-based performance evaluation of the control system reduces the time and resources needed for the implementation of the control system into the pilot-plant and increases the chance of success during implementation. In this section, the ability of the control system to track the set point and reject the disturbances is evaluated. For set point tracking, the step changes in the set point are applied, and the controller is allowed to track those changes. For disturbance rejection, structural disturbances (*e.g.*, sinusoidal) and random disturbances are introduced during closed-loop operation.

A step change in the hardness set point has been introduced while keeping the tablet weight set point at a constant value. The blend density acts as the disturbance and can affect the tablet weight and hardness significantly. The blend density can change because of many factors. For example, the particle size of the powder, lubrication level and shear applied in blender can change the blend density. Since the powder is filled in the tablet press die by volume (not by weight), different blend densities can lead to different tablet weight and hardness. Random disturbances have been added to the blend density throughout the operation. A step change in the blend density has been also introduced at $t = 2500$ s in order to analyze its effects.

Figure 5 shows a closed-loop response of the tablet hardness under the combined feed-forward/feed-back control strategy and feed-back only control strategy. As shown in the figure, under the combined control strategy, the feed-forward controller rejects the variation in the blend density before it can affect the hardness. Therefore, under the combined control scheme, the tablet hardness has been controlled more effectively. A small oscillation can be seen at process startup and at the point where the step change in the hardness set point has been made, which is acceptable. As desired for a good controller, the rise time is less and decay ratio is high. A significant variation in the achieved hardness can be seen in case of only feed-back control scheme (see Figure 5). This variation is

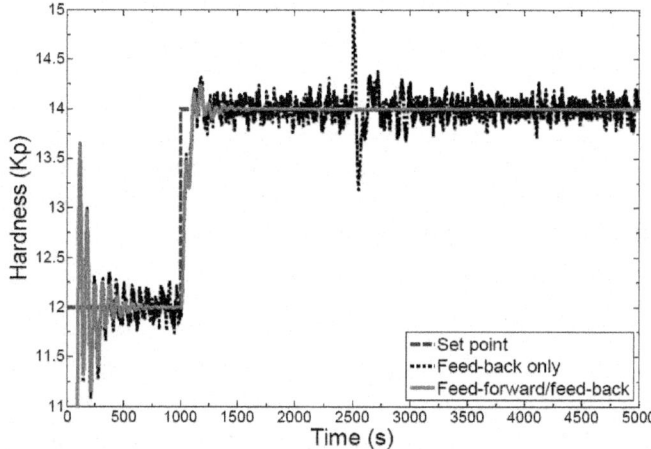

Figure 5. Comparison of combined feed-forward/feed-back control strategy with feed-back only control strategy for hardness control.

because of random variation in the blend density. It should be noted that the hardness variation is more at $t = 2500$ s where the step change in the blend density has been introduced. The results shown in Figure 5 demonstrate the advantages of coupling feed-forward controller with the feed-back control system. It should be noted that only feed-forward control is practically insignificant because there may always be some unknown disturbances that cannot be measured. Feed-back control system is essential to reject the unknown disturbances that can occur during plant operation.

Actuators values obtained from combined feed-forward/feed-back control strategy with that obtained from feed-back only control strategy for tablet hardness control is shown in Figure 6. As shown in the figure, the combined control strategy provides a more consistent and steady actuator response in comparison to the feed-back only control strategy. The oscillations in the actuator obtained by feed-back only control strategy can be attributed to random disturbances introduced in the powder bulk density. In case of a feed-back only control strategy, the hardness is affected by the disturbances introduced in the powder bulk density while, in case of combined control strategy, this effect is compensated proactively. A significant oscillation can be seen at $t = 2500$ s where a step change in the powder bulk density has been introduced.

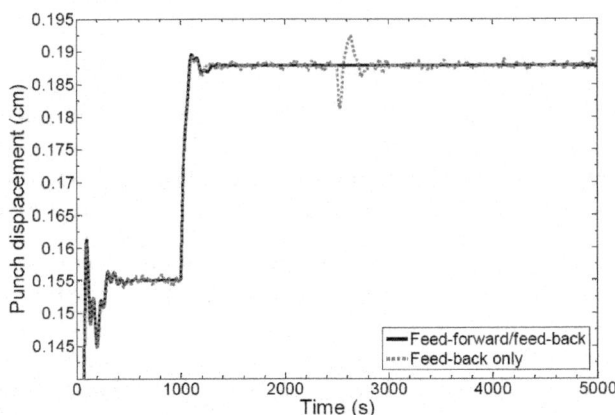

Figure 6. Comparison of actuators obtained from combined feed-forward/feed-back control strategy with feed-back only control strategy for hardness control.

The closed-loop response of the tablet weight under the combined feed-forward/feed-back control strategy and only feed-back control strategy is shown in Figure 7. As shown in the figure, the variation in tablet weight is less under combined control scheme in compare to only feed-back control scheme. Under the combined control scheme, a small oscillation can be seen at startup, which is acceptable because the decay ratio is high. Higher decay ratios reduce the oscillation and overshoot faster. A small oscillation can be seen at $t = 1000$ s where the step change in the hardness set point has been made and this is essentially because the hardness and weight control loops are highly interactive. Note that the step change in the hardness set point has been made for demonstration of the set point tracking ability of the designed control system and in practice for a given formulation hardness and weight set points need to be fixed. At t = 2500 s, a step change in the blend density has been introduced but the combined feed-forward/feed-back control system has efficiently rejected its effect and thereby a consistent desired tablet weight has been achieved. In the case of the feed-back only control scheme, the variation in achieved tablet weight is very high and corresponds to the random variations introduced to the blend density. The deviation of the tablet weight from the desired set point in the interval of 2500-2600 s is significantly higher because of the introduced step change in the blend density. The tablets produced in this range need to be rejected.

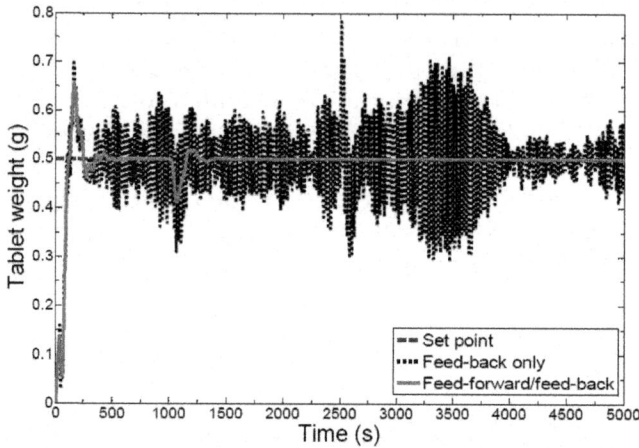

Figure 7. Comparison of the performance of the combined feed-forward/feed-back control scheme with feed-back only control scheme for tablet weight control.

The main compression force has been controlled in real time through a slave controller. The performance of the tablet weight and hardness master controller depends significantly on the performance of the slave 'main compression force' controller. The closed-loop response of the main compression force under the combined and feed-back only control strategy is shown in Figure 8. In each case, the set point of the main compression force has been obtained by master controller specifically designed to control the tablet weight. As shown in Figure 8, in the case of the combined feed-forward/feed-back control scheme, the main compression force is controlled at the set pint generated by master controller. A small oscillation can be seen at startup and at the point where step change in the hardness set point has been made. The decay ratio is however higher as is desired for a well-tuned controller. The step change in the main compression force set point (generated by master controller) can be seen at t = 1000 s. This step change in the main compression force set point is to achieve the step change in the hardness. More compression force is needed to produce the higher hardness tablet. In spite of an increase in compression force, the tablet weight has not been increased meaning that this increase in compression force is because of increased upper punch displacement while keeping the consistent fill depth (see Figures 5–8). In the case of the feed-back only control scheme, first the main compression force set point generated from master controller is oscillatory and second this set point has not been tracked precisely. The variation in the achieved main compression force under the feed-back only control scheme can be seen in Figure 8 which is because of variation in the blend density. The major variation can be seen at t = 2500 s where a step change in the blend density has been introduced.

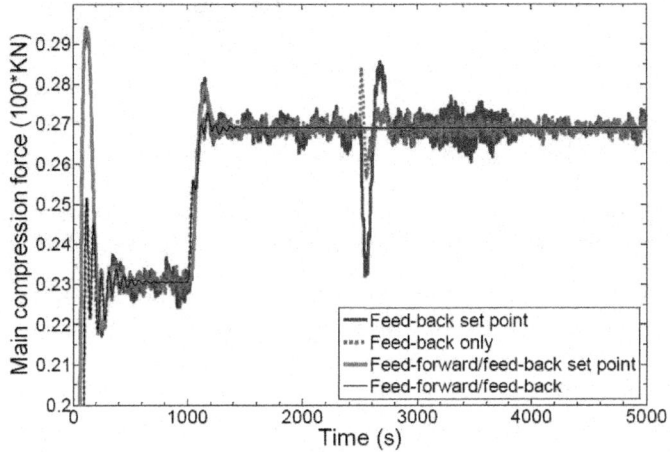

Figure 8. Closed-loop response of main compression force (performance of slave controller).

Results shown in Figures 5–8 demonstrate that the tablet weight and hardness control loops have been decoupled so that both variables can be controlled simultaneously. In order to verify the performance of decoupling control mechanism, a step change in the hardness set point has been made (see Figure 5) while keeping the tablet weight set point at a constant value (see Figure 7). As shown in Figures 5–7, the tablet hardness value has been changed successfully while maintaining a consistent tablet weight. Figure 8 shows that in order to achieve the increased tablet hardness, the set point of the main compression force has been increased. Note that the hardness set point has been changed just for demonstration of the performance of decoupling control mechanism and set point tracking ability of the control system and, for a given formulation, the hardness needs to be consistent throughout the operation.

The response of the fill depth is shown in Figure 9. As shown in the figure under the feed-forward/feed-back control scheme, the control system changes the fill depth more precisely to compensate for the disturbances introduced in the powder bulk density. More change can be seen at t = 1000 s where the step change in hardness set point has been introduced. A step change in the fill depth can be seen at t = 2500 s where step change in the powder bulk density has been introduced. Note that, there are also random disturbances throughout the operation. The fill depth was within the acceptable range during closed-loop operation indicating that the proposed control system is feasible.

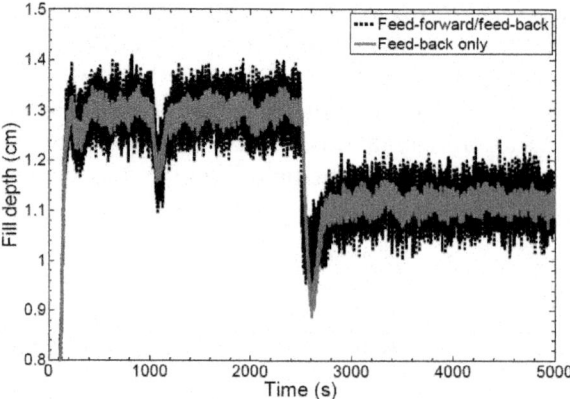

Figure 9. Comparison of actuators obtained from combined feed-forward/feed-back control strategy with feed-back only control strategy for main compression force control.

The set point of the proposed control system can be changed within a feasible operating range of a tablet press. This range depends on many factors such as tablet press configuration and the properties of the raw materials. The actuators should be within the feasible operating range in order to maintain the control variables within the control limits. For example, in case of Kikusui rotary tablet press, the powder filling depth range is 1–15 mm. Maximum tablet diameter is 16 mm and tablet thickness range is 0–5 mm. The range of tablet weight and hardness that can be produced from a tablet press also depends on the raw materials (API, Excipient, Lubricant) and blend properties along with tablet press specifications. Some of the ranges mentioned above can be changed by changing the tooling used in the tablet press. The control architecture will be same for different tablet presses and raw materials but the controller parameters need to be returned.

5. CONCLUSIONS

A well-designed control system incorporating the feed-forward and feed-back control algorithm is developed in this work to obtain a precise pre-defined end-product quality for a pharmaceutical product as mandated by regulatory authorities. The combined control strategy incorporates the advantages of both feed-forward and feed-back control strategies. The feed-forward control strategy rejects the effects of the measured process disturbances proactively before they can influence the product quality and, thereby, minimizes the rejection, while the feed-back control strategy takes into account the effects of other unknown process disturbances in real time. The performance of the combined control strategy is found to be better than only feed-back control. The proposed strategy is complementary to the classical feed-back control approach and should thus have a broad application range for QbD-based manufacturing. The tablet weight and hardness control loops have been also decoupled so that these variables can be controlled simultaneously. For this study, it was assumed that the process model is perfect, which is an ideal scenario, and in practice, there could be some model-plant mismatch. Future work includes the implementation of the designed combined feed-forward/feed-back control system in our continuous tablet manufacturing pilot-plant through commercially available hardware (*e.g.*, Delta V) and the control interface (*e.g.*, OPC).

This work is supported by the National Science Foundation Engineering Research Center on Structured Organic Particulate Systems, through Grant NSF-ECC 0540855.

Author Contributions

The research work reported in this manuscript is conducted by Ravendra Singh and supervised by Fernando J. Muzzio, Marianthi Ierapetritou and Rohit Ramachandran.

Nomenclature

Abbreviations

API	Active Pharmaceutical Ingredient
APAP	Acetyl-Para-Aminophenol
CPP	Critical Process Parameter
CQA	Critical Quality Attribute
DT	Dead Time
MgSt	Magnesium Stearate
MCF	Main Compression Force
NIR	Near Infrared
OPC	OLE (Object linked and embedding) for process control
PAT	Process Analytical Technology
PID	Proportional Integral Derivative
QbD	Quality by Design
QbT	Quality by Testing
RSD	Relative Standard Deviation
SMCC	Silicified Microcrystalline Cellulose
SP	Set point
Symbols	Variables
$Gd(s)$	Disturbance transfer function model
$Gp(s)$	Process transfer function model
$Gc(s)$	Controller transfer function model
d	Disturbances
u	Actuator
y	Control variable
Subscript	Description
d	disturbance
p	process
c	controller
1,2,3,4	Process or controller numbers

Conflicts of Interest

The authors declare no conflict of interest.

REFERENCES

1. FDA. *PAT – A Framework for Innovative Pharmaceutical Development, Manufacturing, and Quality Assurance*; US Food and Drug Administration: Silver Spring, MD, USA, 2004. Available online: http://www.fda.gov/downloads/Drugs/Guidances/ucm070305.pdf (accessed on 19 March 2015).

2. FDA. *Guidance for Industry, Q8 (R2) Pharmaceutical Development*; US Food and Drug Administration: Silver Spring, MD, USA, 2007. Available online: http://www.fda.gov/ downloads/ Drugs/Guidances/ucm073507.pdf (accessed on 19 March 2015).

3. Singh, R.; Sahay, A.; Karry, K.M.; Muzzio, F.; Ierapetritou, M.; Ramachandran, R. Implementation of a hybrid MPC-PID control strategy using PAT tools into a direct compaction continuous pharmaceutical tablet manufacturing pilot-plant. *Int. J. Pharm.* **2014**, *473*, 38–54.

4. Igne, B.; Juan, A.D.; Jaumot, J.; Lallemand, J.; Preys, S.; Drennen, J.K.; Anderson, C.A. Modeling strategies for pharmaceutical blend monitoring and end-point determination by near-infrared spectroscopy. *Int. J. Pharm.* **2014**, *473*, 219–231.

5. Sen, M.; Singh, R.; Vanarase, A.; John, J.; Ramachandran, R. Multi-dimensional population balance modeling and experimental validation of continuous powder mixing processes. *Chem. Eng. Sci.* **2012**, *80*, 349–360.

6. Boukouvala, F.; Niotis, V.; Ramachandran, R.; Muzzio, F.; Ierapetritou, M. An integrated approach for dynamic flowsheet modeling and sensitivity analysis of a continuous tablet manufacturing process: An integrated approach. *Comput. Chem. Eng.* **2012**, *42*, 30–47.

7. Vanarase, A.; Alcal, M.; Rozo, J.; Muzzio, F.; Romaach, R. Real-time monitoring of drug concentration in a continuous powder mixing process using NIR spectroscopy. *Chem. Eng. Sci.* **2010**, *65*, 5728–5733.

8. García-Munoz, S.; Dolph, S.; Ward, H.W., II. Handling uncertainty in the establishment of a design space for the manufacture of a pharmaceutical product. *Comput. Chem. Eng.* **2010**, *34*, 1098–1107.

9. Myerson, A.S.; Krumme, M.; Nasr, M.; Thomas, H.; Braatz, R.D. Control Systems Engineering in Continuous Pharmaceutical Manufacturing. *J. Pharm. Sci.* **2015**, *104*, 832–839.

10. Hsu, S.; Reklaitis, G.V.; Venkatasubramanian, V. Modeling and control of roller compaction for pharmaceutical manufacturing. Part I: Process dynamics and control framework. *J. Pharm. Innov.* **2010**, *5*, 14–23.

11. Hsu, S.; Reklaitis, G.V.; Venkatasubramanian, V. Modeling and control of roller compaction for pharmaceutical manufacturing. Part II: Control and system design. *J. Pharm. Innov.* **2010**, *5*, 24–36.

12. Burggraeve, A.; Tavares da Silva, A.; Van den Kerkhof, T.; Hellings, M.; Vervaet, C.; Remon, J.P.; Vander Heyden, Y.; Beer, T.D. Development of a fluid bed granulation process control strategy based on real-time process and product measurements. *TALANTA* **2012**, *100*, 293–302.

13. Bardin, M.; Knight, P.C.; Seville, J.P.K. On control of particle size distribution in granulation using high-shear mixers. *Powder Technol.* **2004**, *140*, 169–175.

14. Sanders, C.F.W.; Hounslow, M.J.; Doyle, F.J., III. Identification of models for control of we granulation. *Powder Technol.* **2009**, *188*, 255–263.

15. Singh, R.; Sahay, A.; Muzzio, F.; Ierapetritou, M.; Ramachandran, R. Systematic framework for onsite design and implementation of the control system in continuous tablet manufacturing process. *Comput. Chem. Eng. J.* **2014**, *66*, 186–200.

16. Singh, R.; Ierapetritou, M.; Ramachandran, R. System-wide hybrid model predictive control of a continuous pharmaceutical tablet manufacturing process via direct compaction. *Eur. J. Pharm. Biopharm.* **2013**, *85*, 1164–1182.

17. Kawakita, K.; Ludde, K.H. Some considerations on powder compression equations. *Powder Technol.* **1971**, *4*, 61–68.

18. Kuentz, M.; Leuenberger, H. A new model for the hardness of a compacted particle system, applied to tablets of pharmaceutical polymers. *Powder Technol.* **2000**, *111*, 143–145.

19. Kimber, J.A.; Kazarian, S.G.; Stepánek, F. Microstructure-based mathematical modelling and spectroscopic imaging of tablet dissolution. *Comput. Chem. Eng.* **2011**, *35*, 1328–1339.

20. Seborg, D.E.; Edgar, T.F.; Mellichamp, D.A. *Process Dynamics and Control*, 2nd ed.; John Wiley & Sons, Inc.: Hoboken, NJ, USA, 2004.

21. Davies, C.E.; Lankshear, R.C.; Webster, E.S. Direct measurement of the bulk density of cohesive particulate materials by a quasicontinuous in-line weighing method. In Proceedings of *Chemeca 2011*, Sydney, Australia, 18–21 September 2011. Available online: http://search.informit.com. au/ documentSummary;dn=173928395929049;res=IELENG (accessed on 19 March 2015).

22. Ziegler, J.G.; Nichols, B. Optimum settings for automatic controllers. *Trans. ASME* **1942**, *64*, 759–765.

23. Ogunnaike, B.A.; Ray, W.H. *Process Dynamics, Modeling, and Control*; Oxford University Press Inc.: New York, NY, USA, 1994.

24. Marlin, T.E. *Process Control*; McGraw-Hill, Inc.: Blacklick, OH, USA, 2000.

This page left intentionally blank.

INDEX

This page left intentionally blank.